AF389333

Advanced Biometrics with
Deep Learning

Advanced Biometrics with Deep Learning

Editors

Andrew Teoh Beng Jin
Lu Leng

MDPI • Basel • Beijing • Wuhan • Barcelona • Belgrade • Manchester • Tokyo • Cluj • Tianjin

Editors
Andrew Teoh Beng Jin
Yonsei University
Korea

Lu Leng
Nanchang Hangkong University
China

Editorial Office
MDPI
St. Alban-Anlage 66
4052 Basel, Switzerland

This is a reprint of articles from the Special Issue published online in the open access journal *Applied Sciences* (ISSN 2076-3417) (available at: https://www.mdpi.com/journal/applsci/special_issues/Biometrics_Deep_Learning).

For citation purposes, cite each article independently as indicated on the article page online and as indicated below:

LastName, A.A.; LastName, B.B.; LastName, C.C. Article Title. *Journal Name* **Year**, *Article Number*, Page Range.

ISBN 978-3-03936-698-9 (Hbk)
ISBN 978-3-03936-699-6 (PDF)

Contents

About the Editors

Andrew Teoh Beng Jin (Professor) obtained his BEng (Electronic) in 1999 and Ph.D. degree in 2003 from the National University of Malaysia. He is currently a full professor in the Electrical and Electronic Engineering Department, College Engineering of Yonsei University, South Korea. His research, for which he has received funding, focuses on biometric applications and biometric security. His current research interests are Machine Learning and Information Security. He has published more than 300 international refereed journal papers, conference articles, edited several book chapters, and edited book volumes. He served and is serving as a guest editor of IEEE Signal Processing Magazine, associate editor of IEEE Transaction of Information Forensic and Security, IEEE Biometrics Compendium, Machine Learning, and editor-in-chief of the IEEE Biometrics Council Newsletter.

Lu Leng (Associate Professor) received his Ph.D. degree from Southwest Jiaotong University, Chengdu, P. R. China, in 2012. He performed his post-doctoral research at Yonsei University, Seoul, Republic of Korea, and Nanjing University of Aeronautics and Astronautics, Nanjing, P. R. China from 2012 to 2015. He was a visiting scholar at West Virginia University, USA from 2015 to 2016. Currently, he is an associate professor at Nanchang Hangkong University, and also a visiting scholar at Yonsei University, Seoul, Republic of Korea. He has published more than 70 international journal and conference papers. He has been granted several scholarships and funding projects for his academic research. He is the reviewer of more than 50 international journals and conferences. His research interests include computer vision, biometric template protection, and biometric recognition. He is a member of the Institute of Electrical and Electronics Engineers (IEEE), Association for Computing Machinery (ACM), China Society of Image and Graphics (CSIG), and China Computer Federation (CCF).

Editorial

Special Issue on Advanced Biometrics with Deep Learning

Andrew Beng Jin Teoh [1],* and Lu Leng [2]

[1] School of Electrical and Electronic Engineering, College of Engineering, Yonsei University, Seoul 120749, Korea
[2] School of Software, Nanchang Hangkong University, Nanchang 330063, China; leng@nchu.edu.cn
*   Correspondence: bjteoh@yonsei.ac.kr

Received: 16 June 2020; Accepted: 24 June 2020; Published: 28 June 2020

1. Introduction

Biometrics, such as fingerprint, iris, face, hand print, hand vein, speech and gait recognition, etc., as a means of identity management has become commonplace nowadays for various applications. Traditional authentication methods, including possession-based and knowledge-based methods, typically suffer from various problems. It is probable that possessions, such as an ID card or key, can be stolen, broken, or lost; while it is probable that knowledge, such as a password or PIN, can be forgotten or guessed. Compared with the traditional authentication methods, biometric recognition methods are more convenient and secure [1].

Biometric systems follow a typical pipeline that is composed of separate acquisition, preprocessing, feature extraction and classification. Deep learning as a data-driven representation learning approach has been shown to be a promising alternative to conventional data-agnostic and handcrafted pre-processing and feature extraction for biometric systems. Furthermore, deep learning offers an end-to-end learning paradigm to unify preprocessing, feature extraction and recognition based solely on biometric data [2]. The main advantages of deep learning include a strong learning ability, wide coverage, good adaptability, data-driven, good transferability, etc.

2. Advanced Biometrics with Deep Learning

In light of the above, this Special Issue collected high-quality, state-of-the-art research papers that deal with challenging issues in advanced biometric systems based on deep learning. A total of 32 papers were submitted, 12 of which were accepted and published (i.e., 37.5% acceptance rate). The 12 papers can be briefly divided into 4 categories as follows according to biometric modality.

2.1. Face

The paper, authored by H. Tu, G. Duoji, Q. Zhao and S. Wu, extracted invariant features from a single sample per subject for face recognition. The authors generated additional samples to enrich the intra-variation and eliminate external factors [3]. Another paper by S. Zhou, C. Chen, G. Han and X. Hou, proposed a novel loss function, termed as double additive margin Softmax loss (DAM-Softmax) for convolutional neural networks (CNNs) in face recognition [4]. The presented loss has a clearer geometrical explanation and can produce highly discriminative features. B. Ríos-Sánchez, D. Costa-da-Silva, N. Martín-Yuste and C. Sánchez-Ávila, described and evaluated two deep learning models for face recognition in terms of accuracy and size, which were designed for the applications in mobile devices and resource saving environments [5]. The 4th paper, authored by Z. Yang, J. Li, W. Min and Q. Wang, presented real-time pre-identification and cascaded detection for tiny faces to reduce background and other irrelevant information [6]. The cascade detector consisted of a two-stage convolutional neuron network to detect tiny faces in a coarse-to-fine manner. The face-area candidates were pre-identified as a region of interest (ROI) based on a real-time pedestrian detector, while the set of ROI candidates was the input of the second sub-network instead of the whole image.

2.2. Medical Electronic Signal

Medical electronic signals referring to electrocardiogram (ECG) and electroencephalogram (EEG) signals have been identified as a type of behavioral biometrics. The paper, authored by E. Ihsanto, K. Ramli, D. Sudiana and T. S. Gunawan, focused on the accuracy and processing speed of ECG recognition [7]. They proposed a fast and accurate two-stage framework that consisted of ECG beat detection and classification. Hamilton's method and Residual Depthwise Separable CNN (RDSCNN) were used for ECG beat detection and classification, respectively. Another paper by F. Li, X. Li, F. Wang, D. Zhang, Y. Xia and F. He, aimed at enhancing the classification accuracy of P300 EEG signals in a non-invasive brain-computer interface system [8]. They employed principal component analysis (PCA) to remove the noise and artifacts in the data as well as to increase the data processing speed. Furthermore, the parallel convolution method was used for P300 classification, which increased the network depth and improved the accuracy. The third paper, authored by D. Wang, Y. Si, W. Yang, G. Zhang and T. Liu, proposed a novel method suitable for short-term ECG signal identification [9]. An improved heart-rate-free resampling strategy was employed to minimize the influence of heart-rate variability during processing. The PCA Network (PCANet) for feature extraction was implemented to determine the potential difference between subjects.

2.3. Voice Print

Most deep learning-based speaker variability embedding is trained in a supervised manner and requires massive labeled data. To address this issue, W. H. Kang and N. S. Kim proposed a novel technique to extract an i-vector-like feature based on the variational auto-encoder, which was trained in an unsupervised manner to obtain a latent variable, which was represented by a Gaussian mixture model distribution [10]. Another paper, authored by J. Li, X. Zhang, M. Sun, X. Zou and C. Zheng, introduced attention-based long short-term memory (LSTM) to extract representative frames for spoofing detection in noisy environments [11]. Since the selection and weighting of features can improve the discrimination [12,13], the specific and representative frame-level features were automatically selected by adjusting their weights in the framework of attention-based LSTM.

2.4. Other Modalities

In addition to the above three biometric modalities, the remaining three papers are based on other biometric modalities, namely periocular, person re-identification and finger-vein. The paper, authored by L. C. O. Tiong, Y. Lee and A. B. J. Teoh, studied the periocular recognition under unconstrained environments and proposed a dual-stream CNN, which employed the Orthogonal Combination-Local Binary Coded Pattern (OCLBCP) as a color-based texture descriptor [14]. Their network aggregated the RGB image and OCLBCP by using two distinct late-fusion layers. Another paper by Y. Liu, H. Yang and Q. Zhao, focused on the misalignment problem in person re-identification [15]. They presented a two-branch deep joint learning network, where the local branch generated misalignment robust representations by pooling the features around the body parts, while the global branch generated representations from a holistic view. A hierarchical feature aggregation mechanism aggregated different levels of visual patterns assigned learned optimal weights within body part regions. The third paper, authored by H. Qin and P. Wang, proposed an approach to extract robust finger-vein patterns for verification and a supervised learning scheme for vein pattern encoding [16]. Stacked CNN (SCNN) and LSTM were utilized to predict the probability of a vein pixel belonging to a vein pattern.

The accepted papers contain the latest scientific research progress and remarkable achievements, which have important reference significance and values for the research in the fields of biometric recognition, deep learning and computer vision.

3. Technical Challenges and Future Development Trends

Although deep learning methods commonly outperform traditional handcrafted methods for biometric recognition, there are still several technical challenges and open problems, including the availability of high-quality labeled training samples, high computation and storage cost, hardware requirements, poor portability, complicated model design, low interpretability, etc.

Thus, the efforts of solving the aforementioned challenges compose the future trends of deep learning. Many new types of machine learning problems, such as weakly-supervised, semi-supervised and self-supervised learning have been explored to reduce the dependence on the labelled training samples. Compression technologies, such as pruning, quantization and knowledge distillation are employed to reduce the computation/storage cost. Light-weight deep learning models such as MobileNet, ShuffleNet are developed for mobile environments or to improve portability. In addition, many researchers are trying to improve the interpretability of deep learning models.

To sum up, deep learning technologies will definitely play an increasingly important role in biometric recognition with their rapid development.

Author Contributions: Writing—original draft preparation, L.L.; writing—review and editing, A.B.J.T. Both authors have read and agreed to the published version of the manuscript.

Funding: This work was supported by the National Research Foundation of Korea (NRF) grant funded by the Korea government (MSIP) (NO. NRF-2019R1A2C1003306) and National Natural Science Foundation of China (61866028).

Conflicts of Interest: The authors declare no conflict of interest.

References

1. Jain, A.K.; Nandakumar, K.; Ross, A. 50 years of biometric research: Accomplishments, challenges, and opportunities. *Pattern Recognit. Lett.* **2016**, *79*, 80–105. [CrossRef]
2. LeCun, Y.; Bengio, Y.; Hinton, G. Deep learning. *Nature* **2015**, *521*, 436–444. [CrossRef] [PubMed]
3. Tu, H.; Duoji, G.; Zhao, Q.; Wu, S. Improved Single Sample Per Person Face Recognition via Enriching Intra-Variation and Invariant Features. *Appl. Sci.* **2020**, *10*, 601. [CrossRef]
4. Zhou, S.; Chen, C.; Han, G.; Hou, X. Double Additive Margin Softmax Loss for Face Recognition. *Appl. Sci.* **2019**, *10*, 60. [CrossRef]
5. Ríos-Sánchez, B.; Costa-Da-Silva, D.; Martín-Yuste, N.; Avila, M.D.C.S. Deep Learning for Facial Recognition on Single Sample per Person Scenarios with Varied Capturing Conditions. *Appl. Sci.* **2019**, *9*, 5474. [CrossRef]
6. Yang, Z.; Li, J.; Min, W.; Wang, Q. Real-Time Pre-Identification and Cascaded Detection for Tiny Faces. *Appl. Sci.* **2019**, *9*, 4344. [CrossRef]
7. Ihsanto, E.; Ramli, K.; Sudiana, D.; Gunawan, T.S. Fast and Accurate Algorithm for ECG Authentication Using Residual Depthwise Separable Convolutional Neural Networks. *Appl. Sci.* **2020**, *10*, 3304. [CrossRef]
8. Li, F.; Li, X.; Wang, F.; Zhang, D.; Xia, Y.; He, F. A Novel P300 Classification Algorithm Based on a Principal Component Analysis Convolutional Neural Network. *Appl. Sci.* **2020**, *10*, 1546. [CrossRef]
9. Wang, D.; Si, Y.; Yang, W.; Zhang, G.; Liu, T. A Novel Heart Rate Robust Method for Short-Term Electrocardiogram Biometric Identification. *Appl. Sci.* **2019**, *9*, 201. [CrossRef]
10. Kang, W.H.; Kim, N.S. Unsupervised Learning of Total Variability Embedding for Speaker Verification with Random Digit Strings. *Appl. Sci.* **2019**, *9*, 1597. [CrossRef]
11. Li, J.; Zhang, X.; Sun, M.; Zou, X.; Zheng, C. Attention-Based LSTM Algorithm for Audio Replay Detection in Noisy Environments. *Appl. Sci.* **2019**, *9*, 1539. [CrossRef]
12. Leng, L.; Zhang, J.; Khan, M.K.; Chen, X.; Alghathbar, K. Dynamic weighted discrimination power analysis: A novel approach for face and palmprint recognition in DCT domain International. *J. Phys. Sci.* **2010**, *5*, 2543–2554.
13. Leng, L.; Li, M.; Kim, C.; Bi, X. Dual-source discrimination power analysis for multi-instance contactless palmprint recognition. *Multimed. Tools Appl.* **2015**, *76*, 333–354. [CrossRef]
14. Tiong, L.C.O.; Lee, Y.; Teoh, A.B.J. Periocular Recognition in the Wild: Implementation of RGB-OCLBCP Dual-Stream CNN. *Appl. Sci.* **2019**, *9*, 2709. [CrossRef]

15. Liu, Y.; Yang, H.; Zhao, Q. Hierarchical Feature Aggregation from Body Parts for Misalignment Robust Person Re-Identification. *Appl. Sci.* **2019**, *9*, 2255. [CrossRef]
16. Qin, H.; Wang, P. Finger-Vein Verification Based on LSTM Recurrent Neural Networks. *Appl. Sci.* **2019**, *9*, 1687. [CrossRef]

 *applied sciences*

Article

Improved Single Sample Per Person Face Recognition via Enriching Intra-Variation and Invariant Features

Huan Tu [1], Gesang Duoji [2,*], Qijun Zhao [1,2,*] and Shuang Wu [1]

[1] College of Computer Science, Sichuan University, Chengdu 610065, China; tuhuan722@outlook.com (H.T.); ws981117@gmail.com (S.W.)
[2] School of Information Science and Technology, Tibet University, Lhasa 850000, China
* Correspondence: gsdj@utibet.edu.cn (G.D.); qjzhao@scu.edu.cn (Q.Z.)

Received date: 8 December 2019; Accepted date: 3 January 2020; Published: 14 January 2020

Abstract: Face recognition using a single sample per person is a challenging problem in computer vision. In this scenario, due to the lack of training samples, it is difficult to distinguish between inter-class variations caused by identity and intra-class variations caused by external factors such as illumination, pose, etc. To address this problem, we propose a scheme to improve the recognition rate by both generating additional samples to enrich the intra-variation and eliminating external factors to extract invariant features. Firstly, a 3D face modeling module is proposed to recover the intrinsic properties of the input image, i.e., 3D face shape and albedo. To obtain the complete albedo, we come up with an end-to-end network to estimate the full albedo UV map from incomplete textures. The obtained albedo UV map not only eliminates the influence of the illumination, pose, and expression, but also retains the identity information. With the help of the recovered intrinsic properties, we then generate images under various illuminations, expressions, and poses. Finally, the albedo and the generated images are used to assist single sample per person face recognition. The experimental results on Face Recognition Technology (FERET), Labeled Faces in the Wild (LFW), Celebrities in Frontal-Profile (CFP) and other face databases demonstrate the effectiveness of the proposed method.

Keywords: face recognition; single sample per person; sample enriching; intrinsic decomposition

1. Introduction

Face recognition has been an active topic and attracted extensive attention due to its wide potential applications in many areas [1–3]. There are multiple modalities of face data that can be used in face recognition, such as near infrared images, depth images, Red Green Blue (RGB) images, etc. Compared with near infrared and depth images [4], RGB images include more information and have broader application scenarios. In the past decades, many RGB-based face recognition methods have been proposed and great progress has been made, especially with the development of deep learning [5–8]. However, there are still many problems to be solved. Face recognition with single sample per person, i.e., SSPP FR, proposed in 1995 by Beymer and Poggio [9], is one of the most important issues. In SSPP FR, there is only one training sample per person but various testing samples with appearance different from training samples. This situation could appear in many actual scenarios such as criminal tracing, ID card identification, video surveillance, etc. In SSPP FR, the limited training samples provide insufficient information of intra-class variations, which significantly decreases the performance of most existing face recognition methods. Tan et al. [10] showed that the performance of face recognition drops with the decreasing number of training samples per person and a 30% drop of recognition rate happens when only one training sample per person is available. In recent years, many methods have been suggested to solve the SSPP FR problem. These methods can be roughly divided into three categories: robust feature extraction, generic learning, and synthetic face generation.

Algorithms in the first category extract features that are robust to various variations. Some of them extract more discriminative features from single samples based on variants of improved principal component analysis (PCA) [11–13]. Others focus on capturing multiple face representations [14–16], mostly by dividing face image into a set of patches and applying various feature extraction techniques to get face representations. For instance, Lu et al. [14] proposed a novel discriminative multi-manifold analysis (DMMA) method to learn features from patches. They constructed a manifold from patches for every individual and formulated face recognition as a manifold–manifold matching problem to identify the unlabeled subjects. Dadi et al. [17] proposed to represent human faces by Histogram of Oriented Gradients (HOG), which captures edge or gradient structure and is invariant to local geometric and photometric transformations [18]. Local Binary Pattern (LBP) texture features extractor proposed by Ahonen et al. [19] has also been explored for face recognition thanks to its computational efficiency and invariance to monotonic gray-level changes. With the development of deep learning, there are many other methods that utilize the deep learning ability to extract more robust features, such as VGGNet [20], GoogleNet [21], FaceNet [22], ResNet [23], and SENet [24].

Generic learning attempts to utilize a generic set, in which each person has more than one training samples, to enhance the generalization ability of model. An implicit assumption of this kind of algorithms is that the intra-class variations for different datasets are similar and can be employed to share useful information to learn more robust model. The idea of sharing information has been widely used in [25–29] and achieved promising results. Sparse-representation-based classification (SRC) [30] is often used for face recognition, but its performance depends on adequate samples for each subject. Deng et al. [25] extended SRC framework by constructing an intra-class variation dictionary from generic training set together with gallery dictionary to recognize query samples. A sparse variation dictionary learning (SVDL) technique was introduced by Yang et al. [27], which learns a projection from both gallery and generic set and rebuilds a sparse dictionary to perform SSPP FR.

For the last category, some researchers synthesize some samples for each individual from the single sample to compensate the limited intra-class variations [31–37]. Mohammadzade and Hatzinakos [32] constructed expression subspaces and used them to synthesize new expression images. Cuculo et al. [36] extracted features from images that are augmented by standard augmentation techniques, such as cropping, translation, and filtering, and then applied sparsity-driven sub-dictionary learning and k-LIMAPS for face identification. To solve the lighting effect, Choi et al. [37] proposed a coupled bilinear model that generates virtual images under various illuminations using a single input image, and learned feature space based on these synthesized images to recognize a face image. Zeng et al. [33] proposed an expanding sample method based on traditional approach and used the expanded training samples to fine-tune a well-trained deep convolutional neural network model. 3D face morphable model (3DMM) is widely applied to face modeling and face image synthesis [35,38–40]. Zhu et al. [38] fitted 3DMM to face images via cascaded convolutional neural networks (CNN) and generated new images across large poses, which compose a new augmented database, namely 300W-LP. Feng et al. [39] presented a supervised cascaded collaborative regression (CCR) algorithm that exploits 3DMM-based synthesized faces for robust 2D facial landmark detection. SSPP-DAN introduced in [35] combines face synthesis and domain-adversarial network. It first generates synthetic images with varying poses using 3DMM and then eliminates the gap between source domain (synthetic data) and target domain (real data) by domain-adversarial network. Song et al. [40] explored the use of 3DMM in generating virtual training samples for pose-invariant CRC-based face classification.

The core idea of all the aforementioned methods is to train a model that can extract the identity features of face images, and ensure that the features are discriminative enough to find a suitable classification hyperplane that can accurately divide the features of different individuals. Unfortunately, many external factors, such as pose, facial expression, illumination, resolution, etc., heavily affect the appearance of facial images, and the lack of samples with different external factors leads to insufficient learning of feature space and inaccurate classification hyperplane. Being aware of these problems with existing methods, we propose a method to improve SSPP FR from two perspectives: *Normalization* and

Diversification. *Normalization* is to eliminate the external factors so as to extract robust and invariant features, which are helpful for defining more accurate classification hyperplane. *Diversification* means to enrich intra-variation through generating additional samples with various external factors. More diverse samples also enable the model to learn more discriminative features for distinguishing different individuals. To achieve this goal, a 3D face modeling module including 3D shape recovery and albedo recovery is presented at first. For the albedo recovery particularly, we make full use of the physical imaging principle and face symmetry to complete the invisible areas caused by self-occlusion while reserving the identity information. Since we represent albedo in the form of UV map, which is theoretically invariant to pose, illumination and expression (PIE) variations, we can alleviate the influence of these external factors. Based on the recovered shape and albedo, additional face images with varying pose, illumination, and expression are generated to increase intra-variation. Finally, we are able to improve the SSPP face recognition accuracy thanks to the enriched intra-variation and invariant features.

The remaining parts of this paper are organized as follows. Section 2 reviews face recognition with single sample per person and inverse rendering. Section 3 presents the detail of the proposed method. Section 4 reports our experiments and results. Section 5 provides the conclusion of this paper.

2. Related Work

2.1. Face Recognition with Single Sample Per Person

With the unremitting efforts of scholars, face recognition has made great progress. However, the task becomes much more challenging when only one sample per person is available for training the face recognition model. Dadi et al. [17] extracted histogram of oriented gradients (HOG) features and employ support vector machine (SVM) for classification. Li et al. [41] combined Gabor wavelets and feature space transformation (FST) based on fusion feature matrix. They projected the combined features to a low-dimensional subspace and used nearest neighbor classifier (NNc) to complete classification. Pan et al. [42] proposed a locality preserving projection (LPP) feature transfer based algorithm to learn a feature transfer matrix to map source faces and target faces into a common subspace.

In addition to the traditional methods introduced above, there are many other methods that utilize the learning ability of deep learning to extract features. To make up for the lack of data in SSPP FP, some algorithms combine deep learning and sample expanding. In [34], a generalized deep autoencoder (GDA) is firstly trained to generate intra-class variations, and is then separately fine-tuned by the single sample of each subject to learn class-specific DA (CDA). The new samples to be recognized are reconstructed by corresponding CDA so as to complete classification task. Similarly, Zeng et al. [33] used a traditional approach to learn an intra-class variation set and added the variation to single sample to expand the dataset. Then, they fine-tuned a well-trained network using the extended samples. Sample expanding can be done not only in the image space but also in the feature space. Min et al. [43] proposed a sample expansion algorithm in feature space called k class feature transfer (KCFT). Inspired by the fact that similar faces have similar intra-class variations, they trained a deep convolutional neural network on a common multi-sample face dataset at first and extracted features for the training set and a generic set. Then, k classes with similar features in the generic set are selected for each training sample, and the intra-variation of the selected generic data is transferred to the training sample in the feature space. Finally, the expanded features are used to train the last layer of SoftMax classifier.

Unlike these existing methods, this paper simultaneously recovers intrinsic attributes and generates diversified samples. Compared with sample expanding methods mentioned above, our method uses face modeling to decompose intrinsic property and generates images with richer intra-variation via simulating the face image formation process rather than following the idea of intra-variation migration. Our method also takes full advantage of intrinsic properties that can more robustly represent identity information. Deep learning is used as a feature extractor in our method due to its superiority demonstrated in many existing studies.

2.2. Inverse Rendering

The formation of face images is mainly affected by intrinsic face properties and external factors. Intrinsic properties consist of shape (geometry) and albedo (skin properties), while external factors include pose, illumination, expression, camera setting, etc. Inverse rendering refers to reversely decomposing internal and external properties in facial images. Many inverse rendering methods have been proposed. CNN3DMM [44] represents shape and albedo, respectively, as a linear combination of PCA bases and uses a CNN to regress the combination coefficients. SfSNet [45] mimics the process of imaging faces based on physical models and estimates the albedo, light coefficients, and normal of the input face image.

As one of the intrinsic properties, albedo has natural advantage for face recognition owning to its robustness to variations in view angle and illumination. However, most inverse rendering algorithms pay more attention to recovering a more accurate and detailed 3D face shape, and treat the albedo as an ancillary result. As one of the few algorithms using albedo to assist face recognition, Blanz and Vetter [46] captured the personal specific shape and albedo properties by fitting a morphable model of 3D faces to 2D images. The obtained model coefficients that are supposed to be independent of external factors can be used for face recognition. However, due to the limited representation ability of the statistical model, the recovered albedo would lose its discrimination to some extent. To solve this problem, Tu et al. [47] proposed to generate albedo images with frontal pose and neutral expression from face images of arbitrary view, expression, and illumination, and extract robust identity features from the obtained albedo images. They experimentally showed that albedo is beneficial to improving face recognition. However, they only realize the synthesis of normalized albedo images in two-dimensional image space, lacking the exploration on the principle of physical imaging, which leads to a poor performance on a cross-database.

3. Proposed Method

3.1. Overview

3.1.1. Preliminary

In this paper, densely aligned 3D face shapes are used, each containing n vertices. Generally, we denote an n-vertex 3D face shape as point cloud $S \in \mathbb{R}^{3 \times n}$, where each column represents the coordinates of a point. The face normal, represented as $N \in \mathbb{R}^{3 \times n}$, is calculated from the 3D face shape. The texture and albedo are denoted as $T, A \in \mathbb{R}^{3 \times n}$, where each column represents the color and reflectivity of a point on the face.

However, using only a collection of attributes of each point to represent S, N, T, and A misses information about the spatial adjacency between points. Inspired by position maps in [48], we denote albedo as a UV map: $UV_A \in \mathbb{R}^{256 \times 256 \times 3}$ (see Figure 1). Each point in A can find a unique corresponding pixel on UV_A. Different from the traditional UV unwrapping method, each pixel in our UV map will not correspond to multiple points in A. In addition, we also use UV_T and UV_N to represent facial texture and facial normal as UV maps.

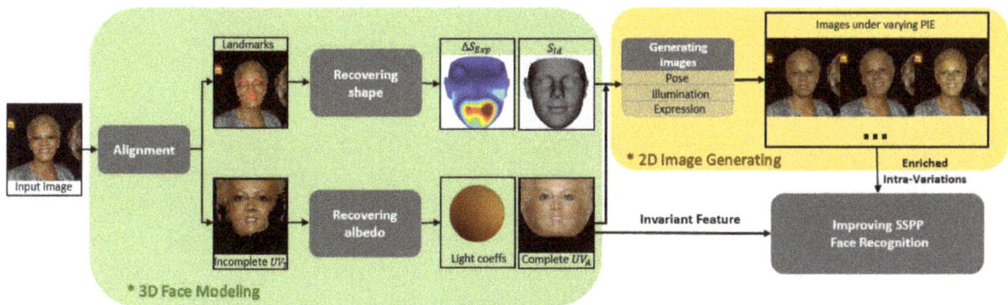

Figure 1. Pipeline of proposed method.

3.1.2. Pipeline

Figure 1 shows the framework of the proposed method for single sample per person face recognition. The method consists of three modules: 3D face modeling, 2D image generation, and improved SSPP FR. Given a face image of a person, we detect 68 landmarks, U, and generate the incomplete UV map of texture (Incomplete UV_T) using PRNet algorithm in [48] at first. We then recover its 3D face shape and complete UV map of albedo (Complete UV_A), respectively, from landmarks and Incomplete UV_T. With the recovered properties, images under varying pose, illumination, and expression are generated in the 2D image generation module. Finally, in the improved SSPP FR module, the reconstructed Complete UV_A and generated images are used to assist SSPP face recognition. Next, we detail: (i) albedo recovery; (ii) shape recovery; (iii) data enrichment; and (iv) SSPP FR.

3.2. Albedo Recovery

3.2.1. Network Structure

We assume that the face is Lambertian and illuminated from the distant. Under the Lambertian assumption, we represent the lighting and reflectance model as second-order Spherical Harmonics (SH) [49,50], which is a natural extension of the Fourier representation to spherical function. In SH, the irradiance at a surface point with normal (n_x, n_y, n_z) is given by

$$B(n_x, n_y, n_z | \Theta^{sh}) = \sum_{k=1}^{b^2} \Theta_k^{sh} H_k(n_x, n_y, n_z), \tag{1}$$

where H_k are the $b^2 = 3^2 = 9$ SH basis functions, and Θ_k^{sh} is the corresponding kth illumination coefficient. Since we consider colored illumination, there are totally $3 \times 9 = 27$ illumination coefficients with nine coefficients for each of the R, G, and B channels. The texture of a surface point can be calculated by multiplying the irradiance and albedo of the point. To sum up, the texture under certain illumination is a function of normal, albedo, and illumination, and can be expressed as

$$\begin{aligned} T(p) &= f_{sh}(A(p), N(p), SHL), \\ UV_T(p) &= f_{sh}(UV_A(p), UV_N(p), SHL), \end{aligned} \tag{2}$$

where p represents a pixel (2D) or point (3D), and SHL denotes the SH illumination coefficients.

Inspired by Sengupta et al. [45], we propose an end-to-end network that can recover the missing part in the incomplete UV_T and generate its complete version UV_A. As can be seen in Figure 2, we concatenate the incomplete UV_T with its horizontally flipped image as input of the network. The proposed network follows an encoder–decoder structure, in which the encoder module extracts a common feature from input image, and the albedo decoder and the normal decoder decode the complete albedo UV_A and the complete normal UV_N from the common feature, respectively, and the light decoder computes the spherical harmonics illumination coefficients SHL from the concatenation of common feature, albedo feature, and normal feature. At last, following Equation (2), a rendering layer is used to recover the texture based on the above decoded attributes.

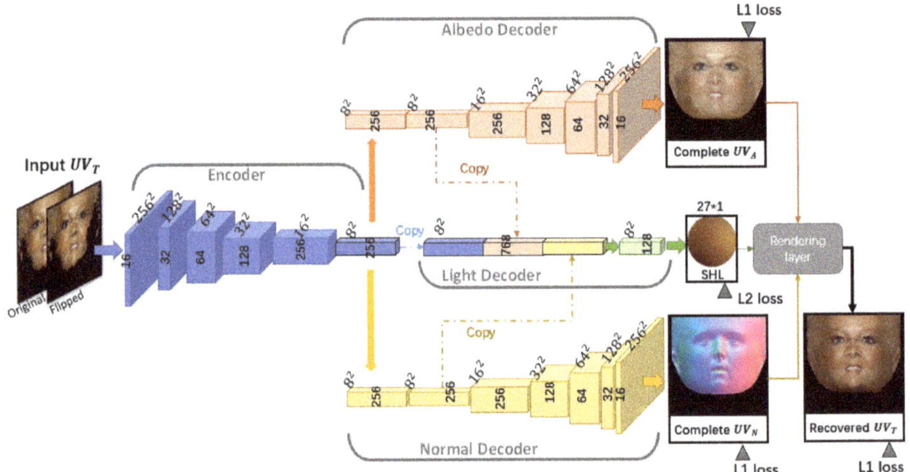

Figure 2. Pipeline of albedo recovery.

3.2.2. Loss Functions

To train the albedo recovery model, we minimize the error between the reconstructed value and the ground truth. However, the ground truth of unseen regions in real face is unavailable. To address the issue, we flip the reconstructed texture horizontally and make it as similar as possible to the input texture image. The loss function for reconstructed texture is defined as

$$
\begin{aligned}
L_{recon} &= \tfrac{1}{t} \sum_{p} \left(|UV_M[p] \left(UV_T^*[p] - U\hat{V}_T[p] \right) \right) \\
&+ \tfrac{\lambda_f}{t} \sum_{p} \left(|UV_M[p] \left(UV_T^*[p] - U\hat{V}_{T_{flip}}[p] \right) \right),
\end{aligned}
\tag{3}
$$

where UV_M is the visibility mask, $[p]$ denotes the pixel spatial location, t is the number of visible pixels, $U\hat{V}_{T_{flip}}$ means horizontally flipping the reconstructed texture $U\hat{V}_T$, λ_f denotes the weight of the reconstruction loss component associated with the horizontally-flipped reconstructed texture with respect to that associated with the original reconstructed texture, and the symbols "*" and "ˆ" indicate the ground truth and reconstructed values, respectively.

The loss functions for SH illumination coefficients, albedo, and normal are formulated, respectively, as

$$
L_l = \| SHL^* - S\hat{H}L \|_2^2,
\tag{4}
$$

$$
L_a = |UV_A^* - U\hat{V}_A|,
\tag{5}
$$

$$
L_n = |UV_N^* - U\hat{V}_N|.
\tag{6}
$$

The total loss is defined as

$$
L_{total} = \lambda_{recon} L_{recon} + \lambda_l L_l + \lambda_a L_a + \lambda_n L_n,
\tag{7}
$$

where λ_{recon}, λ_l, λ_a, and λ_n are the weights to balance different losses.

3.2.3. Implementation Details

Since the ground truth of albedo, normal, and light coefficients of real facial images is not available, synthetic data are firstly used in this paper. Following the definition of 3DMM [46], the shape and the albedo of a face can be represented as

$$S = \bar{S} + \alpha_{id}\, b_{id}^s + \alpha_{exp}\, b_{exp}^s, \tag{8}$$

$$A = \bar{A} + \beta_{id}\, b_{id}^a, \tag{9}$$

where S and A are 3D face shape and albedo; $\bar{S}$ is the mean shape; $\bar{A}$ is the mean albedo; b_{id}^s and b_{id}^a are, respectively, the identity-related shape bases and albedo bases; b_{id}^s is the expression-related shape bases; and α_{id}, α_{exp}, and β_{id} are corresponding shape parameters, expression parameters, and albedo parameters.

Bar et al. [51] provided a method to flatten a 3D shape into 2D embedding and re-sample a 3DMM (including $\bar{S}$, $\bar{A}$, b_{id}^s, b_{exp}^s, and b_{id}^a) from the initial 3DMM over a uniform grid of size $H \times W$ in this flattened space. The re-sampled 3DMM has $n = HW$ vertices and each vertex has an associated UV coordinate. In our work, b_{id}^s and b_{id}^a of initial 3DMM come from BFM 2009 [52] and b_{exp}^s comes from FaceWarehouse [53]. Using the method in [51], we obtain a re-sampled 3DMM with $n = 256^2 = 65{,}536$ vertices. Then, we remove the vertices in the neck area resulting in a 3D face of 43,867 vertices and UV map of 256×256 with 43,867 valid pixels. We use the re-sampled 3DMM to synthesize 2330 subjects with a total of 150,704 facial images.

In view of the difference in data distribution between synthetic images and real images, real training data are necessary. We used real data from the CelebA database [54], including a total of 202,599 facial images of 10,177 subjects, and adopted a two-step training strategy to solve the problem of unavailability of real data labels. (i) We first trained the network on synthetic data. The pre-trained network was applied to real data from CelebA to obtain their albedo, normal, and SH illumination coefficients, which were taken as "pseudo ground truth" of the real data. (ii) We then fine-tuned the network with the combined set of synthetic data with "golden ground truth" and real data with "pseudo ground truth". For both steps, we applied Equation (7) to supervise the learning process and set λ_{recon}, λ_l, λ_a, and λ_n as 1.0 and λ_f as 0.5. Optimization was done by Adaptive Moment Estimation (Adam) with mini-batch of 32 samples, and the exponential decay rate for the first-moment estimates and second-moment estimates were set as 0.5 and 0.999, respectively. When we trained the network with only synthetic data, the learning rate was initialized as 10^{-3} and then decreased by factor of 10 after every four epochs. When training the network with combination of synthetic and real data, we set the learning rate to 10^{-4}.

3.3. Shape Recovery

To recover the 3D face shape, motivated by the method in [55], we train a cascade coupled-regressor $\{R_{Id}^k\}_{k=1}^K$, $\{R_{Exp}^k\}_{k=1}^K$ to reconstruct the 3D shape based on the detected 2D landmarks U^* on the input image. Here, 3D face shape is disentangled as

$$S = \bar{S} + \Delta S_{Id} + \Delta S_{Exp}, \tag{10}$$

where $\bar{S}$ is the mean 3D shape of frontal pose and neutral expression, termed pose-and-express-normalized (PEN) 3D face shape, ΔS_{Id} is the difference between the subject's PEN 3D shape, and $\bar{S}$, ΔS_{Exp} is the deformation in S caused by expression.

Given a 3D face shape and the corresponding 3D landmark vertices D, its 2D landmarks U at the same pose as the input face is obtained through weak perspective projection M. Note that the 3D-to-2D projection matrix is computed via least squares such that the projection of the D is as close as possible to the ground truth landmarks U^*, i.e.,

$$M = ((D)^T D)^{-1}(D)^T U^*. \tag{11}$$

The core idea of this method is to assume that there is a "relationship" between the difference on two 3D face shapes and the difference on their corresponding 2D landmarks, which can be learned from training samples. Thus, the reconstruction can be described as:

$$\begin{aligned} S^k = S^{k-1} \ &+ R_{Id}^k(U^* - U^{k-1})) \\ &+ R_{Exp}^k(U^* - U^{k-1})), \end{aligned} \tag{12}$$

where S^k is the reconstructed 3D face shape after k iteration, $S^0 = \overline{S}$.

To train the coupled-regressors, we synthesized 1000 face images under various poses and expressions based on the resampled 3DMM. The ground truth ΔS_{Id} and ΔS_{Exp} of these images were recorded during the synthesis. In this study, we cascaded five pairs of identity and expression shape regressors, i.e., $K = 5$.

3.4. Data Enrichment

With the obtained identity shape $\overline{S} + \Delta S_{Id}$, in short S_{Id}, and albedo UV_A of a subject, face images of the subject with arbitrary poses, illuminations, and expressions could be generated. Firstly, we obtain shape S with the arbitrary expression via Equation (10), in which random ΔS_{Exp} is generated through the resampled 3DMM. Secondly, we add pose to S and obtained S_{pose}. The process of adding a pose can be expressed as:

$$S_{pose} = f * R * S + t_{3d}, \tag{13}$$

where R is the 3×3 rotation matrix calculated from the pitch, yaw, and roll angles; f is the scale factor; and t_{3d} is the 3×1 translation vector. Thirdly, we generate texture T with random illumination through Equation (2), where N is calculated by the obtained S_{pose}, A is constructed from UV_A, and SHL is randomly generated. Finally, S_{pose} and T are used to render the final image by orthographic projection matrix and z-buffer.

3.5. SSPP FR

In this section, we utilize the decomposed intrinsic facial properties (i.e., albedo) and the generated arbitrary face images to solve SSPP FR problem. This is done from two perspectives: (i) enriching the intra-variation of training samples with the generated arbitrary face images; and (ii) exploring additional invariant features extracted from albedo maps and fusing the match scores of different features.

3.5.1. Face Recognition via Enriching Intra-Variation

In SSPP FR, performance of face matcher drops due to the deficiency of intra-variation. One implementation to improve SSPP face recognition is to enrich intra-variation in training set via 3D-modeling-based image generation. As shown in Figure 3a, 3D face modeling module is used to recover shapes and albedos of SSPP training set, based on which additional face images with varying pose, illumination, and expression are generated. Finally, the enlarged dataset of original and generated images is utilized to train a more discriminative face matcher.

3.5.2. Face Recognition via Enriching both Intra-Variation and Invariant Features

The 3D face modeling proposed in this paper not only makes it possible to enrich intra-variation, but also provides invariant features from its decomposed intrinsic attribute (i.e., albedo). Thus, we come up with another implementation of SSPP face recognition, which is to enrich both intra-variation and invariant features during test. Figure 3b shows the details. As can be seen, this implementation includes three matchers: original face matcher based on features of original input images in gallery and probe, albedo matcher based on invariant features extracted from decomposed albedo of gallery and probe, and enriched face matcher based on features of enriched gallery and the recovered probe. Note that the recovered probe refers to synthetic images with the same PIE and background. The purpose of the enriched face matcher is to exploit the gallery of enriched intra-variation to suppress false rejections. Each matcher computes match scores by calculating cosine similarity of features. The match scores of the three matchers are fused together by a weighted sum rule to obtain the final match score. The identity of a probe image is finally decided as the subject whose gallery sample has the highest final match score with the probe.

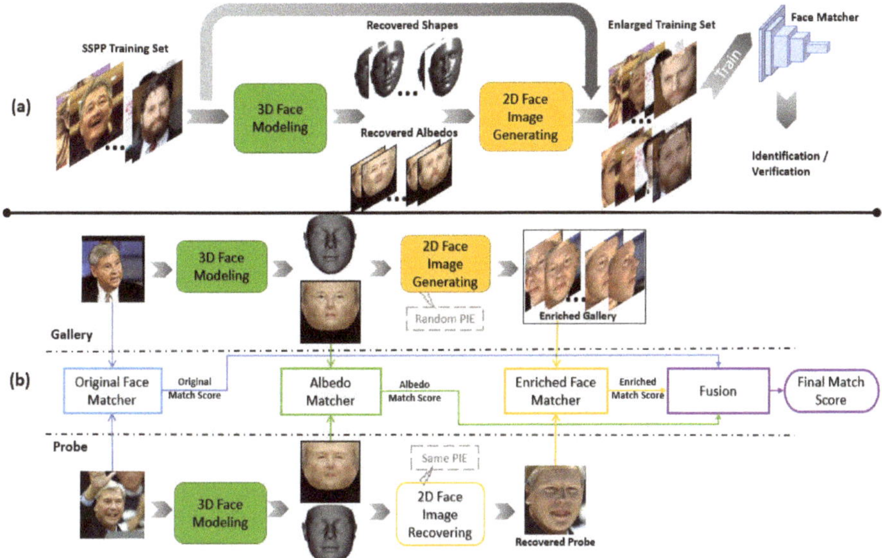

Figure 3. (**a**) Pipeline of SSPP face recognition via enriching intra-variation. (**b**) Pipeline of SSPP face recognition via enriching both intra-variation and invariant features.

3.5.3. Implementation Details

SE-Inception network [24] was employed in all the aforementioned matchers, and Stochastic Gradient Descent (SGD) with momentum coefficient of 0.9 and batch size of 64 was used to train the networks. The learning rate was initialized as 0.1, and gradually decreased by factor of 10 every 25 epochs. All face images were first aligned to the predefined template based on the five landmarks (i.e., left eye, right eye, tip of nose, left corner of mouth, and right corner of mouth) on them, and then cropped and resized to 112×112.

4. Experiments

We evaluated the effectiveness of the proposed method from two aspects: (i) visual inspection of the reconstructed albedo and shape, and the generated facial images accuracy as well; and (ii) single sample per person face recognition based on enriching intra-variation and invariant features.

4.1. Datasets and Protocols

Seven datasets were used in our experiments. Below are the details and evaluation protocols of the datasets.

FERET-b: FERET-b database [56], collected in a laboratory environment, contains 1400 images of 200 subjects with different pose, expression, and illumination. The SSPP FR experiments on this database were conducted in both verification and identification modes. In identification, we used the neutral and frontal image of each person as gallery and the remaining images as probe. In verification, we divided the database into 10 non-overlapping subsets. Each subset contained 120 positive pairs (i.e., pairs of images from the same person) and 120 negative pairs (i.e., pairs of images from different people).

LFW: Labeled Faces in the Wild (LFW) database [57] includes more than 13,000 images of 5749 different individuals taken under an unconstrained environment. The evaluation protocol suggests dividing the database into 10 non-overlapping subsets. Each subset contains 300 positive pairs and 300 negative pairs. LFW-a is a version of LFW after alignment using commercial software. For the sake of fair comparison, we selected the first 50 people who have more than 10 samples for evaluation according to the experimental setup described in [43]. We randomly selected one sample of each of the 50 subjects as gallery, and the remaining images were used as probe.

CPLFW: CPLFW [58], a renovation of LFW, constructs a cross-pose LFW database to evaluate the influence of pose variation in face recognition. It provides 10 disjoint subsets of image pairs for face verification. Each subset contains 300 positive pairs and 300 negative pairs.

CALFW: Cross-Age LFW (CALFW) database [59], similar to the CPLFW database, is a renovation of LFW. It emphasizes aging effect in addition to other variations (pose, illumination, etc.) in face recognition. The dataset is separated into 10 non-repeating subsets of image pairs for face verification, each subset containing 300 positive pairs and 300 negative pairs.

AgeDB: AgeDB [60] contains 16,488 images of various famous people, such as actors/actresses, writers, scientists, politicians, etc. There are 568 distinct subjects in AgeDB. AgeDB provides 10 folds of image pairs, with each fold consisting of 300 positive pairs and 300 negative pairs.

CFP: CFP database [61], a challenging dataset to examine the problem of frontal to profile face verification, collects 7000 images of 500 subjects with each subject having 10 frontal and 4 profile face images. Its evaluation protocol defines two separate experiments of Frontal–Profile (CFP_FP) and Frontal–Frontal (CFP_FF) face verification, and divides in each experiment the whole dataset into 10 folds each containing 350 positive pairs and 350 negative pairs.

VGGFace2-FP: The VGGFacce2 database [62] contains 3.31 million images of 9131 subjects. The dataset is divided into training set and evaluation set, in which the training set contains 8631 subjects and the evaluation set contains 500 subjects. Evaluation scenarios can be divided into two categories by pose and age. We considered the scenario of face matching across different poses, i.e., VGGFace2-FP. The evaluation data are divided into 10 folds of image pairs, with each fold consisting of 250 positive pairs and 250 negative pairs.

4.2. Visualization of Reconstructed Image Components and Generated Images

Figure 4a shows recovered 3D face shapes and complete UV_A for four input face images in LFW. It can be found that our method can not only reconstruct the albedo of the visible area but also recover the albedo of missing part. We also compared our 3D face modeling method with two existing methods [44,45]. Figure 4b shows the comparison results for two input images. As can be seen, the 3D face shape and albedo map reconstructed by Sengupta et al. [45] are incomplete and cannot be used for data augmentation, and the results recovered by Tran et al. [44] are poor in reserving identity-related appearance feature due to the limited representation capacity of PCA bases. Figure 4c shows some generated face images of varying illumination (Columns 3 and 8), pose (Columns 4 and 9), and expression (Columns 5 and 10) of four different subjects.

We also show the ability of the 3D modeling approach to address illumination and pose factors in Figure 5. As can be seen, although expressions, illuminations, and poses have influence on the appearance of input images, our method can effectively eliminate the external influence due to the robustness of the albedo image to lighting as well as to the rigorous alignment of the face in the UV diagram representation.

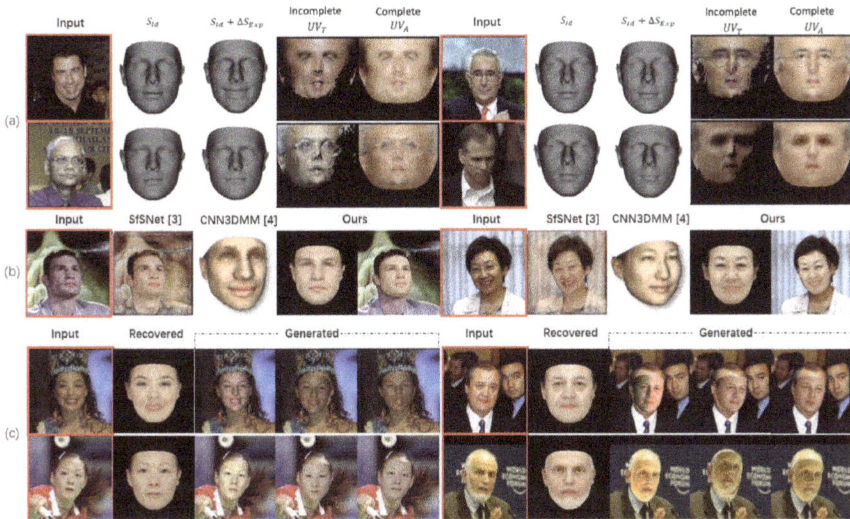

Figure 4. (**a**) Recovered 3D face shapes and albedo UV maps for two input face images by our 3D face modeling module. (**b**) Comparison between the face images reconstructed by our method and two existing methods. (**c**) Example enriched face images of varying illumination (Columns 3 and 8), pose (Columns 4 and 9), and expression (Columns 5 and 10) of four different subjects by our method.

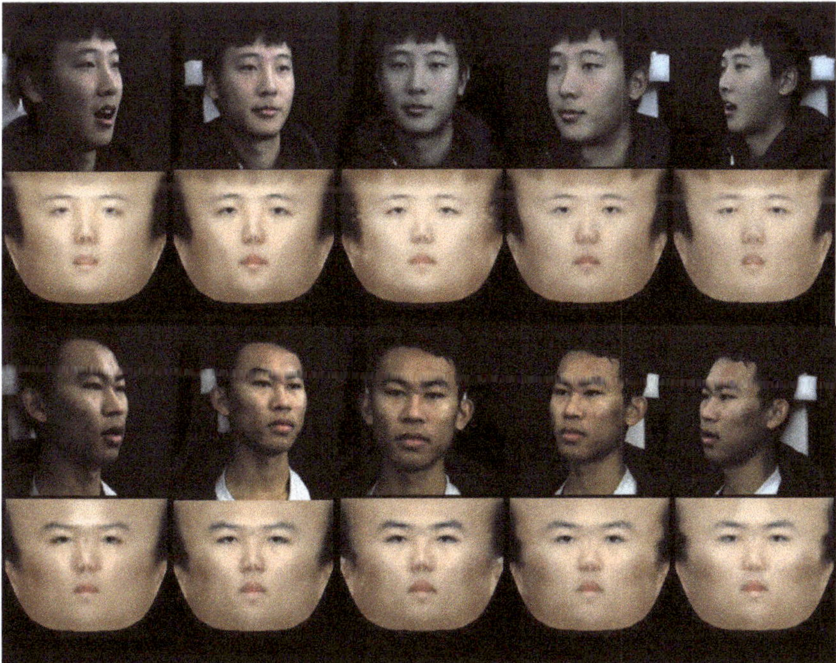

Figure 5. Albedo UV maps recovered by our 3D face modeling module for two subjects in the database constructed by ourselves with different expressions, illuminations, and poses.

4.3. Effectiveness of Enriching Intra-Variation

To validate the effectiveness of our proposed method of enriching intra-variation via 3D-modeling-based image generation, we constructed twelve groups of training data as follows.

- We randomly chose one sample per person in CelebA dataset [54] to form the original SSPP training data (denoted as *Original*). Note that face images in CelebA are in the wild images with varying PIE as well. However, their pose angles are relatively small (mostly within 30 degrees).
- The 3D modeling module was used to recover the intrinsic properties (albedo and shape) and external factors (i.e., PIE) of the face images in *Original*. We then synthesized images with the same PIE and background as the images in *Original* to form another training set, named *Synthetic*.
- With the recovered properties of the images in *Original*, we further generated another seven face images for each person with varying poses, varying illuminations, varying expressions, varying poses and illuminations, or varying poses, illuminations, and expressions. We call the respective databases of generated images as *AugP*, *AugI*, *AugE*, *AugPI*, and *AugPIE*, respectively.
- By combining the above-generated images with the original images, we obtained another five training sets, denoted as *Ori + AugP*, *Ori + AugI*, *Ori + AugE*, *Ori + AugPI*, and *Ori + AugPIE*, respectively.

We trained the implementation of our proposed method with enriching intra-variation (see Figure 3a) by using the above training sets and evaluated its face verification performance on seven of the benchmark datasets, i.e., LFW [57], CPLFW [58], CALFW [59], AgeDB [60], CFP [61], VGGFace2-FP [62], and FERET-b [56]. During the test, the features extracted from the images by the network were used to judge whether two images are from same persons based on the cosine similarity between their features. Here, ten-fold cross validation was employed. Specifically, we divided the dataset into ten subsets, nine of which were used as the training set and the remaining one used as the test set.

The results are reported in Table 1. By comparing the recognition rate obtained by training the network on *Original* and that obtained by training the network on *Synthetic*, it can be found that they are comparable. This indicates that the shape and albedo obtained by using our proposed 3D modeling method can effectively capture the identity information. When the network was trained with the generated data only, the recognition accuracies on most of the datasets were improved. Moreover, the face recognition accuracy was consistently improved when training the network with the combined set of original and generated images. It is worth mentioning that our proposed method of enriching intra-variation improved the recognition accuracy by 8.07% (= 58.82% − 50.72%) and 13.02% (= 82.69% − 69.67%) on the challenging AgeDB and CFP-FF test sets, respectively. It can be found that pose factor is an important factor that affects the verification rate by comparing the results of AugP, AugI, and AugE or the results of Ori+AugP, Ori+AugI, and Ori+AugE. While the results of using merely the augmented data (i.e., *AugP*, *AugI*, *AugE*, *AugPI*, and *AugPIE*) show consistent improvement when more variations are taken into consideration, combining the original and augmented data of expression variations, however, obtained only marginal improvement in some of the test datasets. This is probably because the original data from the CelebA database already have some expression variations and play a major role in the training, and consequently the contribution of augmented expression data becomes marginal.

Table 1. Face verification rates (%) of the first implementation of our proposed method on different benchmark datasets when different enriched face images from the CelebA database were used for training. The best result on each dataset is shown in bold.

TrainData\TestData	AgeDB	LFW	CALFW	CPLFW	CFP_FF	CFP_FP	VGGFACE2_FP	FERET-b
Original	50.75	77.33	63.03	58.7	69.67	60.77	61.46	84.33
Synthetic	53.35	73.77	59.05	57.52	69.09	59.74	60.28	79.96
AugP	52.12	75.95	58.82	60.08	71.46	61.89	61.66	87.54
AugI	50.28	75.45	59.27	57.58	71.7	60.87	61.46	80.17
AugE	50.43	75.05	57.47	58.10	71.09	62.49	61.44	80.04
AugPI	52.47	78.17	60.67	60.83	73.57	62.70	62.80	90.46
AugPIE	55.18	80.08	62.78	61.90	76.59	64.37	62.94	92.08
Ori+AugP	56.62	82.73	69.05	65.68	78.13	66.64	68.54	90.71
Ori+AugI	52.63	77.48	60.93	58.63	70.83	62.07	62.84	81.17
Ori+AugE	50.93	74.85	59.02	57.53	68.36	60.51	60.54	80.20
Ori+AugPI	**58.82**	**86.08**	**72.03**	**67.97**	81.96	**70.44**	**71.02**	94.54
Ori+AugPIE	58.65	85.98	71.02	67.15	**82.69**	70.23	70.22	**95.54**

4.4. Effectiveness of Enriching Both Intra-Variation and Invariant Features

In this experiment, we compared another implementation of our proposed method with both enriched intra-variation and enriched invariant features (see Figure 3b) with the following methods: (i) traditional methods, namely HOG+SVM [17], G-FST [41], and FT-LPP [42]; and (ii) deep learning methods, namely FaceNet [22], CDA [34], TDL [33], and KCFT [43]. For fair comparison, we used multiple samples per person database, CelebA, as the generic data to pretrain three of the matchers. To train albedo matcher and enriched face matcher, we generated the albedo image and the recovered image (synthetic image with same PIE) of CelebA by the proposed 3D face modeling module and 2D face generation module. For our proposed method, the scores from three face matchers were fused by a weighted sum rule. We set the weights for different matchers with respect to their respective recognition accuracy. Specifically, the weights were 0.45, 0.45, and 0.1 for the original matcher (O), albedo matcher (A), and enriched matcher (E), respectively, when fusing the three matchers. In the ablation study, the weights for O and A were both 0.5 when they were fused, while the weights were 0.8 and 0.2 for O and E when O and E were fused.

4.4.1. Results on FERET-b Dataset

We generated six additional images with varying pose, illumination, and expression for each subject in the gallery. Table 2 shows the recognition rates of the proposed method and the counterpart methods. Obviously, our method achieved the best recognition rates.

Table 2. Rank-1 identification rates (%) of different methods on FERET-b and LFW-a. The best result on each database is shown in bold.

Database\Method	HOG+SVM	CDA	G-FST	FT-LPP	TDL	FaceNet	KCFT	Proposed
FERET-b	50.17	77.25	82.08	86.08	89.33	91.42	93.17	**96.41**
LFW-a	40.98	51.12	57.95	62.49	74.01	89.04	**98.83**	97.25

4.4.2. Results on LFW-a Dataset

We generated 20 facial images for each person in the galley of LFW-a. The recognition rates of different methods are shown in Table 2. As can be seen, our method overwhelmedmost of the counterpart methods, and was among the top two methods.

4.4.3. Ablation Study

We further evaluated the contribution of the enriched invariant features and the enriched images on improving SSPP face recognition performance. The comparison results on FERET-b and LFW-a are shown in Table 3. It can be found that the recognition accuracy improved after we fused the match scores of the original images (denoted by O) with the match scores of enriched invariant features extracted from albedo (denoted by A) or the match scores of features extracted from the enriched images (denoted by E), even though the original images already set up relatively high baselines. Figure 6 shows the false rejection rates at different thresholds when fusing different match scores on the FERET-b and LFW-a databases. As can be seen, our proposed matchers A and E significantly reduced the FRR of the original matcher O. It is worth mentioning that both A and E helped in reducing FRR. However, according to the experimental results, A (using albedo-based features) was more effective than E (using augmented images).

Table 3. Rank-1 identification rates (%) of the proposed method when fusing the match scores of the original images (denoted by O) and the match scores of the enriched invariant features (denoted by A) or the match scores of the enriched images (denoted by E) on the FERET-b and LFW-a databases.

O	A	E	FERET-b	LFW-a
✓			92.25	92.46
✓	✓		96.33	97.15
✓		✓	93.17	93.53
✓	✓	✓	**96.41**	**97.25**

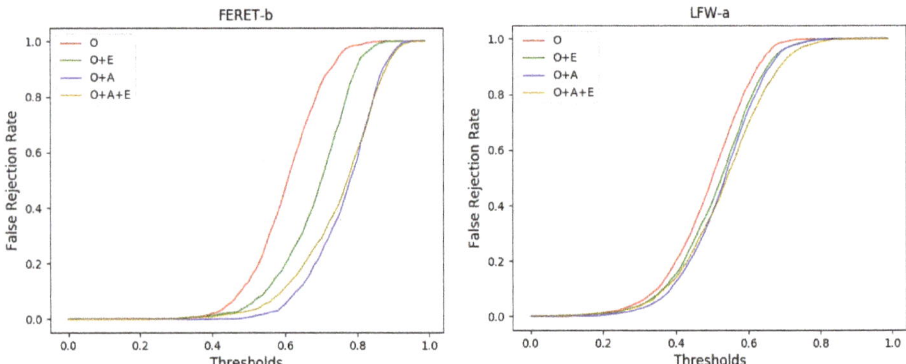

Figure 6. False rejection rates at different thresholds when fusing different match scores on the FERET-b and LFW-a databases.

5. Conclusions

A novel single sample per person face recognition algorithm based on enriching data intra-variation and invariant features is proposed in this paper. The method consists of three modules: 3D face modeling, 2D image generating, and improved SSPP FR. A novel end-to-end network is employed to recover complete albedo from the input image, which not only provides additional invariant identity features, but also can be used with the restored 3D shape to generate images containing richer intra-class variations. While existing SSPP FR methods focus either on generating synthetic images (CDA, KCFT, etc.) or on learning robust features (HOG+SVM, FaceNet, etc.), we improve the SSPP FR accuracy from the perspective of enriching both intra-variation and invariant features. Experiments were performed on multiple databases. The results show that, by using the synthetic images generated by our proposed method to augment the train data, the SSPP FR accuracy is improved significantly by up to 13%. Moreover, using our proposed enriched invariant features boosts the rank 1 identification rate from 92.25% to 96.33% on FERET-b and from 92.46% to 97.15% on LFW-a. In the future, we are going to further improve the modeling and synthesis procedures, for example, by considering more elaborate losses and by integrating the modeling, synthesis, and identification modules in a more unified manner.

Author Contributions: Conceptualization, H.T., G.D., Q.Z., and S.W.; Data curation, H.T.; Formal analysis, H.T.; Funding acquisition, Q.Z.; Investigation, H.T.; Methodology, H.T.; Resources, Q.Z.; Supervision, G.D. and Q.Z.; Validation, H.T. and S.W.; Visualization, H.T.; Writing—original draft, H.T.; and Writing—review and editing, Q.Z. All authors have read and agreed to the published version of the manuscript.

Funding: This work was supported by the National Key Research and Development Program of China (No. 2017YFB0802300) and the National Natural Science Foundation of China (Nos. 61773270, 61971005 and 61961038).

Conflicts of Interest: The authors declare no conflict of interest.

References

1. Abate, A.F.; Nappi, M.; Riccio, D.; Sabatino, G. 2D and 3D face recognition: A survey. *Pattern Recognit. Lett.* **2007**, *28*, 1885–1906. [CrossRef]
2. Hassaballah, M.; Aly, S. Face recognition: Challenges, achievements and future directions. *IET Comput. Vis.* **2015**, *9*, 614–626. [CrossRef]
3. Wang, M.; Deng, W. Deep Face Recognition: A Survey. *arXiv* **2018**, arXiv:1804.06655.
4. Hu, Z.; Gui, P.; Feng, Z.; Zhao, Q.; Fu, K.; Liu, F.; Liu, Z. Boosting Depth-Based Face Recognition from a Quality Perspective. *Sensors* **2019**, *19*, 4124. [CrossRef] [PubMed]
5. Yang, Y.; Wen, C.; Xie, K.; Wen, F.; Sheng, G.; Tang, X. Face Recognition Using the SR-CNN Model. *Sensors* **2018**, *18*, 4237. [CrossRef]
6. Koo, J.H.; Cho, S.W.; Baek, N.R.; Kim, M.; Park, K.R. CNN-Based Multimodal Human Recognition in Surveillance Environments. *Sensors* **2018**, *18*, 3040. [CrossRef]
7. Deng, J.; Guo, J.; Xue, N.; Zafeiriou, S. ArcFace: Additive Angular Margin Loss for Deep Face Recognition. In Proceedings of the IEEE Conference on Computer Vision and Pattern Recognition, CVPR 2019, Long Beach, CA, USA, 16–20 June 2019; pp. 4690–4699.
8. Wang, H.; Wang, Y.; Zhou, Z.; Ji, X.; Gong, D.; Zhou, J.; Li, Z.; Liu, W. CosFace: Large Margin Cosine Loss for Deep Face Recognition. In Proceedings of the 2018 IEEE Conference on Computer Vision and Pattern Recognition, CVPR 2018, Salt Lake City, UT, USA, 18–22 June 2018; pp. 5265–5274. [CrossRef]
9. Beymer, D.; Poggio, T.A. Face Recognition from One Example View. In Procedings of the Fifth International Conference on Computer Vision (ICCV 95), Massachusetts Institute of Technology, Cambridge, MA, USA, 20–23 June 1995; pp. 500–507. [CrossRef]
10. Tan, X.; Chen, S.; Zhou, Z.; Zhang, F. Face recognition from a single image per person: A survey. *Pattern Recognit.* **2006**, *39*, 1725–1745. [CrossRef]
11. Wu, J.; Zhou, Z. Face recognition with one training image per person. *Pattern Recognit. Lett.* **2002**, *23*, 1711–1719. [CrossRef]
12. Yin, H.; Fu, P.; Meng, S. Sampled Two-Dimensional LDA for Face Recognition with One Training Image per Person. In Proceedings of the First International Conference on Innovative Computing, Information and Control (ICICIC 2006), Beijing, China, 30 August–1 September 2006; pp. 113–116. [CrossRef]
13. Lee, S.; Jung, H.; Hwang, B.; Lee, S. Authenticating corrupted photo images based on noise parameter estimation. *Pattern Recognit.* **2006**, *39*, 910–920. [CrossRef]
14. Lu, J.; Tan, Y.; Wang, G. Discriminative multi-manifold analysis for face recognition from a single training sample per person. In Proceedings of the IEEE International Conference on Computer Vision, ICCV 2011, Barcelona, Spain, 6–13 November 2011; pp. 1943–1950. [CrossRef]
15. Abd-Almageed, W.; Wu, Y.; Rawls, S.; Harel, S.; Hassner, T.; Masi, I.; Choi, J.; Leksut, J.T.; Kim, J.; Natarajan, P.; et al. Face recognition using deep multi-pose representations. In Proceedings of the 2016 IEEE Winter Conference on Applications of Computer Vision, WACV 2016, Lake Placid, NY, USA, 7–10 March 2016; pp. 1–9. [CrossRef]

16. Bashbaghi, S.; Granger, E.; Sabourin, R.; Bilodeau, G. Robust watch-list screening using dynamic ensembles of SVMs based on multiple face representations. *Mach. Vis. Appl.* **2017**, *28*, 219–241. [CrossRef]

17. Dadi, H.S.; Pillutla, G.K.M.; Makkena, M.L. Face Recognition and Human Tracking Using GMM, HOG and SVM in Surveillance Videos. *Ann. Data Sci.* **2018**, *5*, 157–179. [CrossRef]

18. Dalal, N.; Triggs, B. Histograms of Oriented Gradients for Human Detection. In Proceedings of the 2005 IEEE Computer Society Conference on Computer Vision and Pattern Recognition (CVPR 2005), San Diego, CA, USA, 20–26 June 2005; pp. 886–893. [CrossRef]

19. Ahonen, T.; Hadid, A.; Pietikäinen, M. Face Description with Local Binary Patterns: Application to Face Recognition. *IEEE Trans. Pattern Anal. Mach. Intell.* **2006**, *28*, 2037–2041. [CrossRef]

20. Simonyan, K.; Zisserman, A. Very Deep Convolutional Networks for Large-Scale Image Recognition. In Proceedings of the 3rd International Conference on Learning Representations, ICLR 2015, San Diego, CA, USA, 7–9 May 2015.

21. Szegedy, C.; Liu, W.; Jia, Y.; Sermanet, P.; Reed, S.E.; Anguelov, D.; Erhan, D.; Vanhoucke, V.; Rabinovich, A. Going deeper with convolutions. In Proceedings of the IEEE Conference on Computer Vision and Pattern Recognition, CVPR 2015, Boston, MA, USA, 7–12 June 2015; pp. 1–9. [CrossRef]

22. Schroff, F.; Kalenichenko, D.; Philbin, J. FaceNet: A unified embedding for face recognition and clustering. In Proceedings of the IEEE Conference on Computer Vision and Pattern Recognition, CVPR 2015, Boston, MA, USA, 7–12 June 2015; pp. 815–823. [CrossRef]

23. He, K.; Zhang, X.; Ren, S.; Sun, J. Deep Residual Learning for Image Recognition. In Proceedings of the 2016 IEEE Conference on Computer Vision and Pattern Recognition, CVPR 2016, Las Vegas, NV, USA, 27–30 June 2016; pp. 770–778. [CrossRef]

24. Hu, J.; Shen, L.; Sun, G. Squeeze-and-Excitation Networks. In Proceedings of the 2018 IEEE Conference on Computer Vision and Pattern Recognition, CVPR 2018, Salt Lake City, UT, USA, 18–22 June 2018; pp. 7132–7141. [CrossRef]

25. Deng, W.; Hu, J.; Guo, J. Extended SRC: Undersampled Face Recognition via Intraclass Variant Dictionary. *IEEE Trans. Pattern Anal. Mach. Intell.* **2012**, *34*, 1864–1870. [CrossRef]

26. Deng, W.; Hu, J.; Zhou, X.; Guo, J. Equidistant prototypes embedding for single sample based face recognition with generic learning and incremental learning. *Pattern Recognit.* **2014**, *47*, 3738–3749. [CrossRef]

27. Yang, M.; Gool, L.V.; Zhang, L. Sparse Variation Dictionary Learning for Face Recognition with a Single Training Sample per Person. In Proceedings of the IEEE International Conference on Computer Vision, ICCV 2013, Sydney, Australia, 1–8 December 2013; pp. 689–696. [CrossRef]

28. Su, Y.; Shan, S.; Chen, X.; Gao, W. Adaptive generic learning for face recognition from a single sample per person. In Proceedings of the Twenty-Third IEEE Conference on Computer Vision and Pattern Recognition, CVPR 2010, San Francisco, CA, USA, 13–18 June 2010; pp. 2699–2706. [CrossRef]

29. Cai, J.; Chen, J.; Liang, X. Single-Sample Face Recognition Based on Intra-Class Differences in a Variation Model. *Sensors* **2015**, *15*, 1071–1087. [CrossRef] [PubMed]

30. Wright, J.; Yang, A.Y.; Ganesh, A.; Sastry, S.S.; Ma, Y. Robust Face Recognition via Sparse Representation. *IEEE Trans. Pattern Anal. Mach. Intell.* **2009**, *31*, 210–227. [CrossRef] [PubMed]

31. Shao, C.; Song, X.; Feng, Z.; Wu, X.; Zheng, Y. Dynamic dictionary optimization for sparse-representation-based face classification using local difference images. *Inf. Sci.* **2017**, *393*, 1–14. [CrossRef]

32. Mohammadzade, H.; Hatzinakos, D. Projection into Expression Subspaces for Face Recognition from Single Sample per Person. *IEEE Trans. Affect. Comput.* **2013**, *4*, 69–82. [CrossRef]

33. Zeng, J.; Zhao, X.; Gan, J.; Mai, C.; Zhai, Y.; Wang, F. Deep Convolutional Neural Network Used in Single Sample per Person Face Recognition. *Comput. Intell. Neurosci.* **2018**, *2018*, 3803627:1–3803627:11. [CrossRef]

34. Zhang, Y.; Peng, H. Sample reconstruction with deep autoencoder for one sample per person face recognition. *IET Comput. Vis.* **2017**, *11*, 471–478. [CrossRef]

35. Hong, S.; Im, W.; Ryu, J.; Yang, H.S. SSPP-DAN: Deep domain adaptation network for face recognition with single sample per person. In Proceedings of the 2017 IEEE International Conference on Image Processing, ICIP 2017, Beijing, China, 17–20 September 2017; pp. 825–829. [CrossRef]

36. Cuculo, V.; D'Amelio, A.; Grossi, G.; Lanzarotti, R.; Lin, J. Robust Single-Sample Face Recognition by Sparsity-Driven Sub-Dictionary Learning Using Deep Features. *Sensors* **2019**, *19*, 146. [CrossRef] [PubMed]

37. Choi, S.; Lee, Y.; Lee, M. Face Recognition in SSPP Problem Using Face Relighting Based on Coupled Bilinear Model. *Sensors* **2019**, *19*, 43. [CrossRef] [PubMed]

38. Zhu, X.; Lei, Z.; Liu, X.; Shi, H.; Li, S.Z. Face Alignment Across Large Poses: A 3D Solution. In Proceedings of the 2016 IEEE Conference on Computer Vision and Pattern Recognition, CVPR 2016, Las Vegas, NV, USA, 27–30 June 2016; pp. 146–155. [CrossRef]

39. Feng, Z.; Hu, G.; Kittler, J.; Christmas, W.J.; Wu, X. Cascaded Collaborative Regression for Robust Facial Landmark Detection Trained Using a Mixture of Synthetic and Real Images With Dynamic Weighting. *IEEE Trans. Image Process.* **2015**, *24*, 3425–3440. [CrossRef] [PubMed]

40. Song, X.; Feng, Z.; Hu, G.; Kittler, J.; Wu, X. Dictionary Integration Using 3D Morphable Face Models for Pose-Invariant Collaborative-Representation-Based Classification. *IEEE Trans. Inf. Forensics Secur.* **2018**, *13*, 2734–2745. [CrossRef]

41. Li, L.; Ge, H.; Tong, Y.; Zhang, Y. Face Recognition Using Gabor-Based Feature Extraction and Feature Space Transformation Fusion Method for Single Image per Person Problem. *Neural Process. Lett.* **2018**, *47*, 1197–1217. [CrossRef]

42. Pan, J.; Wang, X.; Cheng, Y. Single-Sample Face Recognition Based on LPP Feature Transfer. *IEEE Access* **2016**, *4*, 2873–2884. [CrossRef]

43. Min, R.; Xu, S.; Cui, Z. Single-Sample Face Recognition Based on Feature Expansion. *IEEE Access* **2019**, *7*, 45219–45229. [CrossRef]

44. Tran, A.T.; Hassner, T.; Masi, I.; Medioni, G.G. Regressing Robust and Discriminative 3D Morphable Models with a Very Deep Neural Network. In Proceedings of the 2017 IEEE Conference on Computer Vision and Pattern Recognition, CVPR 2017, Honolulu, HI, USA, 21–26 July 2017; pp. 1493–1502. [CrossRef]

45. Sengupta, S.; Kanazawa, A.; Castillo, C.D.; Jacobs, D.W. SfSNet: Learning Shape, Reflectance and Illuminance of Faces 'in the Wild'. In Proceedings of the 2018 IEEE Conference on Computer Vision and Pattern Recognition, CVPR 2018, Salt Lake City, UT, USA, 18–22 June 2018; pp. 6296–6305. [CrossRef]

46. Blanz, V.; Vetter, T. Face Recognition Based on Fitting a 3D Morphable Model. *IEEE Trans. Pattern Anal. Mach. Intell.* **2003**, *25*, 1063–1074. [CrossRef]

47. Tu, H.; Li, K.; Zhao, Q. Robust Face Recognition with Assistance of Pose and Expression Normalized Albedo Images. In Proceedings of the 2019 5th International Conference on Computing and Artificial Intelligence, ICCAI 2019, Bali, Indonesia, 19–22 April 2019; pp. 93–99. [CrossRef]

48. Feng, Y.; Wu, F.; Shao, X.; Wang, Y.; Zhou, X. Joint 3D Face Reconstruction and Dense Alignment with Position Map Regression Network. In Proceedings of the Computer Vision—ECCV 2018—15th European Conference, Munich, Germany, 8–14 September 2018; pp. 557–574. [CrossRef]

49. Ramamoorthi, R. Modeling illumination variation with spherical harmonics—Chapter 12. *Face Process. Adv. Model. Methods* **2006**, 385–424.

50. Ramamoorthi, R.; Hanrahan, P. A signal-processing framework for reflection. *ACM Trans. Graph.* **2004**, *23*, 1004–1042. [CrossRef]

51. Bas, A.; Huber, P.; Smith, W.A.P.; Awais, M.; Kittler, J. 3D Morphable Models as Spatial Transformer Networks. In Proceedings of the 2017 IEEE International Conference on Computer Vision Workshops, ICCV Workshops 2017, Venice, Italy, 22–29 October 2017; pp. 895–903. [CrossRef]

52. Paysan, P.; Knothe, R.; Amberg, B.; Romdhani, S.; Vetter, T. A 3D Face Model for Pose and Illumination Invariant Face Recognition. In Proceedings of the Sixth IEEE International Conference on Advanced Video and Signal Based Surveillance, AVSS 2009, Genova, Italy, 2–4 September 2009; pp. 296–301. [CrossRef]

53. Cao, C.; Weng, Y.; Zhou, S.; Tong, Y.; Zhou, K. FaceWarehouse: A 3D Facial Expression Database for Visual Computing. *IEEE Trans. Vis. Comput. Graph.* **2014**, *20*, 413–425. [CrossRef] [PubMed]

54. Liu, Z.; Luo, P.; Wang, X.; Tang, X. Deep Learning Face Attributes in the Wild. In Proceedings of the 2015 IEEE International Conference on Computer Vision, ICCV 2015, Santiago, Chile, 7–13 December 2015; pp. 3730–3738. [CrossRef]

55. Liu, F.; Zhao, Q.; Liu, X.; Zeng, D. Joint Face Alignment and 3D Face Reconstruction with Application to Face Recognition. *arXiv* **2017**, arXiv:1708.02734.

56. Phillips, P.J.; Moon, H.; Rizvi, S.A.; Rauss, P.J. The FERET Evaluation Methodology for Face-Recognition Algorithms. *IEEE Trans. Pattern Anal. Mach. Intell.* **2000**, *22*, 1090–1104. [CrossRef]

57. Huang, G.B.; Mattar, M.; Berg, T.L.; Learned-Miller, E.G. *Labeled Faces in the Wild: A Database for Studying Face Recognition in Unconstrained Environments*; University of Massachusetts: Amherst, MA, USA, 2007.

58. Zheng, T.; Deng, W. *Cross-pose LFW: A Database for Studying Cross-pose Face Recognition in Unconstrained Environments*; Technical Report 18-01; Beijing University of Posts and Telecommunications: Beijing, China, 2018.
59. Zheng, T.; Deng, W.; Hu, J. Cross-Age LFW: A Database for Studying Cross-Age Face Recognition in Unconstrained Environments. *arXiv* **2017**, arXiv:1708.08197.
60. Moschoglou, S.; Papaioannou, A.; Sagonas, C.; Deng, J.; Kotsia, I.; Zafeiriou, S. AgeDB: The First Manually Collected, In-the-Wild Age Database. In Proceedings of the 2017 IEEE Conference on Computer Vision and Pattern Recognition Workshops, CVPR Workshops 2017, Honolulu, HI, USA, 21–26 July 2017; pp. 1997–2005. [CrossRef]
61. Sengupta, S.; Chen, J.; Castillo, C.D.; Patel, V.M.; Chellappa, R.; Jacobs, D.W. Frontal to profile face verification in the wild. In Proceedings of the 2016 IEEE Winter Conference on Applications of Computer Vision, WACV 2016, Lake Placid, NY, USA, 7–10 March 2016; pp. 1–9. [CrossRef]
62. Cao, Q.; Shen, L.; Xie, W.; Parkhi, O.M.; Zisserman, A. VGGFace2: A Dataset for Recognising Faces across Pose and Age. In Proceedings of the 13th IEEE International Conference on Automatic Face & Gesture Recognition, FG 2018, Xi'an, China, 15–19 May 2018; pp. 67–74. [CrossRef]

Article

Double Additive Margin Softmax Loss for Face Recognition

Shengwei Zhou, Caikou Chen *, Guojiang Han and Xielian Hou

College of Information Engineering, Yangzhou University, Yangzhou 225127, China;
zhoushengwei326@163.com (S.Z.); hanguojiang0821@163.com (G.H.); xielhou@sina.com (X.H.)
* Correspondence: yzcck@126.com; Tel.: +86-139-5278-0010

Received: 30 October 2019; Accepted: 17 December 2019; Published: 19 December 2019

Abstract: Learning large-margin face features whose intra-class variance is small and inter-class diversity is one of important challenges in feature learning applying Deep Convolutional Neural Networks (DCNNs) for face recognition. Recently, an appealing line of research is to incorporate an angular margin in the original softmax loss functions for obtaining discriminative deep features during the training of DCNNs. In this paper we propose a novel loss function, termed as double additive margin Softmax loss (DAM-Softmax). The presented loss has a clearer geometrical explanation and can obtain highly discriminative features for face recognition. Extensive experimental evaluation of several recent state-of-the-art softmax loss functions are conducted on the relevant face recognition benchmarks, CASIA-Webface, LFW, CALFW, CPLFW, and CFP-FP. We show that the proposed loss function consistently outperforms the state-of-the-art.

Keywords: Softmax; angular margin; ResNet; face recognition

---

1. Introduction

Face recogniton problems are ubiquitous in the computer vision domain. In the past few years, Deep Convolutional Neural Networks (DCNNs) have set the community of face recognition (FR) on fire [1]. Due to effectively layered end-to-end learning network frameworks and carefully deep feature extracting techniques from local to global, which are the most important ingredients for their success, DCNNs has immensely improved the state of the art in real-world face recognition scenarios. Numerous layered network architectures for face recognition tasks such as AlexNet [2], VGG [3], InceptionNet [4], ResNet [5], and DenseNet [6], have been proposed. Among them, the most representative one is AlexNet, originally proposed by Krizhevsky et al. AlexNet has become a pioneered architecture developed for image classification and was the winner of the ImageNet Large Scale Challenge in 2012.

It is well known that the effective feature representation for face images plays an important role in FR. Over the recent years, a hot research trend towards DCNNs has been devoted to learning with more discriminative deep features. Intuitively, the learnt deep features for FR are desired if the maximal within-class variance is less than the minimal between-class variance. However, learning deep features satisfying this condition is generally not easy owing to the inherently large intra-class variation and high inter-class similarity [7] in many FR applications. Despite the softmax function with the cross-entropy loss (called softmax loss) is popularly used in the training of a DCNN, recent studies [8,9] made it clear that the current softmax loss is insufficient to encourage the desired deep features meeting the above condition. To boost the discriminative ability of DCNNs and inspired by the previous idea, the center loss [10], pairwise loss [11] and triplet loss [12] were proposed. They unanimously proposed the enhancement of discrimination power of deep features by minimizing winth-class variance and maximizing between-class variance in the Euclidean space of features. While these methods is superior to the traditional softmax loss over classification performance, they usually suffer from some drawbacks. The center loss only explicitly enhanced the intra-class compactness while disregarding the inter-class separability. For the pairwise loss and triplet loss, They require the careful mining of pairs or triplets of samples, which is highly time-consuming.

Due to the fact that few existing softmax losses can effectively achieve the discriminative condition that the maximal within-class variance is less than the minimal between-class variance under the conventional Euclidean metric space, more recently, approaches have been proposed to address this problem by transforming the original Euclidean space of features to an corresponding angular space [10,13–15]. Specifically, both Large-Margin Softmax Loss [13] and A-Softmax Loss [14] are an angular softmax loss that enables DCNNs to learn angular deep features by imposing an angular margin constraint for larger inter-class variance. Compared to the Euclidean margin suggested by [2,16], the learnt angular features is more discriminative with the angular margin because the angular metric with cosine similarity is intrinsically more suitable to the softmax loss. During training of A-Softmax Loss, the orginal Softmax loss must be combined to ensure the convergence. To overcome the optimisation problem of A-Softmax Loss, the Additive Margin Softmax loass (AM-Softmax) [15] is proposed. The loss integrates a angular margin to the softmax loss in an additve manner. Its implementation and optimisation are much easier than A-Softmax Loss since A-Softmax Loss integrates the angular margin in a multiplicative way. AM-Softmax is easily reproducible and achieves state-of-the-art performance.

Motivated by AM-Softmax loss, This paper propose a new additive angular margin Loss, namely double additive margin softmax loss (DAM-Softmax). The idea behind the proposed loss is to impose an additive margin m to both the intra-class angular variation and inter-class angular variation simultaneously to enhance the intra-class compactness and inter-class discrepancy of the learned features. Compared to AM-Softmax loss, our loss has a stronger geometrical significance and will lead to obtain more discriminative features. Experimental results on some relevant face recognition benchmarks show that the proposed loss achieves better classification performance than the current state-of-the-art losses.

The rest of this paper is organized as follows. In Section 2, We will briefly introduce the related works such as the original softmax Loss, L-Softmax Loss, A-Softmax Loss and AM-Softmax Loss. Then we discuss the proposed loss, Double Additive Margin Softmax Loss in detail in Section 3. Finally, extensive experiments are presented in Section 4.

2. Preliminaries

In order to clearly understand the proposed DAM-Softmax loss, we will briefly review the classical softmax loss and AM-softmax loss. The classical softmax loss is formulated by

$$L_1 = -\frac{1}{N}\sum_{i=1}^{N} log \frac{e^{\mathbf{w}_{y_i}^{\top}\mathbf{f}_i}}{\sum_{c=1}^{C} e^{\mathbf{w}_c^{\top}\mathbf{f}_i}} \tag{1}$$

where $\mathbf{w}_c$ ($c = 1, \ldots, C$, C is the number of classes) denotes the weight vector of the last fully connected classifier layer, $\mathbf{f}_i$ is the learned deep feature input vector of the last fully connected classifier layer corresponding to the original input $\mathbf{x}_i$ with the label y_i, and N is the number of training samples in a minibatch. The inner product, $\mathbf{w}_c^{\top}\mathbf{f}_i$, between $\mathbf{w}_c$ and $\mathbf{f}_i$ can be also factorized into $\|\mathbf{w}_c\|\|\mathbf{f}_i\|cos(\theta_c)$ where θ_c is the angle between $\mathbf{w}_c$ and $\mathbf{f}_i$, the loss can thus be rewritten as

$$L_2 = -\frac{1}{N}\sum_{i=1}^{N} log \frac{e^{\|\mathbf{w}_{y_i}\|\|\mathbf{f}_i\|cos(\theta_{y_i})}}{\sum_{c=1}^{C} e^{\|\mathbf{w}_c\|\|\mathbf{f}_i\|cos(\theta_c)}} \tag{2}$$

The A-Softmax is a new loss function derived from the classical Softmax loss which proposed to impose a constraint to make $\|\mathbf{w}_c\| = 1$ and generalize the modified softmax loss to angular softmax (A-Softmax) loss by replacing $\|\mathbf{f}_i\|cos(\theta_{y_i})$ with $\|\mathbf{f}_i\|\psi(\theta_{y_i})$,

$$L_3 = -\frac{1}{N}\sum_{i=1}^{N} log \frac{e^{\|\mathbf{f}_i\|\psi(\theta_{y_i})}}{e^{\|\mathbf{f}_i\|\psi(\theta_{y_i})} + \sum_{c=1,c\neq y_i}^{C} e^{\|\mathbf{f}_c\|cos(\theta_c)}} \tag{3}$$

where the authors proposed to define $\psi(\theta)$ as $\psi(\theta) = (-1)^k cos(m\theta) - 2k$, $\theta \in [\frac{k\pi}{m}, \frac{(k+1)\pi}{m}]$ and $k \in [0, m-1]$ for removing the restriction which θ must be in the range of $[0, \frac{\pi}{m}]$.

In the AM-Softmax loss, the authors suggested to introduce an additive margin to its decision boundary by defining $\psi(\theta)$ as $cos(\theta) - m$. In addition, both the deep feature vector $\mathbf{f}_i$ and weight vectors $\mathbf{w}_c$ are normalized during the implementation. Thus, The AM-Softmax loss is given by

$$\begin{aligned} L_4 &= -\frac{1}{N}\sum_{i=1}^{N} log \frac{e^{s\cdot(cos(\theta_{y_i})-m)}}{e^{s\cdot(cos(\theta_{y_i})-m)} + \sum_{c=1,c\neq y_i}^{C} e^{s\cdot cos(\theta_c)}} \\ &= -\frac{1}{N}\sum_{i=1}^{N} log \frac{e^{s\cdot(\mathbf{w}_{y_i}^{\top}\mathbf{f}_i-m)}}{e^{s\cdot(\mathbf{w}_{y_i}^{\top}\mathbf{f}_i-m)} + \sum_{c=1,c\neq y_i}^{C} e^{s\cdot \mathbf{w}_c^{\top}\mathbf{f}_i}} \end{aligned} \tag{4}$$

where s is a hyper-parameter for scaling the cosine values.

In order to simultaneously enlarge the between-class angular margin and compress the within-class angular variation, It is clear to learn that both the A-Softmax loss and AM-Softmax loss share a common idea to generalize the original softmax loss to angular softmax loss by introducing an integer m to quantitatively control the decision boundary. specifically, In binary class case, a learned feature $\mathbf{f}$ from class 1 is given and θ_i is the angle between $\mathbf{f}$ and $\mathbf{w}_i$. A-softmax loss requires $cos(m\theta_1) > cos(\theta_2)$ to correctly classify $\mathbf{f}$. AM-softmax loss instead proposes $cos(\theta_1) - m > cos(\theta_2)$ to correctly classify $\mathbf{f}$. Both of them explicitly enforce the intra-class compactness to achieve more discriminative deep features by imposing an intra-class angular margin in the multiplicative manner and in the additive manner, respectively. Compared with the A-Softmax loss, AM-Softmax loss is simpler which is simpler and reaches better performance. In addition, It is much easier to implement because the computation of the gradient for back-propagation is no longer required.

3. Double Additive Margin Softmax Loss

One can obviously learn that AM-Softmax loss can obtain better performance by incorporating a single additive margin to its intra-class angular variation. Inspired by that, we propose to impose an additive margin to both the intra-class angular variation and inter-class angular distribution simultaneously to enhance the intra-class compactness and inter-class discrepancy. To give a formal formulation for the idea, we first define a function $g(\theta) = cos(\theta)$. The Equation (4) can be rewritten as

$$L_5 = -\frac{1}{N}\sum_{i=1}^{N} log \frac{e^{s\cdot\psi(\theta_{y_i})}}{e^{s\cdot\psi(\theta_{y_i})} + \sum_{c=1,c\neq y_i}^{C} e^{s\cdot g(\theta_c)}} \tag{5}$$

where $\psi(\theta_{y_i}) = cos(\theta_{y_i})$ and $g(\theta_c) = cos(\theta_c)$.

As analyzed above, we impose an additive margin m to both the intra-class and inter-class angular variation angular distribution simultaneously. Then we have the formulations:

$$\begin{aligned} \psi(\theta_{y_i}) &= cos(\theta_{y_i}) - m \\ g(\theta_c) &= cos(\theta_c) + m \end{aligned} \tag{6}$$

Compared to AM-Softmax loss, our formulation is also simple while explicitly encourages intra-class compactness and inter-class separability simultaneously, we thus term the loss as Double Additive Margin Softmax loss (DAM). Finally, the proposed loss function can be formulated by

$$
\begin{aligned}
L_6 &= -\frac{1}{N} \sum_{i=1}^{N} log \frac{e^{s \cdot (cos(\theta_{y_i}) - m)}}{e^{s \cdot (cos(\theta_{y_i}) - m)} + \sum_{c=1, c \neq y_i}^{C} e^{s \cdot (cos(\theta_c) + m)}} \\
&= -\frac{1}{N} \sum_{i=1}^{N} log \frac{e^{s \cdot (\mathbf{w}_{y_i}^T \mathbf{f}_i - m)}}{e^{s \cdot (\mathbf{w}_{y_i}^T \mathbf{f}_i - m)} + \sum_{c=1, c \neq y_i}^{C} e^{s \cdot (\mathbf{w}_c^T \mathbf{f}_c + m)}}
\end{aligned}
\tag{7}
$$

3.1. Geometric Interpretation

Our double additive margin has a more explicit geometric interpretation on the hypersphere manifold. To simplify the geometric interpretation, we project the features onto two dimensional space and discuss the binary classification case on the hypersphere manifold where there are only $\mathbf{w}_1$ and $\mathbf{w}_2$, and $\|\mathbf{w}_1\| = \|\mathbf{w}_2\| = 1$. Thus, the classification performance depends totally on the angles θ_1 between $\mathbf{f}$ and $\mathbf{w}_1$, and θ_2 between $\mathbf{f}$ and $\mathbf{w}_2$.

Classification Stringency. In order to correctly classify the learned feature vector $\mathbf{f}$ as class 1, the traditional softmax loss requires $\theta_1 < \theta_2$ which easily has $\theta_1 - \theta_2 < 0$, the AM-Softmax loss requires $cos(\theta_1) - m < cos(\theta_2)$ which can be rewriten as $cos(\theta_1) - cos(\theta_2) < m$, while our DAM-Softmax loss needs $cos(\theta_1) - m < cos(\theta_2) + m$ which can be reformulated into $cos(\theta_1) - cos(\theta_2) < 2m$. It is obvious to see that the DAM-Softmax loss is more stringent than the orginal softmax loss and AM-Softmax loss for satisfying the corresponding classification criteria.

Classification Boundary. In Figure 1, we draw a schematic diagram to show the classification boundary of the classical softmax loss, AM-Softmax loss and the proposed DAM-Softmax loss. The classification boundary of the traditional softmax loss is denoted as the vector $\mathbf{p}_0$. In this case, we have $\mathbf{w}_1^T \mathbf{p}_0 = \mathbf{w}_2^T \mathbf{p}_0$ at the decision boundary $\mathbf{p}_0$ ($\mathbf{w}_1 \in class1, \mathbf{w}_2 \in class2$). For the AM-Softmax, the boundary becomes a marginal region instead of a single vector. At the new boundary $\mathbf{p}_1$ for class 1, one has $\mathbf{w}_1^T \mathbf{p}_1 - m = \mathbf{w}_2^T \mathbf{p}_1$, which gives $m = (\mathbf{w}_1^T - \mathbf{w}_2^T)\mathbf{p}_1 = cos(\theta_{\mathbf{w}_1, \mathbf{p}_1}) - cos(\theta_{\mathbf{w}_2, \mathbf{p}_1})$. If we further assume that all the classes have the same intra-class variance and the boundary for class 2 is at $\mathbf{p}_2$, we can get $cos(\theta_{\mathbf{w}_2, \mathbf{p}_1}) = cos(\theta_{\mathbf{w}_1, \mathbf{p}_2})$. Thus, $m = cos(\theta_{\mathbf{w}_1, \mathbf{p}_1}) - cos(\theta_{\mathbf{w}_1, \mathbf{p}_2})$, which is the difference of the cosine scores for class 1 between the two sides of the margin region. For our DAM-Softmax loss, the boundary becomes a wider marginal region than the one of AM-Softmax loss. At the new boundary $\mathbf{p}_3$ for class 1, one has $\mathbf{w}_3^T \mathbf{p}_3 - m = \mathbf{w}_4^T \mathbf{p}_3 + m$ ($\mathbf{w}_3 \in class1, \mathbf{w}_4 \in class2$), which gives $2m = (\mathbf{w}_3^T - \mathbf{w}_4^T)\mathbf{p}_3 = cos(\theta_{\mathbf{w}_3, \mathbf{p}_3}) - cos(\theta_{\mathbf{w}_4, \mathbf{p}_3})$. If we further assume that all the classes have the same intra-class variance and the boundary for class 2 is at $\mathbf{p}_4$, we can get $cos(\theta_{\mathbf{w}_4, \mathbf{p}_3}) = cos(\theta_{\mathbf{w}_3, \mathbf{p}_4})$. Thus, $2m = cos(\theta_{\mathbf{w}_3, \mathbf{p}_3}) - cos(\theta_{\mathbf{w}_3, \mathbf{p}_4})$, which is the difference of the cosine scores for class 1 between the two sides of the margin region. Obviously, the DAM-Softmax loss leads to a larger classification margin between class 1 and class 2.

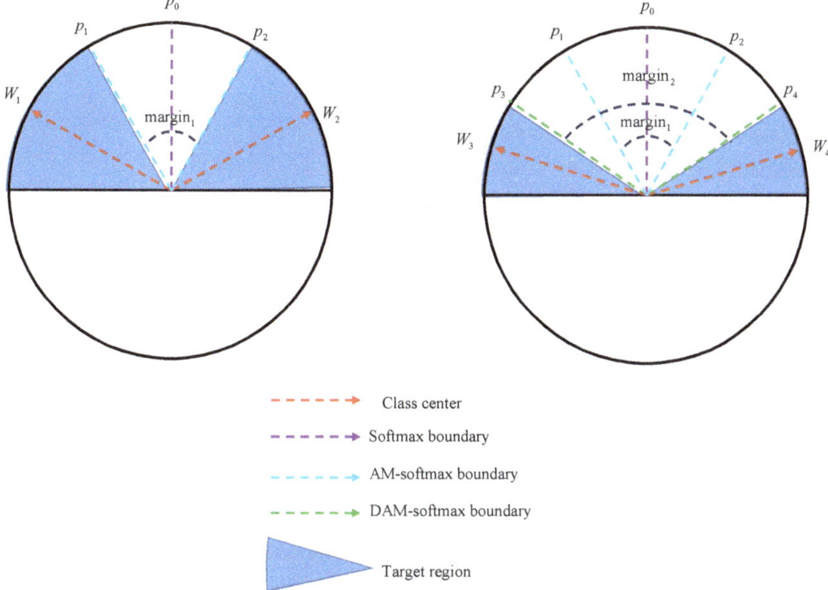

Figure 1. Geometric difference.

3.2. Feature Distribution Visualization on MNIST Dataset

In order to better study and verify the effectiveness of the proposed DAM-Softmax loss function, we conducted an experiment on the MNIST dataset [17] to visualize the learned feature distributions. We chose the 7-layer CNN models with the original Softmax loss, AM-Softmax loss and DAM-Softmax loss for training and required to output two-dimensional deep features for visualization. After the 2-dimensional features were obtained, we then made the normalization to them and ploted them on a circle in the two dimensional space.

The visualization from Figure 2 can well demonstrate that our DAM-Softmax outperforms AM-Softmax [15] when the heperparameters s and m is 30 and 0.4, respectively. Compared to AM-Softmax [15], the DAM-Softmax loss can lead to the larger inter-class margin and smaller intra-class variance property to the features without tuning too many hyper-parameters.

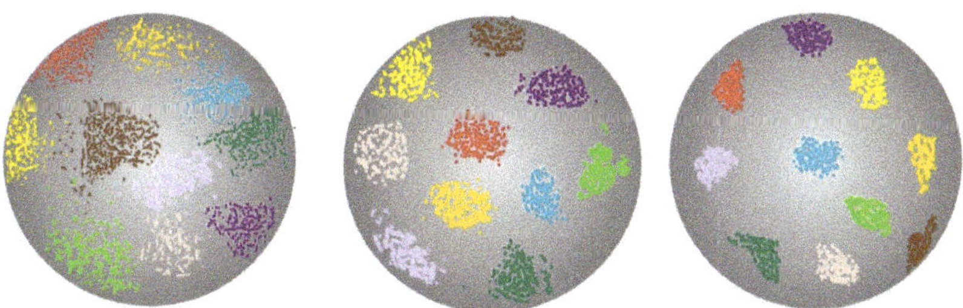

Figure 2. Features visualization in MNIST dataset: Softmax (left), AM-Softmax (middle) and DAM-Softmax (right). Specifically, we set the feature (output of the network) dimension to 2 and then draw the classification result as a scatter plot.

3.3. Algorithm

The proposed DAM-Softmax loss is extremely easy to implement in the popular deep learning frameworks, e.g., Pytorch [18] and Tensorflow [19]. The algorithm for DAM-Softmax loss is given as follow.

Algorithm 1: The steps of the DAM-Softmax Loss

Input: Feature Scale s, Margin Parameter m in Equation (7), Randomly initialized weights $\mathbf{w}$, Input images $\mathbf{f}$, Batch size N

1. Normalize the input image $\mathbf{f}$ ($\hat{\mathbf{f}} = \frac{\mathbf{f}}{|\mathbf{f}|}$), and make the new $\mathbf{f}=\hat{\mathbf{f}}$
2. Normalize the weight $\mathbf{w}$ ($\hat{\mathbf{w}} = \frac{\mathbf{w}}{|\mathbf{w}|}$), and make the new $\mathbf{w}=\hat{\mathbf{w}}$
3. According to the Equation (7), introducing the variable substitutions (the new $\mathbf{f}_i=\hat{\mathbf{f}}_i$ and the new $\mathbf{w}_{y_i}=\hat{\mathbf{w}}_{y_i}$) and get the $cos(\theta_{y_i}) = \mathbf{w}_{y_i}^\top \mathbf{f}_i$
4. According to the Equation (7), introducing the variable substitutions (the new $\mathbf{f}_c=\hat{\mathbf{f}}_c$ and the new $\mathbf{w}_c=\hat{\mathbf{w}}_c$) and get $cos(\theta_c) = \mathbf{w}_c^\top \mathbf{f}_c$
5. Calculate "$cos(\theta_{y_i}) - m$", and get "$s \cdot (cos(\theta_{y_i}) - m)$" in the Equation (6) and the Equation (7)
6. Calculate "$cos(\theta_c) + m$", and get "$s \cdot (cos(\theta_c) + m)$" in the Equation (6) and the Equation (7)
7. Construct loss functioin: $L = -\frac{1}{N} \sum_{i=1}^{N} log \frac{e^{s \cdot (\mathbf{w}_{y_i}^\top \mathbf{f}_i - m)}}{e^{s \cdot (\mathbf{w}_{y_i}^\top \mathbf{f}_i - m)} + \sum_{j \neq y_i} e^{s \cdot (\mathbf{w}_c^\top \mathbf{f}_c + m)}}$

Output: Loss function L

---

4. Experiment

In this section, we first introduce the experiment settings. Then, we will discuss the effect of the hyperparameters. Finally, we will evaluate the performance of our loss function with several existing state-of-the-art loss functions on the benchmark datasets.

4.1. Implementation Settings

4.1.1. Datasets

Training Datasets. The CASIA-WebFace [7] dataset used for training consists of 49,4414 color face images from 10,575 classes.

Test Datasets. LFW dataset [17] contains 13,233 web-collected images from 5749 different identities, with large variations in pose, expression and illuminations. CFP dataset [18] consists of 500 subjects, each with 10 frontal and 4 profile images. The evaluation protocol includes frontal-frontal (FF) and frontal-profile (FP) face verification, each having 10 folders with 350 same-person pairs and 350 different-person pairs. In our experiments, we employ to test the performance the most challenging subsets CFP-FP which contains images of celebrities in frontal and profile views, CPLFW [20] and CALFW [21] which have higher pose and age variations with same identities from LFW dataset. Specific details of the three datasets above are shown in Table 1 and some example images from CFP-FP, CPLFW and CALFW are given in Figure 3–5, respectively.

Table 1. Face datasets for training and testing.

Datasets	Identity	Image
CPLFW [20]	5749	11625
CALFW [21]	5749	12174
CFP-FP [18]	500	7000

Figure 3. Samples of the same people in the CFP-FP.

Figure 4. Samples of different people in the CALFW.

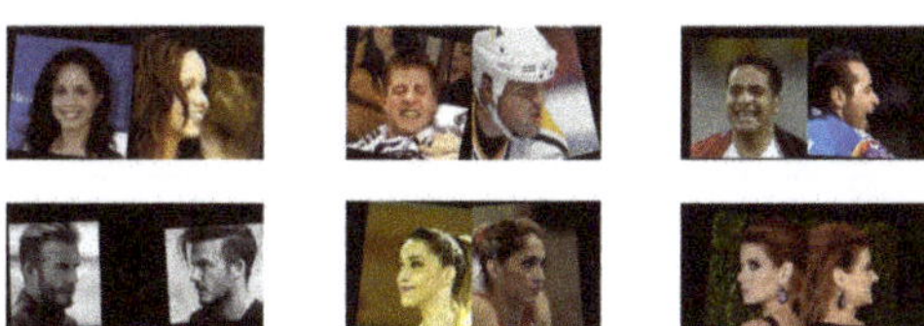

Figure 5. Samples of different people in the CPLFW.

Data Prepossessing. We adopt the data preprocessing method used in [14,22] to detect faces and facial landmarks in images and align them. Then, we crop the aligned face images and resize them to 112×112, and proceed to perform the normalization for the cropped face images by subtracting 128 and dividing 128.

Dataset Overlap Removal. To develop open-set evaluations, we use the overlap checking code provided by F. Wang [22] to get ride of the overlapped subjects between the training dataset of CASIA-WebFace and the testing dataset of LFW.

4.2. Network Architecture and Parameter Settings

For the fair comparison, the CNN architecture used in all experiments of this paper is the ResNet-face18 model specially designed for the training of face recognition, which is a modified ResNet [5]. The model has an improved residual block of BN-Conv-BN-PReLu-Conv-BN structure in which the kernel size and stride in the first convolutional layer is 3×3 and 1 instead of the original 7×7 and 2, and the stride in the second convolutional layer is set to be 2 instead of 1. In addition, PReLu [23] is used to replace the original ReLu. All implementations in the paper are conducted by Pytorch [18]. We set the batch size to be 256 and the weight decay parameter 5×10^{-4}. The initial learning rate is set as 10^{-1}. We set the learning decay rate to be 0.05 which means that the learning rate will be reduced by 5% when the loss value increases. In addition, the total epoch is set as 110. The SGD [9] is used in the optimization process of ResNet-face18.

4.3. Effect of Hyperparameter m

According the discussion in Section 3, our proposed DAM-Softmax loss has two hyperparameters which are the scale s and the margin m. More importantly, the two hyperparameters plays an key role for the performance of the proposed loss. Several recent works [15,24] have already discussed the scale s, we thus follow [15,24] to directly set it to 30 and will no longer discuss it in this paper. In this case, we can focus on the other hyperparameter, margin m. We train the ResNet-face18 model with the DAM-Softmax loss on CASIA-Webface dataset to conduct experiments to seek the best angular margin. For comparison, we train the same network model with the AM-Softmax loss on the same dataset. In Tables 2 and 3, we list the performance of AM-Softmax loss and the proposed DAM-Softmax loss. under the variation of m from 0.1 to 0.25. As illustrated in Tables 2 and 3, we can see that the recognition rate increases gradually from $m = 0.25$ to 0.3 and arrive at the saturated at $m = 0.4$, the performance then begins to drop from 0.4 to 0.5. For DAM-Softmax loss, the classification accuracy improves significantly from $m = 0.1$ to 0.17 and reaches the best at $m = 0.18$, from 0.18 to 0.25, the performance turns to decrease. Therefore, we fix the margins of AM-Softmax loss and DAM-Softmax loss as 0.4 and 0.18, respectively. The experiments for both AM-Softmax loss and DAM-Softmax loss can result in excellent performance without observing any difficulty in convergence. The proposed loss get to the best verification accuracy on CASIA-Webface training dataset.

Table 2. Experimental results of different values of m with AM-Softmax.

Parameter m	0.25	0.3	0.35	0.4	0.45	0.5
Accuracy Rate	97.21%	97.34%	97.45%	**97.68%**	97.57%	97.59%

Table 3. Experimental results of different values of m (110 epoch) with DAM-Softmax.

Parameter m	0.1	0.13	0.15	0.17	0.18	0.2	0.22	0.25
Accuracy Rate	97.68%	97.74%	97.82%	97.94%	**97.97%**	97.89%	97.81%	97.65%

4.4. Comparison with State of the Art Loss Functions on LFW Dataset

In this part, we evaluate the performance of the proposed DAM-Softmax loss and the state-of-the-art loss functions. Following the previous experimental setting, we train a ResNet-Face18 model under the guidance of the original softmax, L-Softmax, A-Softmax, AM-Softmax and DAM-Softmax on the training dataset of CAISAWebFace. The experimental results on the test dataset of LFW are shown in Table 4.

Table 4. Some results of comparative testing experiment.

Model	Accuracy Rate
Softmax (resnet-face18, 110 epoch)	97.08%
L-Softmax (resnet-face18, 110 epoch) [10]	97.33%
A-Softmax (resnet-face18, 110 epoch) [14]	97.52%
AM-Softmax (resnet-face18, 110 epoch) [15]	97.68%
DAM-Softmax (resnet-face18, 47 epoch)	**97.83**%
DAM-Softmax (resnet-face18, 110 epoch)	**97.97**%

From Figure 6, It can be seen that the verification accuracy of DAM-Softmax loss is over 80% after one epoch while AM-Softmax loss requires 20 epoches to achieve the similar accuracy. In the 40th epoch, DAM-Softmax loss reaches the best performance which is still superior to the one of AM-Softmax loss. Figure 7 reports the training loss under the variation of epoch. When epoch = 75, the training loss of the original Softmax loss approaches stabilization with the value of about 13 by using softmax, the AM-softmax's training loss reach stabilization at around 10 when epoch is 55, while DAM-softmax get to stabilization at the 40th epoch and has a lower training loss. Therefore, This can demonstrate that the proposed loss has a faster convergence speed than AM-Softmax loss. As can be seen in Table 4, our proposed DAM-Softmax loss consistently arrives at competitive results compared to the other losses, which demonstrates the effectiveness of DAM-Softmax loss.

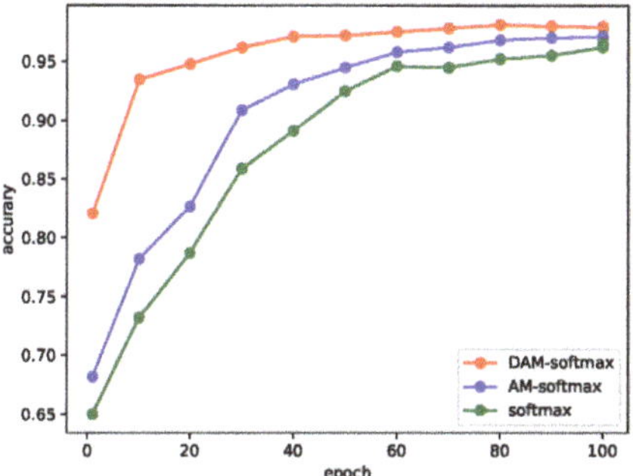

Figure 6. Accuracy rate of the test phase.

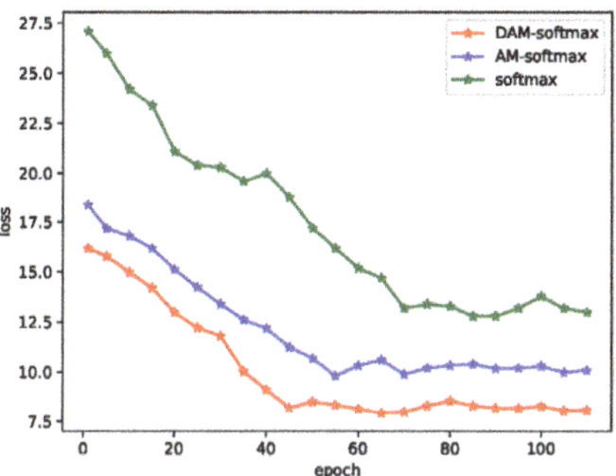

Figure 7. Training loss *v.s* Epoch.

4.5. Comparison with State of the Art Loss Functions on CFP-FP, CPLFW and CALFW Datasets

In order to further verify the effectiveness and robustness of DAM-Softmax, we compare the performance of the proposed losse with related baseline methods, e.g., the original softmax, L-Softmax, A-Softmax, AM-Softmax and DAM-Softmax on three datasets which have large-pose, large-age and different-angle. The experimental results are listed in the Table 5. The details of CFP-FP, CPLFW and CALFW datasets are listed in the Table 1.

Table 5. Verification results of different datasets based on different algorithms.

Method	CALFW	CPLFW	CFP-FP
Softmax	88.21%	77.54%	89.54%
AM-Softmax	89.72%	80.21%	92.12%
DAM-softmax	**90.17%**	**82.08%**	**93.26%**

As seen in Table 5, The proposed DAM-Softmax loss obtains the best performance. From Table 5, we can see that DAM-Softmax works much better on three datasets than AM-Softmax loss. Thus, we further demonstrate that our DAM-Softmax loss has stronger robustness.

5. Conclusions and Future Work

In this paper, we present a novel Double Additive Angular Margin Loss function for face recognition. specifically, we propose to simultaneously impose a angular margin to the intra-class and inter-class variation on the hypersphere manifold, which can effectively enhance the discriminative power of learned deep features. Competitive performance on several popular face benchmarks verify the superiority and robustness of our approach.

Author Contributions: Conceiving the idea, S.Z., and C.C.; writing original draft, S.Z., and C.C.; writing final original manuscript, S.Z., and C.C.; supervision, C.C.; data curation, S.Z., and G.H.; All authors discussed and revised the results, and have read and approved the published version of the manuscript.

Funding: This work was supported by the National Natural Science Foundation of China (60875004).

References

1. Schroff, F.; Kalenichenko, D.; Philbin, J. Facenet: A unified embedding for face recognition and clustering. In Proceedings of the IEEE Conference on Computer Vision and Pattern Recognition, Boston, MA, USA, 7–12 June 2015; pp. 815–823.

2. Krizhevsky, A.; Sutskever, I.; Hinton, G.E. Imagenet classification with deep convolutional neural networks. In *Advances in Neural Information Processing Systems*; Curran Associates, Inc.: Nice, France, 2012; pp. 1097–1105.

3. Simonyan, K.; Zisserman, A. Very deep convolutional networks for large-scale image recognition. *arXiv* **2014**, arXiv:1409.1556.

4. Szegedy, C.; Liu, W.; Jia, Y.; Sermanet, P.; Reed, S.; Anguelov, D.; Erhan, D.; Vanhoucke, V.; Rabinovich, A. Going deeper with convolutions. In Proceedings of the IEEE Conference on cOmputer Vision and Pattern Recognition, Boston, MA, USA, 7–12 June 2015; pp. 1–9.

5. He, K.; Zhang, X.; Ren, S.; Sun, J. Deep residual learning for image recognition. In Proceedings of the IEEE Conference on Computer Vision and Pattern Recognition, Las Vegas, NV, USA, 27–30 June 2016; pp. 770–778.

6. Huang, G.; Liu, Z.; Van Der Maaten, L.; Weinberger, K.Q. Densely connected convolutional networks. In Proceedings of the IEEE Conference on Computer Vision and Pattern Recognition, Honolulu, HI, USA, 21–26 July 2017; pp. 4700–4708.

7. Yi, D.; Lei, Z.; Liao, S.; Li, S.Z. Learning face representation from scratch. *arXiv* **2014**, arXiv:1411.7923 .

8. Russakovsky, O.; Deng, J.; Su, H.; Krause, J.; Satheesh, S.; Ma, S.; Huang, Z.; Karpathy, A.; Khosla, A.; Bernstein, M. ImageNet Large Scale Visual Recognition Challenge. *Int. J. Comput. Vis.* **2015**, *115*, 211–252. [CrossRef]

9. Bottou, L. Large-scale machine learning with stochastic gradient descent. In Proceedings of the COMPSTAT'2010, Paris, France, 22–27 August 2010; pp.177–186.

10. Wen, Y.; Zhang, K.; Li, Z.; Qiao, Y. A discriminative feature learning approach for deep face recognition. In Proceedings of the European Conference on Computer Vision, Amsterdam, The Netherlands, 11–14 October 2016; pp. 499–515.

11. Hadsell, R.; Chopra, S.; LeCun, Y. Dimensionality reduction by learning an invariant mapping. In Proceedings of the 2006 IEEE Computer Society Conference on Computer Vision and Pattern Recognition (CVPR'06). IEEE, Santa Barbara, CA, USA, 26–27 October 2006; Volume 2, pp. 1735–1742.

12. Cheng, D.; Gong, Y.; Zhou, S.; Wang, J.; Zheng, N. Person re-identification by multi-channel parts-based cnn with improved triplet loss function. In Proceedings of the IEEE Conference on Computer Vision and Pattern Recognition, Las Vegas, NV, USA, 26 June–1 July 2016; pp. 1335–1344.

13. Liu, W.; Wen, Y.; Yu, Z.; Yang, M. Large-margin softmax loss for convolutional neural networks. *arXiv* **2016**, arXiv:1612.02295.

14. Liu, W.; Wen, Y.; Yu, Z.; Li, M.; Raj, B.; Song, L. Sphereface: Deep hypersphere embedding for face recognition. In Proceedings of the IEEE Conference on Computer Vision and Pattern Recognition, Honolulu, HI, USA, 21–26 July 2017; pp. 212–220.

15. Wang, F.; Cheng, J.; Liu, W.; Liu, H. Additive margin softmax for face verification. *IEEE Signal Process. Lett.* **2018**, *25*, 926–930. [CrossRef]

16. Sun, Y.; Chen, Y.; Wang, X.; Tang, X. Deep learning face representation by joint identification-verification. *arXiv* **2014**, arXiv:1406.4773.

17. Huang, G.B.; Learned-Miller, E. *Labeled Faces in the Wild: Updates and New Reporting Procedures*; Dept. Comput. Sci., Univ.: Amherst, MA, USA, 2014; pp. 14–003.

18. Sengupta, S.; Chen, J.C.; Castillo, C.; Patel, V.M.; Chellappa, R.; Jacobs, D.W. Frontal to profile face verification in the wild. In Proceedings of the 2016 IEEE Winter Conference on Applications of Computer Vision (WACV). IEEE, Lake Placid, NY, USA, 7–9 March 2016; pp. 1–9.

19. Abadi, M.; Agarwal, A.; Barham, P.; Brevdo, E.; Chen, Z.; Citro, C.; Corrado, D.; Greg S.; Davis, A.; Dean, J.; Devin, M. Tensorflow: Large-scale machine learning on heterogeneous distributed systems. *arXiv* **2016**, arXiv:1603.04467.

20. Kafai, M.; Eshghi, K.; Le, A.; Bhanu, B. A Reference-Based Framework for Pose Invariant Face Recognition. In Proceedings of the IEEE International Conference Workshops on Automatic Face and Gesture Recognition, Ljubljana, Slovenia, 4–8 May 2015.

21. Zheng, T.; Deng, W.; Hu, J. Cross-Age LFW: A Database for Studying Cross-Age Face Recognition in Unconstrained Environments. *arXiv* **2017**, arXiv:1708.08197 .

22. Deng, J.; Guo, J.; Xue, N.; Zafeiriou, S. Arcface: Additive angular margin loss for deep face recognition. *arXiv* **2018**, arXiv:1801.07698.

23. Xu, B.; Wang, N.; Chen, T.; Li, M. Empirical evaluation of rectified activations in convolutional network. *arXiv* **2015**, arXiv:1505.00853.

24. Wang, F.; Xiang, X.; Cheng, J.; Yuille, A.L. Normface: l 2 hypersphere embedding for face verification. In Proceedings of the 25th ACM International Conference on Multimedia, Mountain View, CA, USA, 23–27 October 2017; pp. 1041–1049.

 *applied sciences*

Article

Deep Learning for Facial Recognition on Single Sample per Person Scenarios with Varied Capturing Conditions

Belén Ríos-Sánchez [†,*], David Costa-da-Silva [†], Natalia Martín-Yuste [†] and Carmen Sánchez-Ávila [†]

Group of Biometrics, Biosignals, Security and Smart Mobility, Universidad Politécnica de Madrid, 28040 Madrid, Spain; dcosta@cedint.upm.es (D.C.-d.-S.); nmartin@cedint.upm.es (N.M.-Y.); csa@cedint.upm.es (C.S.-Á.)
* Correspondence: brios@cedint.upm.es; Tel.: +34-91-067-9630
† Current address: Edif. CeDInt-UPM, Campus de Montegancedo, Pozuelo de Alarcón, 28223 Madrid, Spain.

Received: 22 October 2019; Accepted: 5 December 2019; Published: 13 December 2019

Abstract: Single sample per person verification has received considerable attention because of its relevance in security, surveillance and border crossing applications. Nowadays, e-voting and bank of the future solutions also join this scenario, opening this field of research to mobile and low resources devices. These scenarios are characterised by the availability of a single image during the enrolment of the users into the system, so they require a solution able to extract knowledge from previous experiences and similar environments. In this study, two deep learning models for face recognition, which were specially designed for applications on mobile devices and resources saving environments, were described and evaluated together with two publicly available models. This evaluation aimed not only to provide a fair comparison between the models but also to measure to what extent a progressive reduction of the model size influences the obtained results.The models were assessed in terms of accuracy and size with the aim of providing a detailed evaluation which covers as many environmental conditions and application requirements as possible. To this end, a well-defined evaluation protocol and a great number of varied databases, public and private, were used.

Keywords: biometrics; face; deep learning; single sample per person; knowledge generalisation

---

1. Introduction

Facial biometrics is widely extended nowadays because of its great number of applications, the maturity of the technology and the users acceptance. Typical uses include data, devices or facilities access control, surveillance, border crossing, entertainment and human–computer interaction As might be expected, the variety of these application scenarios implies different needs regarding security, storage of information, computing capacity or samples collection. For instance, certain applications cannot store user's templates, are require to verify the identity of the user against a smart card (ID card, driving license or passport) or present difficulties to collect multiple samples of each user.

Focussing on the availability of one or more samples during the training or the enrolment of the users into the system, applications can be divided in two groups, which are known as single sample per person (SSPP) and multiple samples per person (MSPP) problems. SSPP problem has received considerable attention during the last decades because of the relevance of its applications, particularly those related to security, surveillance and border crossing. Nowadays, it is increasingly popular between e-voting and bank of the future service providers. These applications need to guarantee the security of transactions but also to offer an attractive, useful and comfortable user experience, including the access from mobile devices, so they typically avoid to store any information about the user and compare a user's photograph and the image in his/her ID card. SSPP scenarios present lower costs of collecting, storing and processing samples but also add new challenges to typical face recognition difficulties (sensitivity to capturing conditions, facial expressions and changes in the appearance of the users among others). The reduction of the number of images implies a severe reduction on the recognition accuracy of most of the methods developed up to the moment, which strongly rely on the number of samples and their quality to generate a good facial model that generalises inter- and intra-person variability. Accordingly, a solution able to extract and generalise knowledge from previous experiences and similar environments is required.

On the other hand, deep learning solutions for face recognition have received great attention during the last years. In fact, the introduction of convolutional neural networks (CNNs) for facial features extraction marked a turning point when Deep Face [1] and DeepID [2] were presented in 2014, as can be seen in the survey presented by Wang and Deng [3]. At the beginning, CNNs were mostly applied in MSPP scenarios, although they were recently also applied for SSPP problems achieving very promising results. Table 1 summarises the latest works on deep learning for face recognition on SSPP scenarios. Information about the dataset (name and main variations), the number of people and images used during the test, whether an additional dataset is used for training and the match rate are provided for each method to describe their performance. Most of these works provide an evaluation of the methods in terms of accuracy for identification scenarios. However, identity verification is a widely extended scenario. In addition, the specific requirements of the varied applications make necessary a complimentary evaluation of the architectures and the size of the models similarly to the work presented by Hu et al. [4], which quantitatively compare the architectures of CNNs and evaluate the effect of different implementation choices using the public database LFW to train the models. Moreover, the use of public datasets for testing is crucial and the direct comparison against publicly available models seems to be necessary to state a base line.

For these reasons, in this work, two publicly available models, FaceNet [5] and OpenFace [6], were evaluated for SSPP verification scenarios together with two small-size proprietary models, which were specially designed for mobile devices applications and resources saving environments. This evaluation aimed not only to provide a fair comparison between the models but also to measure to what extent a progressive reduction of the model size influences the obtained results. The models were assessed in terms of accuracy and size with the aim of evaluating their applicability to scenarios with different environmental conditions and requirements, including a test where the image of the ID Card is compared against a usual facial image. Many varied databases, public and private, were used with the aim of covering as many scenarios as possible. A distance based classifier was used for matching purposes due to the SSPP nature of the scenarios and to guarantee the straightforward portability of the solution to mobile devices.

The remainder of this article is organised as follows. First, the methods involved in the evaluation are presented in Section 2. Then, the evaluation protocol, the involved databases and the obtained results are presented in Section 3. Finally, conclusions are provided in Section 4.

Table 1. Summary of different works relating to deep learning for face recognition from a single image. ATD column shows if an additional dataset is used for training. E, I, T, LT, O, S, B, V, A and P in Variations column stand for Expression, Illumination, Time, Long Time, Orientation, Scale, Blurring, View, Accessories and Pose, respectively.

Method	Database	Main Variations	# People	# Images	ATD	Match Rate (%)
Face Identity Preserving Features + DNN ([7])	Multi-PIE	P, I, E, T	337	128,940	No	90.46
Face Recovery + FCDN ([8])	LFW	P, I, S, A, E, T	5749	13,233	CelebFaces	97.27
Deep Supervised Autoencoders ([9])	Ext. Yale B	I	38	2432	No	82.22
	AR	I, A	100	2600	No	85.21
	PIE	I, P, E	68	5508	No	72.36
	Multi-PIE	LT, I, V	337	20,209	No	76.12
Virtual image synthesis + DDAN ([10])	EK-LFH	I, B, P, V, S	30	15,930	No	72.08
	LFW	E, I, LT, O, V, A, P	158	>1580	No	97.91
JCR-ACF ([11])	AR	I, E, A	100	2600	CASIA-WF	96.10
	Multi-PIE	P, I, E	249	14,940	CASIA-WF	70.40
	LFW	E, I, LT, O, V, A, P	158	>1580	CASIA-WF	86.00
	CASIA-WF	I, P, E, A	9175	406,423	No	15.00
TLFL framework ([12])	Ext. Yale B	E	38	2404	No	34.86
	AR	E, I	100	1400	No	55.50
	PIE	P, I, E	68	11,630	No	55.89
	LFW	E, I, LT, O, V, A, P	158	>1422	No	15.21
	CAS-PEAL	E	284	1420	No	72.95
	JAFFE	E	10	213	No	89.47
Mean search + LSH + DNN ([13])	Msceleb	-	>10,000	>100,000	No	95.00
	CASIA-WF	I, P, E, A	10,408	>492,744	No	52.43
DCNN ([14])	AR	I, E, A, T	100	2600	CASIA-WF	99.76
	Ext. Yale B	P, I	38	2432	CASIA-WF	88.30
	FERET	P, E, I	200	1400	CASIA-WF	93.90
	LFW	E, I, LT, O, V, A, P	50	>500	CASIA-WF	74.00
NDRDF ([15])	AR	E, I, A, T	116	3016	No	98.00
Transfer Learning + KCFT ([16])	ORL	E, I, A, T	40	400	CASIA-WF	98.14
	FERET	P, E, I	200	1400	CASIA-WF	93.04
	LFW	E, I, LT, O, V, A, P	50	>500	No	97.49

2. Methods

2.1. Face Detection and Alignment

Before extracting the facial biometric features, it is required to locate the face area within the image. It is recommendable that the face area does not contain face surroundings, such as hair, clothes or background. Given the tendency towards deep neural networks to solve object recognition tasks and its good performance, a detector based on cascaded convolutional networks was used [17]. It is composed of three carefully designed deep convolutional networks fed an image pyramid to estimate face position and landmark locations in a coarse-to-fine manner, and exploits the inherent correlation between detection and alignment to increase their performance. In particular, the implementation provided by FaceNet [5] was employed. When multiple faces are present in the image, the biggest face is selected assuming that the user whose identity is validated is closer to the camera.

Since deep learning models are sensitive to the position of the face elements within the images, once the face is detected, it is aligned to a common reference framework. To this end, some reference points belonging to eyes, nose and mouth are detected and transformed to a fixed position.

Finally, face images must have the same size to improve the comparison performance, so all the images are resized to a common size of 160×160 px.

2.2. Feature Extraction

As commented above, face recognition schemes changed since the presentation of Deep Face [1] and DeepID [2] in 2014. Convolutional neural networks learn a mapping from facial images to a compact space where distances directly correspond to a measure of face similarity.

In this study, four models for facial features extraction based on the GoogleNet [18] architecture were compared: FaceNet [5], OpenFace [6], gb2s_Model1 and gb2s_Model2. These models were trained using a deep convolutional network that directly optimises the embedding itself using a triplet-based loss function based on large margin nearest neighbour [19]. These triplets consist of two matching and a non-matching roughly aligned face patches and the loss aims to separate the positive pair from the negative by a squared L2 distance margin. This way, faces of the same person have small distances and faces of different people have large distances.

Given the relevance of triplets selection, FaceNet presents a novel online negative exemplar mining strategy, which ensures consistently increasing difficulty of triplets as the network trains, and explores hard-positive mining techniquess which encourage spherical clusters for the embeddings of a single person to improve clustering accuracy. As a result, FaceNet surpassed state-of-the-art face recognition performance using only 128 bytes per face, getting a classification accuracy of 99.63% $\pm$ 0.09 on the Labeled Faces in the Wild (LFW) dataset [20]. FaceNet was trained with a private dataset containing 100M–200M images and the model size is 90 MB. Details about the methodology, system architecture and network structure can be found in [5]. The TensorFlow implementation provided by Sandberg [21] was used in this comparative.

The good results achieved by FaceNet contributed to increase the accuracy gap between state-of-the-art publicly available and private face recognition systems. Aimed to bridge this gap, a general-purpose face recognition library oriented to mobile scenarios was developed: OpenFace [6]. This real-time face recognition system was specially designed to provide high accuracy with low training and prediction times and adapts depending on the context. The model was trained using a modified version of FaceNet's nn4 network, nn4.small2, which reduces the number of parameters and the number of training images. In this case, 500 k images coming from CASIA-WebFace [22] and FaceScrub [23] datasets were used to train the model after a preprocessing stage where they were aligned. Details about the system architecture, network structure and implementation can be found in [6]. Among the four models offered by OpenFace, the nn4.small2 was selected for this comparative as it gets the best performance (92.92% $\pm$ 1.34% on the LFW dataset).

OpenFace is oriented to mobile applications and considerably reduces the size of the model (30 MB) compared to FaceNet's (90 MB). However, it could still be too large for being embedded in some mobile applications or devices with limited resources. Thus, in this work, two smaller models were trained, viz. gb2s_Model1 and gb2s_Model2, following a similar reduction process than OpenFace. These models were trained using a network inspired by FaceNet but composed of a smaller number of layers and filters. As the idea is to measure the influence of a progressive reduction of the model size on the results, the same structure of layers as OpenFace was used as an starting point but varying the number of filters to reduce the final number of parameters of the model. The architectures of the proposed models are presented in Tables 2 and 3. In addition, the training process followed the same triplet-loss architecture as FaceNet and OpenFace, thus it also provided an embedding on the unit hypersphere and Euclidean distance represented similarity. A subset of the LFW database was used to train the network and the size of the resulting gb2s_Model1 and gb2s_Model2 are 22 MB and 12.5 MB, respectively.

Table 2. gb2s_Model1 network architecture.

Type	Output Size	#1×1	#3×3 Reduce	#3×3	#5×5 Reduce	#5×5	Pool Proj
conv1 (7×7×3,2)	48×48×64						
max pool + norm	24×24×64						m 3×3,2
inception (2)	24×24×192		64	192			
norm + max pool	12×12×192						m3×3,2
inception (3a)	12×12×256	64	96	128	16	32	m, 32p
inception (3b)	12×12×320	64	96	128	32	64	ℓ_2, 64p
inception (3c)	6×6×640		128	256,2	32	64,2	m 3×3,2
inception (4a)	6×6×640	256	96	192	32	64	ℓ_2, 128p
inception (4e)	3×3×1024		160	256,2	64	128,2	m 3×3,2
inception (5a)	3×3×384	128	64	192			ℓ_2, 64p
inception (5b)	3×3×384	128	64	192			m, 64p
avg pool	384						
linear	128						
ℓ_2 normalisation	128						

Table 3. gb2s_Model2 network architecture.

Type	Output Size	#1×1	#3×3 Reduce	#3×3	#5×5 Reduce	#5×5	Pool Proj
conv1 (7×7×3,2)	48×48×64						
max pool + norm	24×24×64						m 3×3,2
inception (2)	24×24×192		64	192			
norm + max pool	12×12×192						m3×3,2
inception (3a)	12×12×256	64	96	128	16	32	m, 32p
inception (3b)	12×12×320	64	96	128	32	64	ℓ_2, 64p
inception (3c)	6×6×640		128	256,2	32	64,2	m 3×3,2
inception (4a)	6×6×312	128	48	96	16	32	ℓ_2,56p
inception (4e)	6×6×504		80	128,2	32	64,2	m 3×3,2
inception (5a)	3×3×256	96	32	128			ℓ_2, 32p
inception (5b)	3×3×256	96	32	128			m, 32p
avg pool	256						
linear	128						
ℓ_2 normalisation	128						

2.3. Matching

A distance based classifier was used to compare the biometric features given the lack of multiple samples during the enrolment in single sample per person scenarios, which are necessary to train more complex classifiers. Its simplicity and low computational requirements also guarantee a straightforward portability of the solution to any environment, including mobile devices. This classifier provides a numeric value as a result, which represents the difference between two feature vectors. Accordingly, the decision policy established in the system is to consider the compared vectors as belonging to the same person if the computed distance is lower than a previously established threshold. Since deep learning approaches evaluated in this study map the images to a compact Euclidean space, Euclidean distance was applied in this study.

3. Evaluation

3.1. Databases

Many images coming from different databases both, public and private, were used in this evaluation. The choice of databases aimed to cover as many of the most realistic cases of use as possible. Accordingly, the images present different degrees of difficulty regarding to pose, scale, background, lighting conditions, appearance, accessories and expressions. In particular, BioID ([24]),

EUCFI ([25]), ORL ([26]), Extended Yale B ([27]), Print-Attack ([28]) and gb2sμMOD_Face_Dataset ([29]) were used together with three proprietary datasets:

- gb2sTablet: A set of 250 frames coming from the gb2sTablet_Face_Dataset which is part of the proprietary gb2sTablet_Database. It contains images from 60 people captured in an indoor environment using artificial lighting and simple but uncontrolled backgrounds. Images were recorded in a frontal position but without restrictions about the distance to the camera, appearance or accessories. The size of the images is 320 × 240 px.
- gb2s_Selfies: A proprietary database oriented to emulate real daily-life scenarios. Accordingly, this dataset shows different illumination conditions and backgrounds, possible presence and absence of glasses, hear variations, or expressions. Twenty-six individuals participated in the database creation and 10 images per person were recorded in 10 different sessions, one image per day. Each person captured images of his/her face using the frontal camera of his/her own mobile phone the more frontal as possible (Selfie position).
- gb2s_IDCards: A private database oriented to evaluate security applications which require from an official document to verify the identity of the users. In this case, the same 26 individuals participated in the database creation and 10 images per person were recorded in 10 different sessions, one image per day. Each person captured images of his/her ID card using the back camera of his/her own mobile phone.

In general, feature extraction models able to generalise knowledge for recognising new people are required. It is even more necessary in SSPP scenarios, thus feature extraction models were previously trained using images coming from totally different datasets. Since this training is separated from the evaluation of the whole system, training datasets are not referenced at this point.

Table 4 summarises the databases involved in the evaluation.

Table 4. Databases overview. A, B, E, L, P, S and T stand for Accessories, Background, Expressions, Lighting, Pose, Scale and Time, respectively.

Database	Access	#Users	#Images	Image Size	Color	Variations
Ext. Yale B	Public	28	16,128	640×480	gray	L, P
ORL	Public	40	400	92×112	gray	L, T, E, A
BioID	Public	23	1521	384×286	gray	B, L, S
EUCFI	Public	395	7900	180×200	color	B, S, E, A
PrintAttack	Public	38	1400	320×240	color	L
gb2sμMOD	Public	60	4220	640×480	color	B, S, A
gb2sTablet	Private	60	16,593	320×240	color	B, S, A
gb2s_Selfies	Private	26	262	-	color	B, L, T, E, A
gb2s_IDCards	Private	26	261	-	gray	L, T

3.2. Evaluation Protocol

An evaluation protocol based on the definitions suggested by the ISO/IEC 19795 standard ([30,31]) was applied to quantify the performance of the different methods described above, ensuring fair comparison between them and hopefully future research. It is composed of three parts:

1. *Dataset Organisation.* First, each evaluation dataset was divided into validation and test subsets, which contain 70% and 30% of the samples respectively. Next, validation subset is in turn separated into enrolment and access samples (also 70% and 30%, respectively), to generate the biometric profile of the users and to simulate accesses into the system. This way, methods were validated and the acceptance threshold was adjusted. Then, new accesses into the already configured system were made using the test samples, allowing for the calculation of more realistic performance rates.

2. *Computation of Scores*. A list of genuine and zero-effort impostor scores was generated at this stage. To this end, the biometric template of each user was created using only one enrolment sample, and access samples were divided into genuine and impostor, corresponding to authentic and forger users, respectively. Then, genuine scores were computed by comparing the access samples against the reference template of the same user, and impostor scores were obtained by comparing each access sample against the biometric templates of the other users. The same process was repeated for each enrolment sample.
3. *Metrics calculation*. Finally, certain metrics about the performance of the system were obtained from genuine and impostor scores. Concretely, validation results are offered in terms of Equal Error Rate (EER) and test results wer measured in terms of False Match Rate (FMR) and False Non-Match Rate (FNMR).The threshold associated to the EER computed in the validation stage was used in the test stage.

3.3. Results

An evaluation of the complete system in a SSPP scenario with identity verification purposes was carried out for each database separately, allowing for the comparability of the results. In addition, a test where an image coming from the gb2s_Selfies dataset was compared against an image of the same user coming from the gb2s_IDCards dataset was included to evaluate this relevant and complicated scenario. Tables 5 and 6 gather the results obtained during de validation and test stages respectively. Results are also illustrated in Figures 1 and 2.

It can be seen that capturing conditions have a great influence on the results. In fact, the results achieved on databases recorded under quite controlled capturing conditions, which usually present lower variability between images, are quite good for every model. However, as soon as the complexity of the datasets increases, the influence of the model become more evident. It can be observed that the most influencing conditions are lighting, in particular low light levels such as those present on Extended Yale B database, expressions and appearance of the users. The image size and the number of images per person does not seem to be very relevant. As can be expected, stronger differences between enrolment and access samples lead to worse results. It is clearly shown in the Selfies-IDCards scenario.

Focusing on the feature extraction models, it can be seen that the progressive model reduction from 9 MB to 30 MB, 22 MB and 12 MB increases the averaged validation EER in 8.37%, 10.31% and 9.79%, respectively, while the averaged test FMR raises are 8.27%, 10.11% and 9.80%, respectively, and the averaged test FNMR increments are 8.81%, 10.74% and 10.94%, respectively. It is evident that a big and well trained model is required to deal with the most complicated scenarios. Accordingly, FaceNet is able to deal even with the Selfies-IDCards, providing quite acceptable results. Applications associated to this scenario typically have strong security requirements, and the use of a bigger model could be justified. However, more research efforts are required to determine a competitive solution in terms of model size and accuracy for this particular scenario. On the other hand, applications where resources saving is a priority need smaller models such as OpenFace and gb2s model(s). The results achieved by these models do not present a clear pattern and they strongly depend on the dataset used during the evaluation. In one half of the cases, OpenFace surpasses gb2s_models, but, on the other half, the results are similar or even worse. Differences between gb2s Model1 and gb2s Model2 are not relevant, being a little bit better the gb2s Model2 in many cases, which is the smallest. Obtained results highlight that the use of a reduced model for face recognition is viable when they are trained with a sufficiently big dataset containing enough representativeness.

Comparing validation and test results, a great consistency can be observed. The biometric template of each user is composed just by one single sample and the feature extraction models were trained with separated datasets, thus the learnt features were not directly related with the images used in the experiments and their capturing conditions. Therefore, validation and test results obtained for each dataset are pretty similar.

Finally, even when comparison against the state of the art is difficult because different authors follow different evaluation methodologies, it can be seen that the presented results generally match or exceed the state-of-the-art solutions. Focusing on those works which use the same databases for evaluation purposes, it can be observed that FaceNet and OpenFace clearly surpasses the actual state of the art. This is not the case of gb2s_Model(s), which in some cases surpasses the state of the art but in many cases does not. However, it can be observed that most state-of-the-art solutions use the same databases for training and testing purposes, which could influence the results. Thus, it is not possible to provide a fair comparison against the state of the art, and a new study following the same evaluation protocol and using the same databases, which should be different from those involved in the training process, must be carried out to this end.

Table 5. Facial recognition using deep learning and DBC on single sample per person scenarios: validation results (EER (%)).

Feature Extraction	BioID	EUCFI	ORL	Ext. Yale B	Print Attack	gb2s Tablet	gb2s μMOD	gb2s Selfies	gb2s IDCards	Selfies-IDCards
Face Net	1.47	1.16	0.25	3.60	0.05	0.05	4.28	1.91	0.55	4.66
Open Face	6.77	2.39	3.54	9.35	3.46	2.21	10.44	16.04	1.43	46.05
gb2s_Model1	16.85	2.65	6.65	27.76	4.09	0.76	18.94	11.98	2.56	28.86
gb2s_Model2	13.22	2.63	6.25	29.02	6.30	0.74	19.72	9.56	1.79	26.61

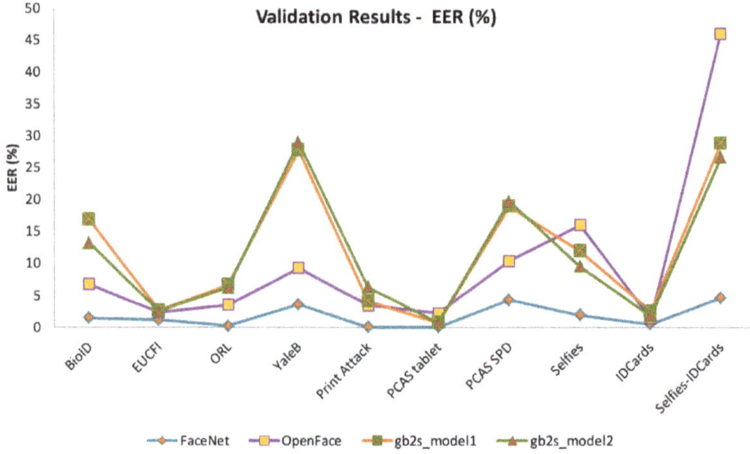

Figure 1. SSPP validation results.

Table 6. Facial recognition using deep learning and DBC on single sample per person scenarios: test results (%).

Database	FaceNet		OpenFace		gb2s_Model1		gb2s_Model2	
	FMR	FNMR	FMR	FNMR	FMR	FNMR		
BioID	1.34	0.07	6.99	7.73	16.17	17.72	13.70	14.30
EUCFI	1.21	1.85	2.44	2.16	2.55	3.71	2.79	3.30
ORL	0.35	1.25	2.62	5.60	3.29	7.25	6.35	8.25
Ext. Yale B	3.85	3.96	9.39	9.63	30.19	27.91	31.68	29.62
Print Attack	0.02	0.09	3.60	2.94	4.15	6.64	6.07	8.61
gb2sTablet	0.05	0.07	2.20	2.19	0.72	1.21	0.68	1.07
gb2sμMOD	3.89	4.77	10.67	9.76	16.18	24.87	16.53	26.91
gb2s_Selfies	1.66	0.00	13.91	17.90	10.60	11.83	8.67	11.45
gb2s_IDCards	0.70	0.76	1.86	3.33	2.47	3.05	1.51	2.29
Selfies - IDCards	4.89	2.29	47.00	42.00	29.77	18.32	28.00	18.70

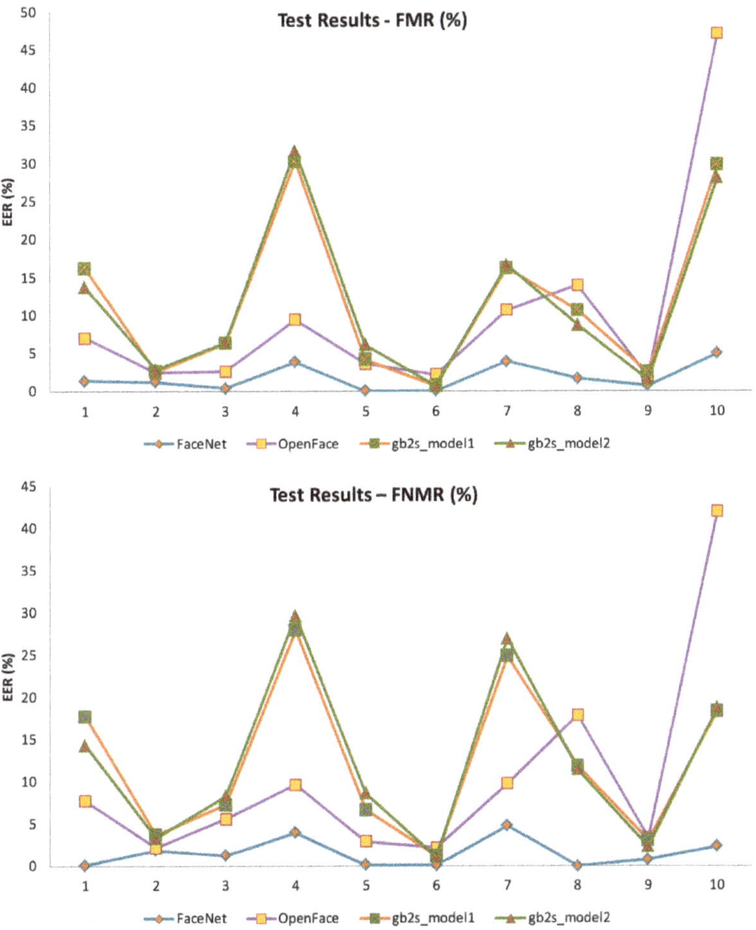

Figure 2. SSPP Validation Results.

4. Conclusions

In this study, two deep learning models for face recognition, which were specially designed for applications on mobile devices and resources saving environments, were described and evaluated together with two publicly available models (FaceNet and OpenFace) for identity verification at single sample per person scenarios. This evaluation aimed not only to provide a fair comparison between the models but also to measure to what extent a progressive reduction of the model size influences the obtained results. The models were assessed in terms of accuracy and size with the aim of evaluating their applicability to scenarios with different environmental conditions and requirements. To this end, a great number of varied databases, public and private, was used. Given that SSPP scenarios imply that there is only one sample per person during the enrolment into the system, a distance based classifier was used to compare the biometric features.

The results show that the influence of the feature extraction model become more evident as the complexity of the images increases. In fact, a big and well trained model is required to deal with really complicated scenarios that present high variability between images used to enrol users into the system and posterior accesses. However, for those scenarios where resources saving is a priority, smaller models are viable if they are trained with a sufficiently big dataset containing enough representativeness.

Author Contributions: B.R.-S. and C.S.-Á. conceived the research and the methodology. B.R.-S., D.C.-d.-S. and N.M.-Y. implemented the software and carried out the experiments. B.R.-S., D.C.-d.-S., N.M.Y. and C.S.-Á. analysed the results. B.R.-S. wrote and edited the paper and C.S.-Á. reviewed it. C.S.-Á. was in consultation at every moment and provided valuable suggestions and criticisms. C.S.-Á. and B.R.-S. acquired the fundings and administered the projects.

Funding: This work was partially funded by the project "Awesome Possum: Data driven authentication", lead by Telenor Digital AS and financed by The Research Council of Norway under project number 256581, and a private project lead by Indra Sistemas S.A.

Acknowledgments: The authors would also like to thank BioID® for the use of BioID Face Database, Libor Spacek for the use of Essex University's Collection of Facial Images (Face Recognition Data), AT&T Laboratories Cambridge for the use of the ORL Database, UCSD for the use of Extended Yale B database and IDIAP for the use of Print-Attack database, as well as all the contributors who have participated in the creation of our proprietary databases for their patience and cooperation.

Conflicts of Interest: The authors declare no conflict of interest.

References

1. Taigman, Y.; Yang, M.; Ranzato, M.; Wolf, L. DeepFace: Closing the Gap to Human-Level Performance in Face Verification. In Proceedings of the 2014 IEEE Conference on Computer Vision and Pattern Recognition, Columbus, OH, USA, 23–28 June 2014; pp. 1701–1708.
2. Sun, Y.; Wang, X.; Tang, X. Deep Learning Face Representation from Predicting 10,000 Classes. In Proceedings of the 2014 IEEE Conference on Computer Vision and Pattern Recognition, Columbus, OH, USA, 23–28 June 2014; IEEE Computer Society: Washington, DC, USA, 2014; pp. 1891–1898.
3. Wang, M.; Deng, W. Deep Face Recognition: A Survey. *arXiv* **2018**, arXiv:1804.06655 .
4. Hu, G.; Yang, Y.; Yi, D.; Kittler, J.; Christmas, W.; Li, S.Z.; Hospedales, T. When Face Recognition Meets with Deep Learning: An Evaluation of Convolutional Neural Networks for Face Recognition. In Proceedings of the 2015 IEEE International Conference on Computer Vision Workshop (ICCVW), Tampa, FL, USA, 5–8 December 2015; pp. 384–392.
5. Schroff, F.; Kalenichenko, D.; Philbin, J. FaceNet: A Unified Embedding for Face Recognition and Clustering. In Proceedings of the IEEE Computer Society Conference on Computer Vision and Pattern Recognition 2015, Boston, MA, USA, 7–12 June 2015.
6. Amos, B.; Bartosz, L.; Satyanarayanan, M. *OpenFace: A General-Purpose Face Recognition Library with Mobile Applications*; Technical Report, CMU-CS-16-118; CMU School of Computer Science: Pittsburgh, PA, USA, 2016.
7. Zhu, Z.; Luo, P.; Wang, X.; Tang, X. Deep Learning Identity-Preserving Face Space. In Proceedings of the 2013 IEEE International Conference on Computer Vision and Pattern Recognition, Sydney, Australia, 1–8 December 2013; IEEE Computer Society: Washington, DC, USA, 2013; pp. 113–120.
8. Zhu, Z.; Luo, P.; Wang, X.; Tang, X. Recover Canonical-View Faces in the Wild with Deep Neural Networks. *arXiv* **2014**, arXiv:1404.3543.
9. Gao, S.; Zhang, Y.; Jia, K.; Lu, J.; Zhang, Y. Single Sample Face Recognition via Learning Deep Supervised Autoencoders. *IEEE Trans. Inf. Forensics Secur.* **2015**, *10*, 2108–2118. [CrossRef]
10. Hong, S.; Im, W.; Ryu, J.; Yang, H.S. SSPP-DAN: Deep domain adaptation network for face recognition with single sample per person. In Proceedings of the 2017 IEEE International Conference on Image Processing (ICIP), Beijing, China, 17–20 September 2017; IEEE: Piscataway, NJ, USA, 2017; pp. 825–829.
11. Yang, M.; Wang, X.; Zeng, G.; Shen, L. Joint and collaborative representation with local adaptive convolution feature for face recognition with single sample per person. *Pattern Recognit.* **2017**, *66*, 117–128. [CrossRef]
12. Guo, Y.; Jiao, L.; Wang, S.; Wang, S.; Liu, F. Fuzzy Sparse Autoencoder Framework for Single Image per Person Face Recognition. *IEEE Trans. Cybern.* **2018**, *48*, 2402–2415. [CrossRef] [PubMed]
13. Xihua, L. Improving Precision and Recall of Face Recognition in Sipp with Combination of Modified Mean Search and Lsh. Ph.D. Thesis, Beihang University, Beijing, China, 2018.

14. Zeng, J.; Zhao, X.; Gan, J.; Mai, C.; Zhai, Y.; Wang, F. Deep Convolutional Neural Network Used in Single Sample per Person Face Recognition. *Comput. Intell. Neurosci.* **2018**, *2018*, 3803627. [CrossRef] [PubMed]

15. Ouanan, H.; Ouanan, M.; Aksasse, B. Non-linear dictionary representation of deep features for face recognition from a single sample per person. *Procedia Comput. Sci.* **2018**, *127*, 114–122. [CrossRef]

16. Min, R.; Xu, S.; Cui, Z. Single-Sample Face Recognition Based on Feature Expansion. *IEEE Access* **2019**, *7*, 45219–45229. [CrossRef]

17. Zhang, K.; Zhang, Z.; Li, Z.; Qiao, Y. Joint Face Detection and Alignment Using Multitask Cascaded Convolutional Networks. *IEEE Signal Process. Lett.* **2016**, *23*, 1499–1503. [CrossRef]

18. Szegedy, C.; Liu, W.; Jia, Y.; Sermanet, P.; Reed, S.; Anguelov, D.; Erhan, D.; Vanhoucke, V.; Rabinovich, A. Going deeper with convolutions. In Proceedings of the 2015 IEEE Conference on Computer Vision and Pattern Recognition (CVPR), Boston, MA, USA, 7–12 June 2015; pp. 1–9.

19. Weinberger, K.Q.; Blitzer, J.; Saul, L.K. Distance Metric Learning for Large Margin Nearest Neighbor Classification. In *Advances in Neural Information Processing Systems 18*; Weiss, Y., Schölkopf, B., Platt, J.C., Eds.; MIT Press: Cambridge, MA, USA, 2006; pp. 1473–1480.

20. Huang, G.B.; Mattar, M.; Berg, T.; Learned-Miller, E. Labeled Faces in the Wild: A Database forStudying Face Recognition in Unconstrained Environments. In Proceedings of the Workshop on Faces in 'Real-Life' Images: Detection, Alignment, and Recognition, Marseille, France, 17 October 2008.

21. Sandberg, D. Face Recognition Using Tensorflow. 2017. Available online: https://github.com/davidsandberg/facenet (accessed on 8 March 2019).

22. Yi, D.; Lei, Z.; Liao, S.; Li, S. Learning Face Representation from Scratch. *arXiv* **2014**, arXiv:1411.7923.

23. Ng, H.W.; Winkler, S. A data-driven approach to cleaning large face datasets. In Proceedings of the 2014 IEEE International Conference on Image Processing (ICIP), Paris, France, 27–30 October 2014; IEEE Xplore Digital Library: Piscataway, NJ, USA, 2014; pp. 343–347.

24. BioID. BioID Face Database. Available online: https://www.bioid.com/facedb/ (accessed on 8 March 2019).

25. EUCFI. The Essex University Collection of Face Images. Available online: https://cswww.essex.ac.uk/mv/allfaces/index.html (accessed on 8 March 2019).

26. ORL. The Database of Faces. Available online: http://www.cl.cam.ac.uk/research/dtg/attarchive/facedatabase.html (accessed on 8 March 2019).

27. YaleB. The Extended Yale Face Database B. Available online: http://vision.ucsd.edu/~iskwak/ExtYaleDatabase/ExtYaleB.html (accessed on 8 March 2019).

28. Anjos, A.; Marcel, S. Counter-Measures to Photo Attacks in Face Recognition: A public database and a baseline. In Proceedings of the International Joint Conference on Biometrics 2011, Washington, DC, USA, 11–13 October 2011.

29. Ríos-Sánchez, B.; Arriaga-Gómez, M.; Guerra-Casanova, J.; de Santos-Sierra, D.; de Mendizábal-Vázquez, I.; Bailador, G.; Sánchez-Ávila, C. gb2sμMOD: A MUltiMODal biometric video database using visible and IR light. *Inf. Fusion* **2016**, *32*, 64–79. [CrossRef]

30. ISO. *ISO/IEC 19795-1:2007: Information Technology—Biometric Performance Testing and Reporting—Part 1: Principles and Framework*; International Organization for Standardization (ISO): Geneva, Switzerland; 2007.

31. ISO. *ISO/IEC 19795-2:2007: Information Technology—Biometric Performance Testing and Reporting—Part 2: Testing Methodologies for Technology and Scenario Evaluation*; International Organization for Standardization (ISO): Geneva, Switzerland; 2007.

Article

Real-Time Pre-Identification and Cascaded Detection for Tiny Faces

Ziyuan Yang [1,†]**, Jing Li** [1,†]**, Weidong Min** [2,3,*] **and Qi Wang** [1]

[1]   School of Information Engineering, Nanchang University, Nanchang 330031, China;
     yangziyuan@email.ncu.edu.cn (Z.Y.); jingli@ncu.edu.cn (J.L.); 351029018003@email.ncu.edu.cn (Q.W.)
[2]   School of Software, Nanchang University, Nanchang 330047, China
[3]   Jiangxi Key Laboratory of Smart City, Nanchang 330047, China
*   Correspondence: minweidong@ncu.edu.cn; Tel.: +86-0791-8830-4080
†   The first two authors contributed equally to this work.

Received: 6 September 2019; Accepted: 6 October 2019; Published: 15 October 2019

Abstract: Although the face detection problem has been studied for decades, searching tiny faces in the whole image is still a challenging task, especially in low-resolution images. Traditional face detection methods are based on hand-crafted features, but the features of tiny faces are different from those of normal-sized faces, and thus the detection robustness cannot be guaranteed. In order to alleviate the problem in existing methods, we propose a pre-identification mechanism and a cascaded detector (PMCD) for tiny-face detection. This pre-identification mechanism can greatly reduce background and other irrelevant information. The cascade detector is designed with two stages of deep convolutional neural network (CNN) to detect tiny faces in a coarse-to-fine manner, i.e., the face-area candidates are pre-identified as region of interest (RoI) based on a real-time pedestrian detector and the pre-identification mechanism, the set of RoI candidates is the input of the second sub-network instead of the whole image. Benefiting from the above mechanism, the second sub-network is designed as a shallow network which can keep high accuracy and real-time performance. The accuracy of PMCD is at least 4% higher than the other state-of-the-art methods on detecting tiny faces, while keeping real-time performance.

Keywords: face detection; tiny faces; pre-identification mechanism; cascaded detector; deep learning; convolutional neural network

---

1. Introduction

Face detection is one of the most hot topics in computer vision as it is a key step for many different applications, such as face recognition [1], facial expression analysis [2], eye-tracking [3], facial performance capture [4], facial expression transformation [5], etc. In fact, the applications are not limited to the traditional areas, there are still some exciting interdisciplinary applications [6–11] in the field of animation. However, many factors such as the illumination, occlusion, and the diversity of faces cause huge challenges in face detection.

Using universal face templates to detect faces is one of the main research fields of traditional methods. Determining whether there is a face is undertaken by calculating the correlation coefficient between the area which is detected and the template [12]. However, facial skin color, different expressions, and occlusion lead to the method being less robust and computationally complex.

In recent years, many researchers shift their attention from the traditional methods to convolutional neural networks (CNNs) [13–15] since they have achieved remarkable success in many important tasks of computer vision, such as classification, detection, and recognition. Lots of approaches have been proposed to solve the problem of tiny-face detection, which aims to search a tiny face in a whole image, especially in a low-resolution image. However, these methods cannot achieve satisfactory performance because the features of tiny faces are different from those of big faces and tiny faces contain limited information available for face detection.

It is challenging to detect small objects in image detection. That is because these networks are designed to propose default boxes and the classification score is calculated from one single deep CNN. For example, faster Region-based Convolutional Neural Networks (R-CNN) [16] extracts features by visual geometry group (VGG)-16 [17], but when the face size is less than 16×16, the output in 'conv5' is less than one pixel. As the convolutional layer is deeper, each pixel in the feature map gathers more information outside of the original input area and lower information of the region of interest (RoI), which means these methods cannot keep the performance when targets are small. However, a single shallow CNN cannot get enough information for object detection, and thus a cascaded structure which is usually divided into a prediction part and a regression part becomes a popular and effective framework for face detection. A cascade framework can generate a large number of bounding box candidates based on a low threshold, and then extracts the regression scores of these candidates. That is, the information loss of shallow convolutional layers can be effectively relieved.

Moreover, the above-mentioned face detection algorithms cannot keep good performance in video surveillance, because they are designed for high-resolution images and big faces. Limited by the cost of large-scale surveillance and the scale of data, the face targets are often small and not clear. Additionally, the head movement tends to be more frequent than the body, making it difficult to detect tiny faces in multiple views. Previous methods are difficult to accurately detect tiny faces in this kind of scenario. If tiny faces can be detected automatically in real time, the telephoto lens would immediately be aimed at the suspicious persons' faces, which can help the police to conduct reconnaissance. This is very useful in public security, since violent incidents and terrorist attacks have occurred in many places in recent years.

In order to alleviate the above problems, this paper presents a cascaded framework named pre-identification mechanism and a cascaded detector (PMCD) to detect tiny faces based on two independent CNNs and a pre-identification mechanism. The two sub-networks can be trained separately, which greatly improves the flexibility in training networks. The first sub-network of PMCD and the pre-identification mechanism generate a set of RoI candidates, defined by the pedestrian area in the image. Then, the RoI candidates are resized to different scales to build an image pyramid as the input of the following network. As the input of the second sub-network has greatly reduced the irrelevant area, a shallow network can learn enough features and guarantee real-time performance.

We tested PMCD on a self-collected dataset and Caltech Pedestrian [18] dataset. Due to the scarcity of the tiny-face detection dataset, we collected 1370 images, 2450 faces in total and most faces were less than 20×20. In order to test in different situations, the size of 562 faces were larger than 20×20, while the rest were all smaller than the size in the self-collected dataset. As the results showed, compared with other state-of-the-art methods, PMCD achieves impressive performance in tiny-face detection and it can achieve real-time detection.

To this end, this paper aims to alleviate the issues discussed above which are tiny-face detection and keeping real-time features.

- We propose a new pre-identification mechanism to obtain a set of face candidates as the input for the face detector, which contains higher face proportion than that in the original image. The mechanism greatly reduces miss rate of tiny-face detection and leads to the robustness of PMCD.
- We propose a novel cascade neural network called PMCD for real-time tiny-face detection. The first sub-network is a deep pedestrian detector as a part of the pre-identification mechanism, and the second sub-network is designed based on a shallow multi-task CNN to detect faces in the RoI candidates.

The rest of this paper is as follows. Section 2 introduces related works on face detection. Section 3 details the framework including two sub-networks and the pre-identification mechanism. Section 4 mainly describes the performance of PMCD and compares it with other state-of-the-art methods. Section 5 concludes and looks forward to future works.

2. Related Works

Face detection is the key step in many different face-related applications and studies [19]. Most of the early work was designed for high-resolution images and large targets by using statistical learning methods to automatically extract features. Yaman et al. [20] proposed a framework to detect faces by utilizing histogram-based feature extraction with random subspace and voting ensemble learners. Luo et al. [21] proposed a face location algorithm which is developed to extract face regions with a high proportion of skin. This type of method increases the speed of the operation. Since the features are not completely selected by humans, the robustness is improved. However, the effect is still poor when the target is small. Mohanty et al. [22] proposed a new feature and combined it with the gray level feature and skin color feature, and this method improves detection speed in complex backgrounds and reduces the computational complexity. Additionally, Ma et al. [23] used the geometric relationship between facial organs to generate four Haar-like features for face detection based on the traditional Adaboost classifier, which greatly reduces the detection time. One of the most impressive traditional methods is Viola-Jones [24] which designed cascade classifiers based on the Haar feature and AdaBoost classifier, but all of these traditional methods focus on improving the performance with more effective hand-crafted features and more powerful classifiers [25–28]. These features or detection structures have certain subjective factors, which leads to the robustness of these frameworks being poor, and the operation being complicated and time consuming.

In recent years, deep learning models offer new ideas in solving many research problems such as classification, object detection, image segmentation, image restoration etc. Convolutional neural networks have achieved remarkable success in many different tasks of computer vision, especially for object detection. The family of Region-based Convolutional Neural Networks (R-CNN) [17,29,30], You Only Look Once (YOLO) [31], and Single Shot Multibox Detector (SSD) [32] are the most efficient and popular approaches for detecting objects these years. Inspired by these brilliant methods, researchers have proposed many effective and robust face detection structures. Jiang et al. [33] investigated applying faster R-CNN to face detection, compared with previous proposed models, and its accuracy has a significant improvement. Contextual multi-scale region-based CNN (CMS-RCNN) [34] combined multi-scale information which consists of the region proposal component and the RoI detection component to detect faces. Wan et al. [35] improved Faster R-CNN with the hard negative mining and significant boosts, and the hard negatives harvested from a large set of background examples. Sun et al. [36] applied some effective strategies including feature concatenation, hard negative mining, multi-scale training, model pre-training, and proper calibration to improve Faster R-CNN in face detection. Zhang et al. [37] light-designed R-CNN and improved the performance by integrating multi-scale training, multi-scale testing, some tricks for inference, and a vote-based ensemble method. Enlightened by SSD, Hsu et al. [38] proposed Multiple Dropout Framework to detect faces. Due to the anchor mechanism, the family of R-CNN can detect objects which occupy the majority of an image which are clear and huge, but the accuracy of these methods drop rapidly when the target is small.

Li et al. [39] proposed a novel cross-level parallel network (CLPNet) to extract low-level features and fuse them in the high-level stage, and CLPNet achieved remarkable performance on crowd counting. Triantafyllidou et al. [40] proposed a novel lightweight deep neural network and a new training method of progressive positive and hard negative sample mining to improve training speed and accuracy. A Fully Convolutional Network (FCN) [41] generated face proposals by the heat map of facial parts which are scored by a new facial parts responses method by their spatial structure and arrangement. As described in Section 1, cascaded structures are more advantageous than single networks in the task of face detection, and thus have been widely used in face detection, CascadedCNN [42] is a cascade framework built on CNNs and a CNN-based calibration stage was introduced to adjust the position of the detection window. Qin et al. [43] proposed joint training to achieve end-to-end optimization for CNN cascade and showed back propagation used in training a single CNN can be naturally used in training the cascade CNN structure. Coarse-to-Fine Auto-encoder Network (CFAN) [44] cascades a few Stacked Auto-encoder Networks to accomplish different tasks including face detection. Multitask Cascaded Convolutional Networks (MTCNN) [45] is a multi-task cascade network to detect faces by three stages in a coarse-to-fine detection structure. Min et al. [46] proposed a multi-scale and multi-channel shallow convolutional network (MMSC) for real-time face detection after the pre-identified method detecting faces in the images based on a traditional pedestrian detection method. Hu et al. [47] proposed a multi-task detector with hybrid resolutions (HR), which detects different face scales from multiple layers of a single neural network.

The methods mentioned above have good performance in face detection when images are of high resolution and the faces are big, but all of these traditional methods and region-based single deep CNNs cannot keep a high accuracy in tiny-face detection, the reason is that the effective information of tiny faces is very limited, as explained in Section 1. Inspired by [46], we designed a novel structure to reduce the unnecessary input of the face detection order to improve the effectiveness and propose a new pre-identification mechanism.

3. Proposed Method

3.1. Overview of Our Method

In this paper, we used two convolutional neural networks (CNNs) for coarse-to-fine face detection. CNN is actually a multi-layer perception, where each layer is composed of multiple feature maps through different convolutional kernels. The most important advantage of this structure is that the parameters of CNN are self-learned through training data, so the structure avoids the generalization weakness caused by hand-crafted features. CNN usually consists of convolutional layers, pooling layers, and fully connected layers. A convolutional layer is used to extract features, a pooling layer aims to reduce the amount of data for calculation, and a fully connected layer is designed to combine the extracted local features into a powerful global feature. In practical operations, convolution layers and pooling layers are often designed as a whole, i.e., the structure would pool the feature maps after the convolutional operation. However, the output of convolutional layers is not immediately pooled, it is passed to the activation function first and then the results are passed to the pooling layers. Activation functions, which are always nonlinear functions, are used for feature mapping in order to help solve nonlinear problems. The parameters in convolutional layers and pooling layers are learned through the back-propagation algorithm, so the self-learned characteristic makes CNN far superior to the traditional algorithms in precision and robustness in many computer vision tasks such as classification, detection, tracking, etc.

Although face detection has received extensive attention in recent years, it still entails many challenges, such as occlusion, complex environments, small targets, etc. This paper aims to solve the problem of tiny-face detection. Due to the low resolution of face images, detecting faces is not an easy task when a pedestrian is far away from the camera. In this situation, the information of a body is much more than the face information. The pedestrians' faces can be roughly calibrated based on body context information. Considering this, we defined the region at the top of the pedestrians' bounding boxes as RoI candidates, obtained the candidate regions by a pedestrian detector and a novel pre-identification algorithm to reduce the search range as pre-processing.

Methods of pedestrian detection include the method based on statistical learning methods, background modeling, and neural networks. The most influential pedestrian detection structure is what Dalal et al. [48] proposed, which uses Support Vector Machine (SVM) as the classifier to detect faces in the images based on the histograms of oriented gradient (HOG) features. Lots of features have been proposed with better robustness and result in better classification accuracy for pedestrian detection [49–52]. However, these features are often specifically designed to a particular situation, hence the robustness of these traditional methods is not good. In order to alleviate the problems mentioned above, more and more researchers have adopted convolutional neural networks (CNNs) to detect objects these years [17,29–32,53]. There are many images from different scenes in the training set, so the robustness of features learned by neural networks is better than hand-crafted features. Therefore, these methods based on CNNs have a huge improvement in accuracy and robustness.

Our method consists of two parts, pre-identification detection and face detection. In the pre-identification detection part, we detected the bounding box of the pedestrian by a deep CNN first, then the RoI candidates would be selected by the self-adaptively algorithm, after that we built an image pyramid, which was used to adapt different sizes of faces, as the input for the face detection network. In the face detection part, image pyramid would be passed to our proposed multitask neural network to detect faces. As the previous steps significantly reduce background interference, the multi-task face detector requires only a shallow structure to perform efficiently and in real-time. What we proposed is a cascaded framework, and we can train the two sub-networks separately. The whole framework is shown in Figure 1.

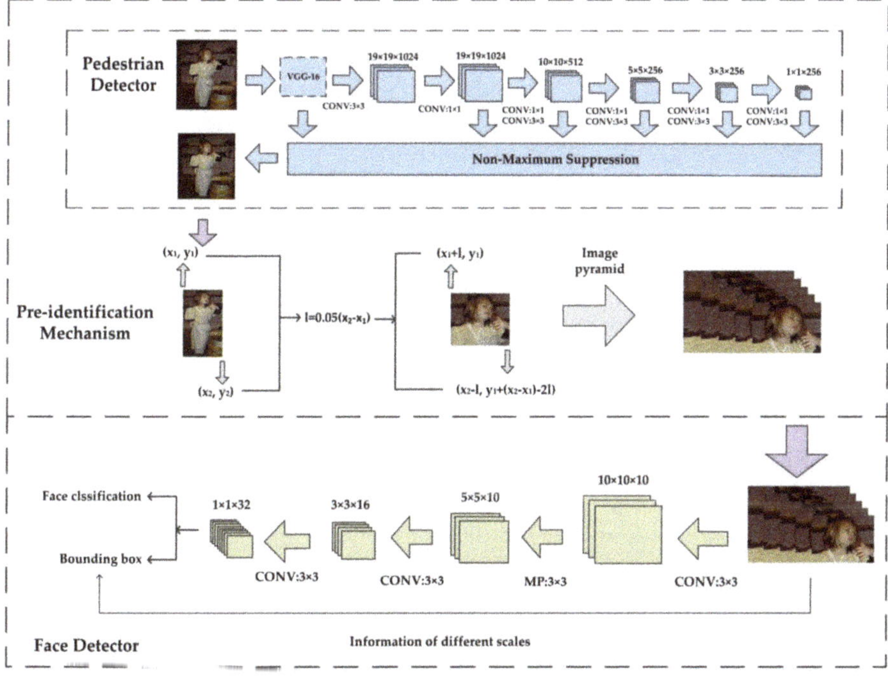

Figure 1. The whole framework.

3.2. Proposed Pre-Identification Mechanism

3.2.1. Pedestrian Detector

The bounding boxes of the pedestrians are detected by a deep CNN [32] in order to narrow the detection range of the face detector, as seen in Figure 1. Firstly, the image size is resized to 300 × 300. Afterwards, the features are extracted by VGG-16 and then fed to six additional convolutional layers for detecting the targets. All default boxes are predicted by combining many feature maps with different scales and ratios, multiple default boxes are set to cover various sizes and shapes of targets. Finally, non-maximum suppression (NMS) is used to obtain the final bounding box which has the highest score in the set of bounding boxes of the target.

Convolutional layers, except VGG-16 in the pedestrian detector, predict bounding boxes and offsets. There are different receptive fields in different levels of feature maps. The scale of the default boxes is computed as:

$$s_k = s_{min} + \frac{s_{max} - s_{min}}{m - 1}(k - 1), \ k \in [1, \, m], \tag{1}$$

where s_{min} is 0.2 and s_{max} is 0.9, the length l_k and the width w_k of the bounding box are calculated based on s_k; m is the number of default boxes, W is the width of the input and L is the length of the input, r is the ratio of the length and the width in each feature map. The detailed calculation formula is shown as:

$$\begin{cases} w_k = W \times s_k \times \sqrt{r} \\ l_k = L \times s_k / \sqrt{r} \end{cases}. \tag{2}$$

3.2.2. Pre-Identification Mechanism

In order to reduce the misclassification rate in the face detector and reduce the interference of other information, we should reduce the input of the face detector. Face areas are adaptively estimated according to the proportion of the pedestrian area. The proportional relationship between the face-region and the pedestrian-region is estimated by our newly proposed mechanism. The face-region selection operator is used to generate a set of face candidate regions which is described in Algorithm 1, and the whole process of face pre-identification is shown in Figure 1. Herein, N_i and D_i are two sets of coordinates, N_i contains the upper-left corner coordinates of the bounding box N_i^1 and the lower-right corner coordinates N_i^2 through the pedestrian detector, N_i^1 and N_i^2 are composed by x and y; D_i contains the upper-left corner coordinates of the bounding box D_i^1 and the lower-right corner coordinates D_i^2 through the pre-identification algorithm, D_i^1 and D_i^2 are composed by x and y, and i means the i-th person in the image.

Algorithm 1. Pre-identification.

Input: Coordinates of the bounding boxes N_i through the pedestrian detector
Output: Coordinates of the RoI D_i
1. $T \leftarrow 0.9$
2. $i \leftarrow 0$
3. $\theta \leftarrow (1 - T)/2$
4. **while** $N_i \neq \varnothing$
5. $l \leftarrow \theta \times \left(N_i^2.x - N_i^1.x\right)$
6. $D_i^1.x \leftarrow N_i^1.x + l$
7. $D_i^1.y \leftarrow N_i^1.y$
8. $D_i^2.x \leftarrow N_i^2.x - l$
9. $D_i^2.y \leftarrow N_i^1.y + \left(N_i^2.x - N_i^1.x\right) - 2 * l$
10. $i \leftarrow i + 1$
11. **end while**

3.2.3. Image Pyramid

The input image size is fixed for the face detector, but the size of the target is not fixed. Therefore, image pyramids can be used to detect different sizes of objects with the fixed input size. We build image pyramids upon the set of RoI candidates. RoIs are resized to different scales which are adaptively computed by the image pyramid method to build an image pyramid, which is the input of the second multi-task face detector. The image pyramid algorithm is described in Algorithm 2. Herein, D_i is the coordinate set of RoI, which is the same as in Section 3.2.2, i means the i-th person in the image, factor is the scaling factor, P_i is an image pyramid of the i-th person, containing many resized images. The threshold of minlin Algorithm 2 is set to 12, since the size of input of the face detector is 12 and the information is pretty limited if the size of face is smaller than 12.

Algorithm 2. Image pyramid.

Input: Coordinates of the RoI D_i
Output: Image pyramid P_i
1. $w \leftarrow D_i^2.x - D_i^1.x$
2. $l \leftarrow D_i^2.y - D_i^1.y$
3. $count \leftarrow 0$
4. $factor \leftarrow 0.7$
5. **if** $w < l$
6. $minl \leftarrow w$
7. **else**
8. $minl \leftarrow l$
9. **end if**
10. **if** $minl < 12$
11. $minl \leftarrow 12$
12. **end if**
13. $m \leftarrow 12/minl$
14. $minl \leftarrow minl * \mathrm{m}$
15. **while** $minl \geq 12$
16. $scales \leftarrow scales + m * factor^{count}$
17. $minl \leftarrow minl * factor$
18. $P_i^{count} \leftarrow Resize(D_i, scales)$
19. $count \leftarrow count + 1$
20. **end while**

3.3. Multitask Face Detector

Before detecting the face, we pre-processed the image to reduce the size of the input. The detector is a shallow neural network, but the depth is enough to learn useful features of tiny faces, because the input of the detector is preprocessed, and background information and other interferences is greatly reduced. Another benefit of shallow neural networks is that the parameters are much smaller than deeper detectors, resulting in fast operation and real-time performance.

Each image pyramid is passed to the face detector to detect the coordinates of the face. The real coordinates are calculated based on the detected coordinates of bounding boxes and the image pyramid scales. Finally, the NMS method is used to eliminate the redundant bounding boxes by getting multiple overlapped bounding boxes and reducing them to only one. The structure of our face detector is shown in Figure 1. The stride of the convolutional layer is 1, and the step size of the pooling layer is 2.

Different convolutional kernel sizes may result in different feature extraction effects. Considering that the detected target is a tiny face with the input size is 12×12, we apply a 3×3 filter for all the convolutional layers. There are two tasks in the face detector, face classification and bounding box regression. Compared to other complex multiclass objection classification and detection methods, the input of our proposed network is the image pyramid generated from the set of RoI candidates and the number of classes is only 2, face and non-face, so this network does not need a deep network structure and a lot of convolutional kernels in each layer. ReLU is applied as nonlinearity activation function in this detector, which is:

$$f(x) = \begin{cases} 0, & x \leq 0 \\ x, & x < 0 \end{cases} . \tag{3}$$

Face classification is a binary classification problem, so we use the cross-entropy logistic regression function as the loss function. The function is designed as follows:

$$L_i = -\left(y_i \log p_i + \left(1 - y_i\right)\left(1 - \log p_i\right)\right), \tag{4}$$

where p_i is the possibility calculated based on the input x_i; y_i is the predicted class, $y_i \in \{1, 0\}$.

For the task of bounding box regression, we employ the Euclidean loss of the offset between the predicted bounding box and the nearest ground truth. Predicted by the network are the upper-left coordinate, width, and length. The loss function is given by:

$$G_i = \| \hat{b_i} - b_i \|^2,$$ (5)

where $\hat{b_i}$ is the class that the network predicts and b_i is the ground-truth coordinate. We use four values to represent a bounding box, i.e., the upper-left coordinate, width, and length. So $\hat{b_i}$ and b_i are four-dimensional vectors.

The total loss of this network is the weighted sum of the loss values of the above two functions. Faces are contained in RoI candidates after we preprocess the input of the multitask detector, therefore we give the face detector a high tolerance for features of tiny faces. This trick makes the tiny faces pass to the classification task of the multitask detector with a low threshold. Therefore, we give a big weight to the loss function of the bounding box. In addition, the total loss function is as follows:

$$Loss_i = t_1 \times L_i + t_2 \times G_i,$$ (6)

where L_i and G_i are the losses of face classification and bounding box regression respectively; t_1 and t_2 are the weights of each task. Here, t_1 is set to 0.3 and t_2 is set to 0.7.

4. Experiments

The experimental environment used in this paper is as follows: Intel Xeon E-2136 CPU @3.30 GHz, 16 GB internal storage, Windows 10 64 bit operating system, Microsoft, US.

We tested the performance of our framework on different datasets composed of a self-collected dataset and Caltech Pedestrian dataset 18. The self-collected dataset was introduced in Section 1. The Caltech dataset was collected by the California Institute of Technology on 2012, and it is often used in the design and testing of pedestrian detection algorithms. It contains a video of a city environment with a duration of about 10 h, and the image resolution is 640×480. The pedestrian targets are divided into different levels according to the size and the occlusion.

Figure 2 shows PMCD could detect the face accurately when the pedestrians are in different environments containing occlusion, incomplete body, and poor light.

Our mechanism consists of SSD [32] and the pre-identification mechanism, which is mentioned in Section 3.2.2. The accuracy of face detection in our framework is directly related to the performance of the mechanism, so we tested our mechanism on the Caltech Pedestrian dataset. In order to prove that our mechanism is superior to other state-of-the-art methods, we drew the curve of false positive per image (FPPI)-miss rate by the evaluation method [54], as shown by:

$$FPPI = m/N,$$ (7)

where m is the number of false positive, N is the number of images.

$$miss\ rate = fa/(fa + tr),$$ (8)

where fa is the number of false negatives, tr is the number of true positives.

We compared our pre-identification mechanism with Viola-Jones (VJ) [24], histogram of oriented gradient (HOG) [48], Scale Aware (SA)-Fast RCNN [55] and region proposal network(RPN)+ boosted forests (BF) [56]. As shown in Figure 3, ours performs better than the other methods, and the detection results of the corresponding method are better when the miss rate is lower.

Figure 2. Face detection in different situations based on pre-identification mechanism and a cascaded detector (PMCD): (**a**) the face detected with occlusion on some part of the body; (**b**) the face detected with occlusion on some part of the body in complex backgrounds; (**c**) the tiny face detected with good light; (**d**) the tiny face detected with poor light.

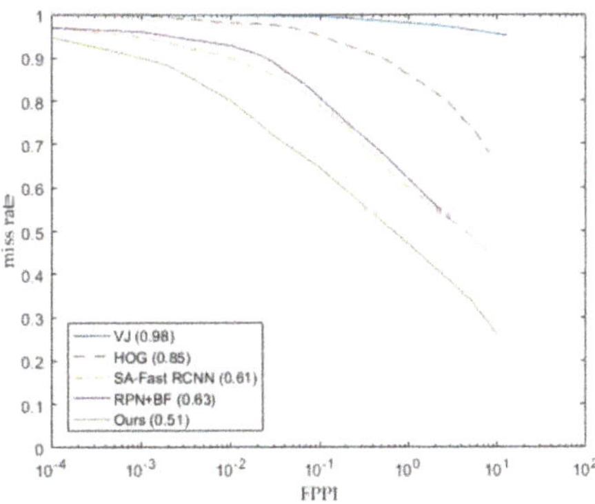

Figure 3. The curve of false positive per image (FPPI)-miss rate.

As shown in Figure 3, SSD achieves the best performance in these methods. Meanwhile, the miss rates of VJ and HOG are high, the main reason is that traditional methods usually train the classifiers by hand-crafted features, resulting in poorer robustness than that of convolutional neural networks.

We compared the whole framework with five of the most important face detection algorithms on the self-collected dataset, which are Viola-Jones(VJ) [24], single shot multibox detector(SSD) [32], multi-task cascaded neural networks (MTCNN) [45], multi-channel shallow convolutional network(MMSC) [46], and hybrid resolutions(HR) [47]. Figure 4 shows the results of the six methods in different situations. VJ and MTCNN cannot detect faces correctly when the face is fairly small. MMSC uses traditional methods to extract pre-processing regions, and thus it cannot keep good performance in detecting faces on incomplete bodies or in relatively complex background. SSD can detect faces roughly, but it cannot correctly obtain the bounding boxes when the faces are small. On one hand, HR can achieve remarkable performance in detecting tiny faces when the environment is relatively simple, but when the light is dim or faces are in a complex environment, the performance of HR would be very bad. On the other hand, PMCD achieves the best performance among the six methods. In the case of poor lighting conditions, only PMCD can achieve remarkable performance.

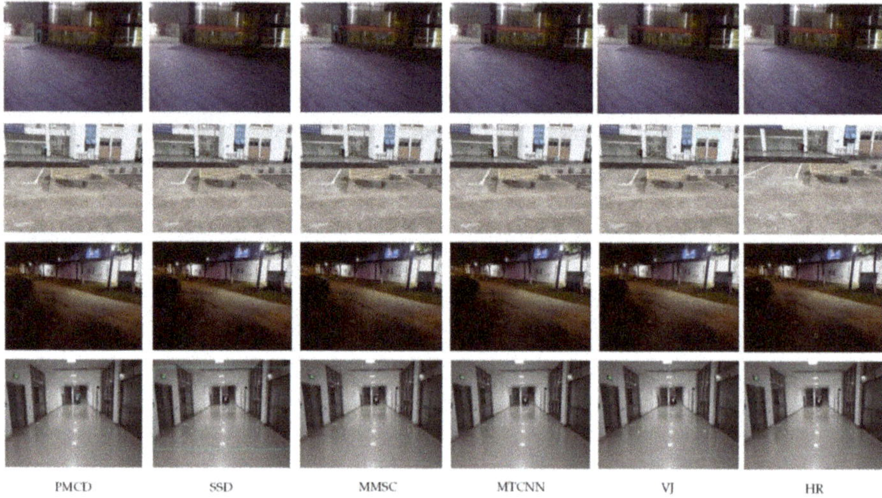

Figure 4. Face detection results.

Intersection over union (IoU) is the ratio between the intersection and the union of the candidate bounding box predicted by the model and the ground truth bounding box. The formula is:

$$\mathrm{IoU} = (\mathrm{area}(C) \cap \mathrm{area}(G)) / (\mathrm{area}(C) \cup \mathrm{area}(G)) \tag{9}$$

where $\mathrm{area}(C)$ is the area of the candidate bounding box; $\mathrm{area}(G)$ is the area of the ground truth bounding box.

We collected the true positives' IoUs which are predicted by the above five methods and calculated the mean IoU (MIoU) to judge the accuracy of the predicted bounding boxes. The formula is shown as follows:

$$\mathrm{MIoU} = \sum_{i=1}^{n} \mathrm{IoU}_i / n, \tag{10}$$

where n is the total number of true positives; IoU_i is the IoU score of the i-th true positive.

The results are shown in Table 1. PMCD achieves the best performance in these six methods. The second best is HR, no matter how small the face is, HR would be detected very precisely when the light is good. However, when the face is in a dark environment, the face's bounding box will be detected with some deviation. While the performance of SSD is the lowest because SSD cannot retain its ability when the target is small even though it can detect tiny faces, because its default anchors are set to different shapes, but the box shape with the highest score is not always the most suitable one. The other three methods can get great scores only if the targets are big.

Table 1. Mean intersection over union (MIoU) on the self-collected dataset.

	PMCD	HR	MMSC	MTCNN	SSD	VJ
MIoU	84.8%	83.4%	76.9%	79.2%	71.2%	73.6%

In order to verify the effectiveness of PMCD, the F1 measures, where precision and recall are equal, are shown in Table 2. The table of our model is higher than benchmark results of the other three important face detection neural networks. The MTCNN, VJ, and MMSC cannot keep high performance when targets are less than 20×20, which could result in many misclassifications. The framework proposed in this paper can solve this problem very well because we use two neural networks for grading detection, which can avoid the subjectiveness of hand-crafted features. In addition, the pre-identification mechanism can avoid the information loss of small targets, which often happens in a deep neural network.

Table 2. The F1 measures of different methods.

	VJ	MTCNN	MMSC	SSD	HR	PMCD
F1 measures	0.32	0.554	0.569	0.602	0.673	0.712

In addition, we tested the detection speed of these methods based on frames per second (FPS), and the results are shown in Table 3. Although PMCD is not the fastest method, it still exceeds 24 FPS, maintaining real-time performance. PMCD is designed as a cascade framework, in which some time is wasted in passing data, that is, the end-to-end detectors can outperform cascade frameworks in detection speed. Nevertheless, for the tiny-face detection task, the advantages of the cascaded detector are obvious.

Table 3. Detection speed of different methods.

	VJ	MTCNN	MMSC	SSD	HR	PMCD
FPS	1.2	32.8	3.9	43.2	2.6	34.6

5. Conclusions and Future Works

A real-time pre-identification mechanism and a cascaded detector were proposed for tiny-face detection in this paper. We combined pedestrian detection with face detection to reduce the search region. After obtaining the bounding box of the pedestrian, we estimated the face area as RoI based on the proportion of the size of the bounding box. The mechanism not only can improve efficiency but also reduce the number of false positives. After that, we built an image pyramid as the input of the face detector. The algorithm can ensure the performance of the face detector on the targets of different sizes. Based on the experimental results, the proposed cascaded neural network is more efficient in tiny-face detection than other state-of-the-art methods.

There are many open studies in the above research. Our method is constrained, but when the body is not completely occluded, PMCD can perform better in tiny-face detection under extreme conditions such as poor light and occlusion, etc. Traditional face detection methods and region-based CNNs cannot keep a high accuracy in tiny-face detection by themselves, but it is possible to combine these unconstrained methods with our proposed framework in the future. The highest priority should be given to this framework to avoid unnecessary false positives and reduce the miss rate of tiny-face detection.

As mentioned in Section 1, cascade frameworks have significant advantages in detecting tiny faces, and the experimental results also support this conclusion in Section 4. This is because the characteristics of tiny faces are gradually lost with convolutional layers going deeper. Therefore, this problem can be solved by finding a larger upper-body to complete the coarse-to-fine detection. PMCD has achieved remarkable results on our self-collected dataset when the targets are small or in extreme environments such as darkness, complex background, occlusion, etc. When the targets are of normal sizes, the end-to-end detection frameworks can obtain a faster speed. Therefore, PMCD still has potential for further improvement. In the future, we will design an end-to-end detection network based on PMCD, which can internally extract hierarchical features for the face detection task. This network can ensure satisfactory performance on tiny-face detection while keeping a faster detection speed than PMCD.

Author Contributions: All authors of the paper made significant contributions to this work. Z.Y. and J.L. contributed equally to this paper, conceived the idea of work, implemented algorithms, analyzed the experiment data, and wrote the manuscript. W.M. led the project, directed and revised the paper writing. Q.W. helped to code and analysis the experiment data.

Funding: This work was supported by the National Natural Science Foundation of China (Grant No. 61762061, 61963027, and 61703198), the Natural Science Foundation of Jiangxi Province, China (Grant No. 20161ACB20004), Natural Science Foundation for Distinguished Young Scholars of Jiangxi Province (Grant No. 2018ACB21014), and Jiangxi Key Laboratory of Smart City (Grant No. 20192BCD40002).

Conflicts of Interest: The authors declare no conflict of interest.

References

1. Dang, L.M.; Hassan, S.I.; Im, S.; Lee, J.; Lee, S.; Moon, H. Deep Learning Based Computer Generated Face Identification Using Convolutional Neural Network. *Appl. Sci.* **2018**, *8*, 2610. [CrossRef]
2. Bai, W.; Quan, C.; Luo, Z. Uncertainty Flow Facilitates Zero-Shot Multi-Label Learning in Affective Facial Analysis. *Appl. Sci.* **2018**, *8*, 300. [CrossRef]
3. Kang, S.J. Multi-user identification-based eye-tracking algorithm using position estimation. *Sensors* **2016**, *17*, 41. [CrossRef] [PubMed]
4. Ma, L.; Deng, Z.G. Real-time hierarchical facial performance capture. In Proceedings of the Symposium on Interactive 3D Graphics and Games (ACM), New York, NY, USA, 21–23 May 2019. [CrossRef]
5. Weise, T.; Li, H.; Gool, L.V.; Pauly, M. Face/off: Live facial puppetry. In Proceedings of the SIGGRAPH/Eurographics ACM Symposium on Computer animation, New Orleans, LA, USA, 1–2 August 2009. [CrossRef]
6. Bouaziz, S.; Li, H.; Pauly, M. Realtime performance-based facial animation. *ACM Trans. Graph.* **2011**, *30*, 77.
7. Li, H.; Weise, T.; Mark, P.X. Example-based facial rigging. *ACM Trans. Graph.* **2010**, *29*, 32. [CrossRef]
8. Li, W.; Deng, Z.G. A practical model for live speech driven lip-sync. *IEEE Comput. Graph. Appl.* **2014**, *35*, 70–78.
9. Li, H.; Yu, J.H.; Ye, Y.T.; Bregler, C. Realtime facial animation with on-the-fly correctives. *ACM Trans. Graph* **2013**, *32*, 42. [CrossRef]
10. Ouzounis, C.; Kilias, A.; Mousas, C. Kernel projection of latent structures regression for facial animation retargeting. *arXiv* **2017**, arXiv:1707.09629.
11. Ma, L.; Deng, Z. Real-time Facial Expression Transormation for Monocular RGB Video. *Comput. Graph. Forum Wiley Online Libr.* **2019**, *38*, 470–481. [CrossRef]
12. Kaewmart, P.; Markus, B. The shape of the face template: Geometric distortions of faces and their detection in natural scenes. *Vis. Res.* **2015**, *109*, 99–106.
13. Ranjan, R.; Patel, V.M.; Chellappa, R. HyperFace: A Deep Multi-task Learning Framework for Face Detection, Landmark Localization, Pose Estimation, and Gender Recognition. *IEEE Trans. Pattern Anal. Mach. Intell.* **2019**, *41*, 121–135. [CrossRef] [PubMed]
14. Liao, Y.Q.; Xiong, P.W.; Min, W.D.; Min, W.Q.; Lu, J.H. Dynamic sign language recognition based on video sequence with BLSTM-3D residual networks. *IEEE Access* **2019**, *7*, 38044–38054. [CrossRef]
15. Min, W.D.; Fan, M.D.; Guo, X.G.; Han, Q. A new approach to track multiple vehicles with the combination of robust detection and two classifiers. *IEEE Trans. Intell. Transp. Syst.* **2018**, *19*, 174–186. [CrossRef]

16. Ren, S.Q.; He, K.M.; Girshick, R.; Sun, J. Faster R-CNN: Towards real-time object detection with region proposal networks. *IEEE Trans. Pattern Anal. Mach. Intell.* **2017**, *39*, 1137–1149. [CrossRef] [PubMed]

17. Simonyan, K.; Zisserman, A. Very Deep Convolutional Networks for Large-Scale Image Recognition. *arXiv* **2014**, arXiv:1409.1556.

18. Wojek, C.; Dollar, P.; Schiele, B.; Perona, P. Pedestrian Detection: An Evaluation of the State of the Art. *IEEE Trans. Pattern Anal. Mach. Intell.* **2012**, *34*, 743–761. Available online: http://www.vision.caltech.edu/Image_Datasets/CaltechPedestrians/ (accessed on 18 March 2009).

19. Zou, F.Y.; Li, J.; Min, W.D. Distributed Face Recognition Based on Load Balancing and Dynamic Prediction. *Appl. Sci.* **2019**, *9*, 94. [CrossRef]

20. Yaman, M.A.; Subasi, A.; Rattay, F. Comparison of Random Subspace and Voting Ensemble Machine Learning Methods for Face Recognition. *Symmetry* **2018**, *10*, 651. [CrossRef]

21. Luo, Y.; Guan, Y.P. Adaptive skin detection using face location and facial structure estimation. *IET Comput. Vis.* **2017**, *11*, 550–559. [CrossRef]

22. Mohanty, R.; Raghunadh, M.V. A new approach to face detection based on YCgCr color model and improved AdaBoost algorithm. In Proceedings of the International Conference on Communication and Signal Processing, Melmaruvathur, India, 6–8 April 2016; pp. 1392–1396.

23. Ma, S.; Bai, L. A face detection algorithm based on Adaboost and new Haar-Like feature. In Proceedings of the IEEE International Conference on Software Engineering and Service Science, Beijing, China, 26–28 August 2016; pp. 651–654.

24. Viola, P.; Jones, M.J. Robust Real-Time Face Detection. *Int. J. Comput. Vis.* **2004**, *57*, 137–154. [CrossRef]

25. Liao, S.C.; Jain, A.K.; Li, S.Z. A fast and accurate unconstrained face detector. *IEEE Trans. Pattern Anal. Mach. Intell.* **2016**, *38*, 211–223. [CrossRef] [PubMed]

26. Yang, B.; Yan, J.J.; Lei, Z.; Li, S.Z. Aggregate channel features for multi-view face detection. In Proceedings of the IEEE International Joint Conference on Biometrics, Clearwater, FL, USA, 29 September–2 October 2014; pp. 1–8.

27. Bilal, M. Algorithmic optimisation of histogram intersection kernel support vector machine-based pedestrian detection using low complexity features. *IET Comput. Vis.* **2017**, *11*, 350–357. [CrossRef]

28. Baek, J.; Kim, J.; Kim, E. Fast and efficient pedestrian detection via the cascade implementation of an additive kernel support vector machine. *IEEE Trans. Intell. Transp. Syst.* **2017**, *18*, 902–916. [CrossRef]

29. Girshick, R.; Donahue, J.; Darrell, T. Region-based Convolutional Networks for accurate object detection and segmentation. *IEEE Trans. Pattern Anal. Mach. Intell.* **2016**, *38*, 142–158. [CrossRef] [PubMed]

30. Girshick, R. Fast R-CNN. In Proceedings of the IEEE International Conference on Computer Science, Wuhan, China, 20–22 November 2015; pp. 1440–1448. [CrossRef]

31. Redmon, J.; Divvala, S.; Girshick, R.; Farhadi, A. You Only Look Once: Unified, real-time object detection. In Proceedings of the IEEE Conference on Computer Vision and Pattern Recognition, Las Vegas, NV, USA, 27–30 June 2016; pp. 779–788.

32. Liu, W.; Anguelov, D.; Erhan, D.; Szegedy, C.; Reed, S.; Fu, C.Y.; Berg, A.C. SSD: Single Shot MultiBox Detector. In Proceedings of the European Conference on Computer Vision, Amsterdam, The Netherlands, 8–16 October 2016; pp. 21–37.

33. Jiang, H.Z.; Learned-Miller, E. Face detection with the Faster R-CNN. In Proceedings of the IEEE International Conference on Automatic Face and Gesture Recognition, Washington, DC, USA, 30 May–3 June 2017; pp. 650–657.

34. Zhu, C.; Zheng, Y.; Luu, K.; Savvides, M. CMS-RCNN: Contextual multi-scale region-based CNN for unconstrained face detection. *arXiv* **2016**, arXiv:1606.054413.

35. Wan, S.; Chen, Z.; Zhang, T.; Zhang, B.; Wong, K.K. Bootstrapping face detection with hard negative examples. *arXiv* **2016**, arXiv:1608.02236.

36. Sun, X.; Wu, P.; Hoi, S.C.H. Face detection using deep learning: An improved faster RCNN approach. *arXiv* **2017**, arXiv:1701.08289. [CrossRef]

37. Zhang, C.; Xu, X.; Tu, D. Face detection using improved Faster RCNN. *Neurocomputing* **2018**, *299*, 42–50.

38. Hsu, G.S.; Hsieh, C.H. Cross-pose landmark localization using multi-dropout framework. In Proceedings of the IEEE International Joint Conference on Biometrics, Denver, CO, USA, 1–4 October 2017; pp. 390–396.

39. Li, J.; Xue, Y.; Wang, W.; Ouyang, G. Cross-level Parallel Network for Crowd Counting. *IEEE Trans. Ind. Inform.* **2019**. [CrossRef]

40. Triantafyllidou, D.; Nousi, P.; Tefas, A. Fast deep convolutional face detection in the wild exploiting hard sample mining. *Big Data Res.* **2018**, *11*, 65–76. [CrossRef]
41. Yang, S.; Luo, P.; Loy, C.C.; Tang, X. From facial parts responses to face detection: A deep learning approach. In Proceedings of the IEEE International Conference on Computer Vision, Santiago, Chile, 11–18 December 2015; pp. 3676–3684.
42. Li, H.; Lin, Z.; Shen, X.; Brandt, J.; Hua, G. A convolutional neural network cascade for face detection. In Proceedings of the IEEE Conference on Computer Vision and Pattern Recognition, Boston, MA, USA, 7–12 June 2015; pp. 5325–5334.
43. Qin, H.; Yan, J.; Li, X. Joint training of cascaded CNN for face detection. In Proceedings of the IEEE Conference on Computer Vision and Pattern Recognition, Las Vegas, NV, USA, 27–30 June 2016; pp. 3456–3465.
44. Zhang, J.; Shan, S.; Kan, M.; Chen, X. Coarse-to-Fine Auto-Encoder Networks (CFAN) for Real-Time Face Alignment. In Proceedings of the European Conference on Computer Vision, Zurich, Switzerland, 6–12 September 2014; pp. 1–16.
45. Zhang, K.; Zhang, Z.; Li, Z.; Qiao, Y. Joint face detection and alignment using multitask cascaded convolutional networks. *IEEE Signal Process. Lett.* **2016**, *23*, 1499–1503. [CrossRef]
46. Min, W.D.; Fan, M.D.; Li, J.; Han, Q. Real-time face recognition based on face pre-identification detection and multi-scale classification. *IET Comput. Vis.* **2018**, *13*, 165–171. [CrossRef]
47. Hu, P.; Ramanan, D. Finding tiny faces. In Proceedings of the IEEE Conference on Computer Vision and Pattern Recognition (CVPR), Honolulu, HI, USA, 21–26 July 2017; pp. 1522–1530.
48. Dalal, N.; Triggs, B. Histograms of Oriented Gradients for human detection. In Proceedings of the IEEE Conference on Computer Vision and Pattern Recognition, Boston, MA, USA, 7–12 June 2015; pp. 886–893.
49. Xie, Y.; Yang, L.; Sun, X.; Wu, D.; Chen, Q.; Zeng, Y.; Liu, G. An auto-adaptive background subtraction method for Raman spectra. *Spectrochim. Part A Mol. Biomol. Spectrosc.* **2016**, *161*, 58–63. [CrossRef] [PubMed]
50. Han, J.; Quan, R.; Zhang, D.; Nie, F. Robust object co-segmentation using background prior. *IEEE Trans. Image Process.* **2018**, *27*, 1639–1651. [CrossRef] [PubMed]
51. Kim, G.; Yang, S.; Sim, J.Y. Saliency-based initialization of Gaussian mixture models for fully-automatic object segmentation. *Electron. Lett.* **2017**, *53*, 1648–1649. [CrossRef]
52. Chan, K. Segmentation of moving objects in image sequence based on perceptual similarity of local texture and photometric features. *EURASIP J. Image Video Process.* **2018**, *62*. [CrossRef]
53. Park, K.; Kim, S.; Sohn, K. Unified multi-spectral pedestrian detection based on probabilistic fusion networks. *Pattern Recognit.* **2018**, *80*, 143–155. [CrossRef]
54. Determe, J.F.; Louveaux, J.; Jacques, L.; Horlin, F. Improving the Correlation Lower Bound for Simultaneous Orthogonal Matching Pursuit. *IEEE Signal Proc. Lett.* **2016**, *23*, 1642–1646. [CrossRef]
55. Li, J.N.; Liang, X.D.; Shen, S.M.; Xu, T.F.; Feng, J.S.; Yan, S.C. Scale-Aware Fast R-CNN for Pedestrian Detection. *IEEE Trans. Multimed.* **2018**, *20*, 985–996. [CrossRef]
56. Zhang, L.; Lin, L.; Liang, X.; He, K. Is faster R-CNN Doing Well for Pedestrian Detection? *arXiv* **2016**, arXiv:1607.07032v2.

 *applied sciences*

Article

Fast and Accurate Algorithm for ECG Authentication Using Residual Depthwise Separable Convolutional Neural Networks

Eko Ihsanto [1], Kalamullah Ramli [1], Dodi Sudiana [1,*] and Teddy Surya Gunawan [2]

[1] Department of Electrical Engineering, Universitas Indonesia, Depok, Jawa Barat 16424, Indonesia; eko.ihsanto@ui.ac.id (E.I.); kalamullah.ramli@ui.ac.id (K.R.)

[2] Department of Electrical and Computer Engineering, International Islamic University Malaysia, Kuala Lumpur 53100, Malaysia; tsgunawan@iium.edu.my

* Correspondence: dodi.sudiana@ui.ac.id

Received: 21 February 2020; Accepted: 15 April 2020; Published: 9 May 2020

Abstract: The electrocardiogram (ECG) is relatively easy to acquire and has been used for reliable biometric authentication. Despite growing interest in ECG authentication, there are still two main problems that need to be tackled, i.e., the accuracy and processing speed. Therefore, this paper proposed a fast and accurate ECG authentication utilizing only two stages, i.e., ECG beat detection and classification. By minimizing time-consuming ECG signal pre-processing and feature extraction, our proposed two-stage algorithm can authenticate the ECG signal around 660 μs. Hamilton's method was used for ECG beat detection, while the Residual Depthwise Separable Convolutional Neural Network (RDSCNN) algorithm was used for classification. It was found that between six and eight ECG beats were required for authentication of different databases. Results showed that our proposed algorithm achieved 100% accuracy when evaluated with 48 patients in the MIT-BIH database and 90 people in the ECG ID database. These results showed that our proposed algorithm outperformed other state-of-the-art methods.

Keywords: electrocardiogram (ECG); biometric authentication; beat detection; depthwise separable convolution (DSC); ECG ID database; MIT-BIH database

---

1. Introduction

As early as 1977, electrocardiogram (ECG) was identified for its potential for biometric authentication [1]. ECG is relatively easy to acquire, for example, using finger sensors [2]. Many methods have been implemented of ECG identification using feature extraction based on time, amplitude, and frequency [3–6], as well as using machine learning [7–9]. Nevertheless, ECG biometrics still cannot be fully implemented in real-time security systems due to their high computational time and low accuracy.

ECG identification using multivariate analysis feature extraction has been reported in [3]. Using principal components analysis (PCA) and soft independent modeling of class analogy (SIMCA), the ECG identification results vary from 90% to 100%. The accuracy of this method depends on two main factors, including the number of features and the type of lead used to obtain an ECG signal. In [3], they used 12 leads and 30 types of time-based ECG features. Around 50 samples were evaluated from 20 ECG records. Although ECG identification can be performed with just one lead, this research needs to use at least three leads to improve identification accuracy [3].

In [5], the number of leads and the number of ECG features were reduced; specifically, the number of leads has been reduced one, and the number of features has been reduced to seven. After preprocessing and feature extraction, two different methods were normally used for identification purposes, i.e., template matching (TM) and the decision-based neural network (DBNN). Evaluation of 400 ECG beats from 20 people revealed that the accuracy of TM and DBNN was 95% and 80%, respectively. Moreover, combining both methods produced 100% identification accuracy.

In 2005, the ECG-ID database became publicly accessible [10]. The database contains ECG recording files of 90 people using a single ECG lead. Recorded ECG signals from the same person are stored in one folder. Each folder contains two or more ECG recording files, except the 74th folder, containing only one record. After preprocessing and time-based feature extraction, ECG identification is conducted using PCA, linear discriminant analysis (LDA), and majority vote classifier (MVC). Moreover, the accuracy of this method is around 96%.

Apart from using the database of normal ECG heartbeats, ECG biometrics have also been tested using arrhythmia ECG signals [11]. Using test data from the ECG records of 47 patients in the MIT-BIH database, 91.1% accuracy was reported. They applied five stages of blind signal processing (BSP), including preprocessing, wavelet transform, autocorrelation, component selection, and parallel one-dimension convolutional neural networks (1-D CNN). A similar identification method can be applied to other databases, including ECG records from the ECG-ID and MIT-BIH databases [8,9]. Bidirectional gated recurrent units (BGRUs) have been used to identify 47 patients from the MIT-BIH database and 90 ECG healthy people from the ECG-ID database with an accuracy of 98.40% and 98.60%, respectively [8]. When, instead, a PCA neural network (PCANet) was used, accuracies of 100% and 97.75% were achieved when tested on ECG-ID and MIT-BIH databases, respectively [9].

Recently, depthwise separable convolution (DSC) has been applied to a CNN model to reduce computation by reducing the number of model parameters and arithmetic calculations. DSC has been utilized to improve CNN model quality in image classification [12] and neural machine translation [13]. For cardiac arrhythmia classification, DSC has been tested on 16 ECG classes of the MIT-BIH database, outperforming other methods [14]. Besides DSC, the CNN model can be further optimized using the residual technique. The residual technique has been used in CNN to simplify the training process and increase accuracy, as is evident in image classification [15] and ECG classification [16].

Residual DSC has been implemented in Xception [12] and PydMobileNet [17]. Similar algorithms with slight modifications will be used for ECG authentication in this paper. The first modification was on the filter dimension, i.e., 2D-CNN to 1D-CNN. The second modification was on the structure of CNN layers. In Xception and PydMobileNet, DSC was applied to single layers while, in our algorithm, DSC was implemented for the group of three layers, as has been explained for a cardiac arrhythmia classifier [14]. A further modification was conducted for ECG authentication, in which a shortcut residual is applied to the three-layer DSC. Furthermore, for ECG authentication, none of the convolutional layers is similar to the ones in Xception and PydMobileNet. Each convolutional layer is an ordinary 1D convolution, not a DSC.

As discussed earlier, the common problems in ECG authentication include time-consuming signal processing and feature extraction. Another problem is accuracy; it is difficult to obtain 100% accuracy, which is required for authentication purposes. Therefore, in this paper, a fast and efficient algorithm will be proposed to overcome these problems using a residual depthwise separable convolution neural network. The training process required high computational resources, which can be alleviated using GPU. However, the testing process required low computational resources; a CPU alone is sufficient to be used during the authentication process.

2. Proposed ECG Authentication Algorithm Using Residual Depthwise Separable CNN

To simplify the ECG authentication process, the proposed method in this paper requires only two stages, including the segmentation stage and the classification stage. The segmentation stage carried out ECG beat detection and segmentation, while the classification stage performed feature extraction and classification. ECG beat detection is performed using Hamilton's methods [18,19]. The segmentation is based on the relative position of the R_{peak}, and occurs $\frac{1}{4}N$ before R_{peak} and $\frac{3}{4}N$ after R_{peak}. N depends on the sampling frequency and is set to 256 samples for both ECG-ID and MIT-BIH databases. Note that preprocessing has been performed on both databases to eliminate baseline drift and to remove powerline and high-frequency noises [10]. Therefore, it can be assumed that the ECG signals from both databases were ready to be used for further processing.

The second stage, or classification stage, is realized by implementing the Residual Depthwise Convolutional Neural Network. The CNN used in this second stage is 1D-CNN that contains 28 layers. The 21 layers consist of seven layers of residual depthwise separable convolution neural network (RDSCNN) repeated three times, as illustrated in Figure 1.

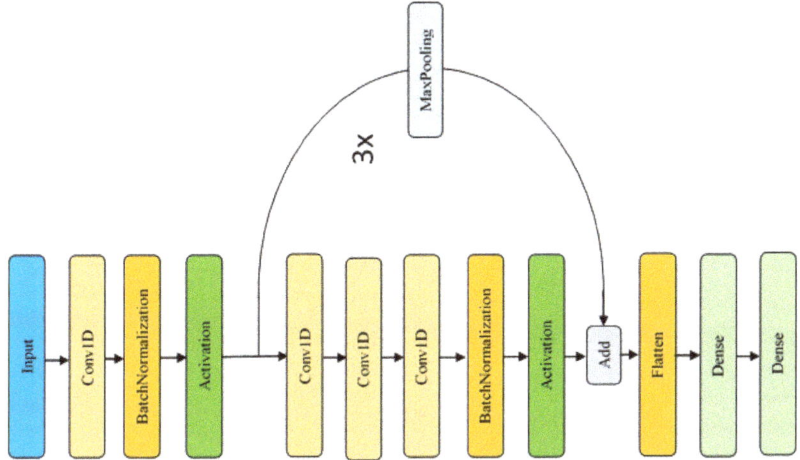

Figure 1. The proposed algorithm using 28 layers of residual depthwise separable convolution neural network.

As shown in Figure 1, there are seven layers repeated three times, including a max pooling layer, three layers of 1D-CNN (Conv1D), a batch normalization layer, an activation layer, and an add layer. This structure is called as Residual Depthwise Separable Convolution implementation. The residual implementation uses a shortcut Max Pooling layer, while depthwise separable convolution implementation uses three Conv1D layers with the filter size [5 1 5], respectively. The summary of the proposed CNN model will be discussed in more detail in Section 3.3.

There are 10 Conv1D layers in the proposed algorithm, which can be calculated using Equation (1) to (4), as follows [20]:

$$x_k^l = \beta_k^l + \sum_{i=1}^{N_{l-1}} conv1D\left(\omega_{ik}^{l-1}, s_i^{l-1}\right) \tag{1}$$

$$\Delta s_k^l = \sum_{i=1}^{N_{l+1}} conv1Dz\left(\Delta_i^{l+1}, rev\left(\omega_{ik}^l\right)\right) \tag{2}$$

$$\frac{\partial E}{\partial \omega_{ik}^l} = conv1D\left(s_k^l, \Delta_i^{l+1}\right) \tag{3}$$

$$\frac{\partial E}{\partial \beta_k^l} = \sum_n \Delta_k^l(n) \tag{4}$$

where l is the layer number, k is the filter number at the particular layer, i is the index coefficient w of the filter, x_k^l is the input and ω_{ik}^{l-1} is the kernel (weight) from the i-th neuron at layer $l-1$ to the k-th neuron at layer l. s_i^{l-1} is the output of the i-th neuron at layer $l-1$ and β_k^l is the bias, which can be assumed to be zero. $conv1Dz(\cdot, \cdot)$ performs full convolution in 1-D with $K-1$ zero padding. Δs_k^l

is the input difference after filtering and $rev(\cdot)$ flips the array. Δ_i^{l+1} is the delta error according to the loss function. $\frac{\partial E}{\partial \omega_{ik}^l}$ denotes error change with respect to the filter coefficients, while $\frac{\partial E}{\partial \beta_k^l}$ denotes error change with respect to the bias.

To maximize the accuracy, the CNN model in this paper is optimized using two methods, i.e., Depthwise Separable [12,13] and Residual [15]. The Depthwise Separable can be explained using Equation (5) to (8), as follows:

$$Conv1D(\omega, y)_{(i)} = \sum_{k.m}^{K.M} \omega_{(k.m)} \cdot y_{(i+k,m)} \tag{5}$$

$$PointwiseConv1D(\omega, y)_{(i)} = \sum_{m}^{M} \omega_m \cdot y_{(i,m)} \tag{6}$$

$$DepthwiseConv1D(\omega, y)_{(i)} = \sum_{k}^{K} \omega_{(k)} \odot y_{(i+k)} \tag{7}$$

$$SepConv1D\left(\omega_p, \omega_d, y\right)_{(i)} = PointwiseConv1D_{(i)}\left(\omega_p, DepthwiseConv1D_{(i)}(\omega_d, y)\right) \tag{8}$$

where i is the element index in layer y. ω_p and ω_d are the filters calculated from Equations (6) and (7). The calculation in Equation (5) can be replaced with Equation (8) through calculations of Equations (6) and (7). Although the output layer from Equation (5) is similar to Equations (6) to (8), the total number of parameters and the training time were decreased, as can be seen in Table 1. Note that the depthwise separable technique will reduce the training time and reduce the accuracy as well. To compensate for the slight decrease in performance, the residual technique can be applied. Therefore, combining these two techniques, i.e., Depthwise Separable and Residual, not only reduces the number of parameters for training but also improves the accuracy.

Table 1. Experimental scenarios using ECG-ID and MIT-BIH databases with various CNN configurations.

Profile	Beat Detection	Depthwise Separable Convolution	Residual
1	Manual	No	Yes
2	Manual	Yes	No
3	Manual	Yes	Yes
4	Hamilton's method	Yes	Yes
5	Hybrid	Yes	Yes

Residual is a technique in CNN, in which one of the branches is a shortcut bypassing one or several other branch layers. Initially, this technique was intended to overcome the saturation of the increasing number of layers. Too many layers will lead to longer training iteration, lower speed, and lower accuracy. With the residual technique, the training iteration can be shorter, and the accuracy will increase. The general formulas for the identity function related to the residual shortcut are as follows [15]:

$$y = \mathcal{F}(x, \{W_i\}) + x \tag{9}$$

$$y = \mathcal{F}(x, \{W_i\}) + W_s x \tag{10}$$

where x is the input, y is the output layer and $\mathcal{F}(x, \{W_i\})$ is a filter or residual mapping which can be optimized. W_i is a collection of overlapped layers. W_s is a linear projection to adjust the x and y dimensions when a shortcut is performed. There is almost no change in the number of parameters and arithmetic operations, except for the addition operation which has a little computational load. Therefore, the application of this residual technique can minimize the training iteration while improving accuracy.

Generally, this residual technique is applied to CNN models with millions of training data, hundreds of layers, and tens of thousands of iteration epochs. However, in this paper, the residual technique can be applied to a smaller number of training data (between 900 and 4800), fewer layers (less than 30), and less iteration (100 epochs). As the number of biometric data to be authenticated increases, the number of layers, the number and size of filters, and the number of iterations can be increased as well.

3. Experimental Setup

This section describes the experimental setup to evaluate our proposed algorithm, including the ECG databases used, various experimental scenarios, and the summary of the CNN model.

3.1. ECG Databases and Computing Platform

Two popular Physionet databases were used to evaluate the performance of our proposed algorithm, i.e., ECG-ID [10] and MIT-BIH [21]. These two databases have been used by several researchers to test the performance of the ECG authentication algorithm. In this paper, each database is used to test five different conditions, with variations in the application of beat detection, as well as DSC and residual configurations.

The ECG-ID database contains 90 folders from 90 different people. Each folder contains several ECG records with a 500 Hz sampling frequency. Each record has been provided with 10 annotations related to the position of R_{peak} in the first 10 beats. The recording duration ranges from 20 s to 5 min. For testing purposes, only 179 files were used, i.e., two files from each folder, except the 74th folder which has only one file. For training purposes, ten beats of the first file in each folder were obtained, so that around 900 beats were used. For validation purposes, ten beats of the second file were acquired. As the 74th folder contains only one file, 10 beats from this file were used for both training and validation. The ECG beat extraction from 179 files from 90 folders produced 900 beats for training and 900 beats for validation. These validation beats were used to validate the CNN weights obtained in the training process.

Besides using the ECG database of healthy people, our proposed algorithm was also tested with ECG records with several cardiac arrhythmia beats taken from the MIT-BIH database. The database contains 48 files, in which each file contains around 30 min of ECG recordings of cardiac arrhythmia patients. Each data file is complemented with the annotation file containing the R_{peak} position in the associated ECG recording. There are two channels provided for each of these ECG records, but in this paper only channel 0 (first channel) was used. For training and testing purposes, each ECG recording is divided into two, in which each half contains 325,000 sample points. The first half is used for training, while the second half is used for testing. For training, only 100 beats of the first half were used, producing a total of 4800 ECG beats. For validation, only 10 beats of the second half were used, producing a total of 480 ECG beats. These validation beats were used to validate the CNN weights obtained in the training process.

The proposed algorithm was implemented in Python with Tensorflow with GPU [22] and Keras libraries [23], as well as other standard libraries, such as Numpy, Matplotlib, and Scipy. The experiments are performed on a computer with Intel Core i7-7700 CPU with a total of eight logical processors, memory of 8 GBytes, and an Nvidia GeForce GTX 1060 6GB DDR5 graphics card, using the Microsoft Windows 10 64-bit operating system.

3.2. Experimental Profiles

Using the two ECG databases, five experimental profiles with various CNN models were carried out as shown in Table 1. In the first profile, DSC was not implemented and only residual was implemented. In the second profile, DSC was implemented and residual was not implemented. The third, fourth, and fifth profiles used both DSC and residual (our main proposed algorithm) with different methods in beat detection and segmentation. Manual beat detection used the annotated file marked by medical experts. Hamilton's automatic beat detection and segmentation used Hamilton's method [18,19]. The hybrid beat detection used manual beat detection using annotated files for the training and Hamilton's automatic beat detection for the testing.

3.3. Residual Depthwise Separable Convolution Neural Network Model (RDSCNN)

The proposed residual depthwise separable convolution neural network (RDSCNN) algorithm is shown in Figure 1 and summarized in Table 2. There are two almost identical models applied to the two ECG databases. The difference is only in the number of neurons in the last layer configured according to the number of classes, i.e., 90 and 48 classes for ECG-ID and MIT-BIH database, respectively. For experimental scenario one (without DSC), the number of filters, size, and stride in layers 6, 13, and 20 should be modified (see column three in Table 2) from [16 1 1] to [32 5 1]. For experimental scenario two (without residual), layers 10, 11, 17, 18, 24, and 25 should be removed. The rest of the experimental scenarios used the complete RDSCNN algorithm.

Table 2. Summary of the proposed RDSCNN algorithm.

#	Layer	Number of Filters, Size, Stride	MIT-BIH		ECG-ID	
			Output Shape	Number of Parameters	Output Shape	Number of Parameters
1	inputlayer		(256, 1)	0	(256, 1)	0
2	convolution1d_1	32, 5, 4	(64, 32)	192	(64, 32)	192
3	batch_normalization_1		(64, 32)	128	(64, 32)	128
4	activation_1		(64, 32)	0	(64, 32)	0
5	convolution1d_2	32, 5, 1	(64, 32)	5152	(64, 32)	5152
6	convolution1d_3	16, 1, 1	(64, 16)	528	(64, 16)	528
7	convolution1d_4	32, 5, 4	(16, 32)	2592	(16, 32)	2592
8	batch_normalization_2		(16, 32)	128	(16, 32)	128
9	activation_2		(16, 32)	0	(16, 32)	0
10	max_pooling1d_1		(16, 32)	0	(16, 32)	0
11	add_1		(16, 32)	0	(16, 32)	0
12	convolution1d_5	32, 5, 1	(16, 32)	5152	(16, 32)	5152
13	convolution1d_6	16, 1, 1	(16, 16)	528	(16, 16)	528
14	convolution1d_7	32, 5, 4	(4, 32)	2592	(4, 32)	2592
15	batch_normalization_3		(4, 32)	128	(4, 32)	128
16	activation_3		(4, 32)	0	(4, 32)	0
17	max_pooling1d_2		(4, 32)	0	(4, 32)	0
18	add_2		(4, 32)	0	(4, 32)	0
19	convolution1d_8	32, 5, 1	(4, 32)	5152	(4, 32)	5152
20	convolution1d_9	16, 1, 1	(4, 16)	528	(4, 16)	528
21	convolution1d_10	32, 5, 4	(1, 32)	2592	(1, 32)	2592
22	batch_normalization_4		(1, 32)	128	(1, 32)	128
23	activation_3		(1, 32)	0	(1, 32)	0
24	max_pooling1d_3		(1, 32)	0	(1, 32)	0
25	add_3		(1, 32)	0	(1, 32)	0
26	flatten_1		(32)	0	(32)	0
27	dense_1		(64)	2112	(64)	2112
28	dense_2		(48)	3120	(90)	5850

Total parameters: 30,752 (MIT-BIH) and 33,482 (ECG-ID). Trainable parameters: 30,496 (MIT-BIH) and 33,226 (ECG-ID). Non-trainable parameters: 256 (MIT-BIH) dan 256 (ECG-ID).

4. Results and Discussion

This section will elaborate on ECG beat segmentation and detection, training and validation, ECG biometric authentication experiments, and comparison with other methods.

4.1. Experiment on ECG Beat Detection and Segmentation

Using ECG-ID and MIT-BIH databases, the ECG record and its R_{peak} positions were obtained, as shown in Figure 2. For the ECG-ID database, three samples were taken from 90 people. The first record was used for training, while the second record was used for validation. There were ten annotated R_{peak} as shown in Figure 2 for each person. In the MIT-BIH database, records 201 and 202 belonged to the same patient with different ECG beats, so they were classified as two classes. The red dots in Figure 2 are an example of the R_{peak} position detected manually by cardiologists. Using this R_{peak} position, the ECG beat was detected and extracted. An example of ECG beat detection and segmentation for both databases is shown in Figure 3.

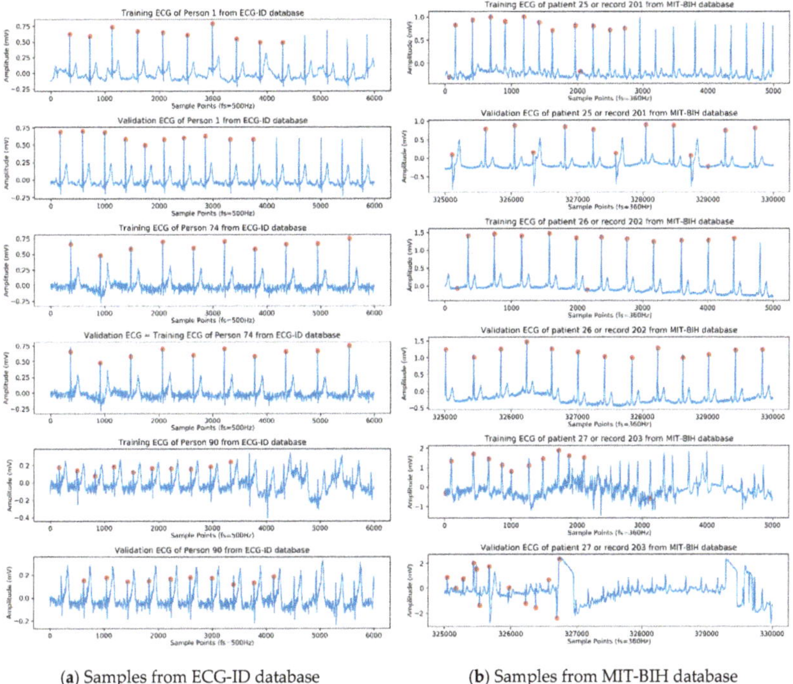

(**a**) Samples from ECG-ID database   (**b**) Samples from MIT-BIH database

Figure 2. Sample of ECG records of three people obtained from the ECG-ID (the 1st, 74th, and 90th person) and MIT-BIH (the 25th, 26th, and 27th person) databases. The red dots were the R_{peak} positions annotated manually by cardiologists.

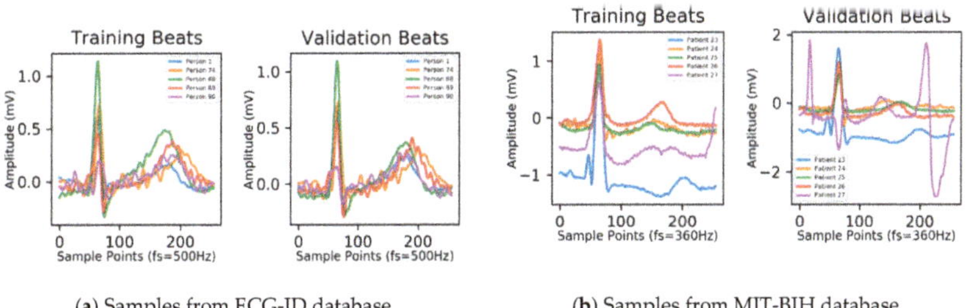

(**a**) Samples from ECG-ID database   (**b**) Samples from MIT-BIH database

Figure 3. Sample of detected and segmented ECG beats for training and validation.

As shown in Figure 3a, ECG records from the ECG-ID database have been preprocessed to remove three types of noises, i.e., baseline drift, powerline noise, and high-frequency noise. Meanwhile, Figure 3b shows that the effect of baseline drift and powerline noise is still visible in the ECG signal.

4.2. Experiment on the Training and Validation

As explained in Section 3.1, the detected and segmented ECG beat was divided into two parts, i.e., training and validation. ECG beats for validation were conducted only after the training process was completed, by producing a weight file with the optimum value of all model parameters. There were a total of 33,226 and 30,496 parameters to be trained and optimized for the ECG-ID and MIT-BIH database, respectively. An example of ECG features extracted using the trained model can be seen in Figure 4. Referring to Table 2, layer 27 (dense_1 layer) produces 64 values which can be used to represent the ECG beat feature of each person.

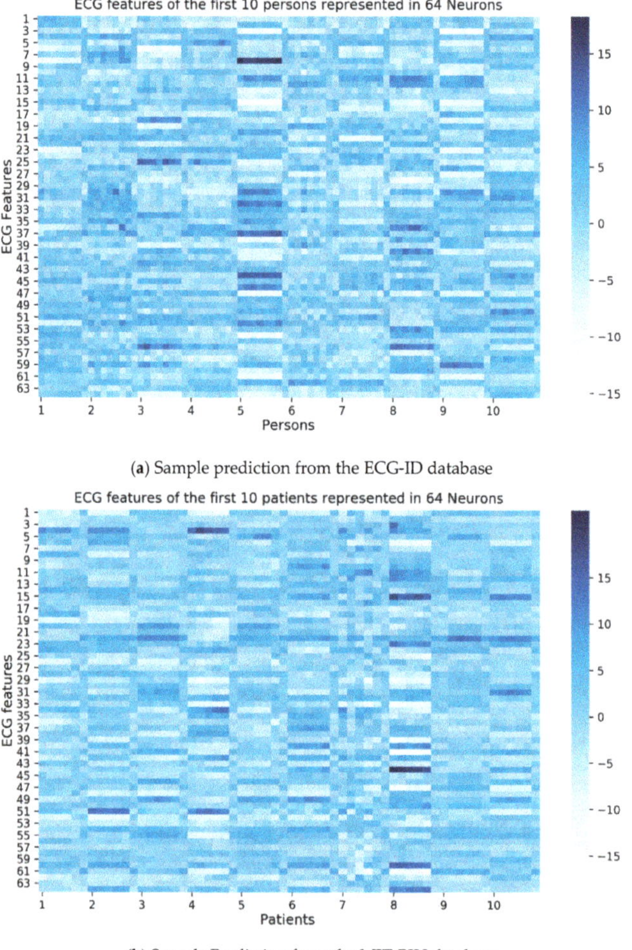

(**a**) Sample prediction from the ECG-ID database

(**b**) Sample Prediction from the MIT-BIH database

Figure 4. Eight ECG beats feature from 10 people represented by 64 neurons of the fully connected layer (layer 27, Table 2).

Figure 4a shows that there were eight columns for each person, while Figure 4b shows that there were six columns for each person. The horizontal lines indicate the regularity of each person's ECG beat feature. Figure 4a shows that ECG beats from persons five, seven, and eight have visible horizontal lines, which indicate they are easier to identify. Figure 4b shows that the regular ECG beat feature was seen for the first until the sixth patient. Meanwhile, the seventh patient does not show clear horizontal lines, indicating that the seventh patient's ECG beats are rather difficult to identify. Table 3 shows the authentication prediction in more detail for eight and six ECG beats of the ECG-ID and MIT-BIH databases, respectively. For ECG-ID, the sixth person could be authenticated as the 6th or 62nd person. Meanwhile, for MIT-BIH, the seventh patient could be authenticated as the 7th or 30th patient.

Table 3. Samples of authentication predictions of eight and six ECG beats for the ECG-ID and MIT-BH databases.

Person	Beats # (ECG-ID)								Predict	Patient	Beats # (MIT-BIH)						Predict
	1	2	3	4	5	6	7	8			1	2	3	4	5	6	
1	28	72	72	1	1	1	16	44	1	1	1	1	1	1	1	9	1
2	2	2	2	2	36	59	2	44	2	2	2	2	2	2	2	9	2
3	3	3	3	24	3	3	3	44	3	3	3	3	3	3	3	9	3
4	4	4	4	4	4	40	4	44	4	4	4	4	4	4	4	9	4
5	5	5	5	5	5	5	5	44	5	5	5	5	5	5	5	9	5
6	6	87	89	62	6	62	14	44	6/62	6	6	6	6	6	6	9	6
7	7	7	7	7	7	7	7	44	7	7	7	24	7	30	30	9	7/30
8	8	8	8	8	8	8	8	44	8	8	8	8	8	8	8	9	8
9	9	9	9	79	79	9	79	44	9	9	45	9	9	9	9	9	9
10	10	76	10	76	10	10	10	44	10	10	10	10	10	10	10	9	10

Based on the previous discussion, the ECG beat features in the 27th layer could be utilized for selection of the training and validation beats. This layer could be further investigated to produce a biometric key. Each biometric key of these 64 features can be quantized into 6 bits only, because the number keys are integer numbers between -20 to 20. Moreover, the number of bits and the number of features can be adjusted according to the number of persons to be authenticated.

The summary of the authentication results of all validation beats can be seen in the confusion matrix shown in Figure 5. Figure 5a,b shows the confusion matrix of the authentication results for the 90 validation beats. These 90 beats were extracted from ECG records in the ECG-ID database. Each person is represented by eight beats or a total of 729 validation beats. Figure 5c,d is the confusion matrix of the authentication results for the 48 validation beats. These 48 patient beats were extracted from ECG records in the MIT-BIH database. Each patient is represented by six beats or a total of 288 validation beats.

Figure 5a,c shows the confusion matrix for a single validation beat, i.e., the first beat of eight or six consecutive beats. Figure 5b,d shows the confusion matrix for multiple validation beats. From Figure 5, it can be shown that 100% accuracy can be obtained for the multiple validation beats, with six beats and eight beats for the ECG-ID and MIT-BIH databases, respectively. The optimum number of multiple beats will be evaluated in the next experiment. Nevertheless, Figure 5 shows that multiple ECG beats produced better accuracy compared to a single ECG beat.

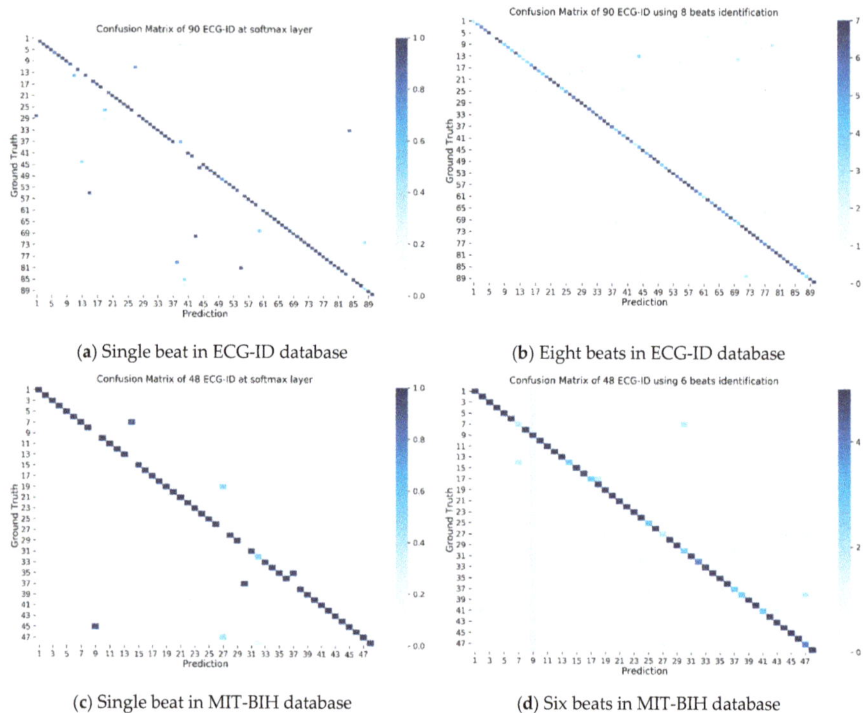

(**a**) Single beat in ECG-ID database

(**b**) Eight beats in ECG-ID database

(**c**) Single beat in MIT-BIH database

(**d**) Six beats in MIT-BIH database

Figure 5. Confusion matrix of the softmax layer (output layer) using single and multiple ECG beats for ECG-ID and MIT-BIH databases.

4.3. Experiment on the Effect of the Number of Consecutive ECG Beats

In this section, the effect of ECG beat detection methods, i.e., manual (using annotated files) vs. automatic Hamilton's method, and the effect of the number of ECG beats on the accuracy, will be elaborated. Table 4 shows the results for various numbers of ECG beats for two ECG beat detections and two ECG databases.

Table 4. Experimental Results for Various CNN Models.

P#	Beat Detection	DSC	Res	Database	#Layer	#Floating Points	t_{train} (ms)	$\frac{t_{val}}{beat}$ (µs)	# Beats Per ID	Accuracy (%)
1	Manual	No	Yes	ECG-ID	28	55,034	29,301	75	8	100
				MIT-BIH	28	52,304	104,519	70	8	100
2	Manual	Yes	No	ECG-ID	22	33,482	29,808	75	8	97.78
				MIT-BIH	22	30,752	99,236	75	5	100
3	Manual	Yes	Yes	ECG-ID	28	33,482	27,001	70	8	100
				MIT-BIH	28	30,752	102,295	73	6	100
4	Hamilton	Yes	Yes	ECG-ID	28	33,482	33,110	75	7	**98.89**
				MIT-BIH	28	30,752	102,464	71	5	**97.92**
5	Hybrid	Yes	Yes	ECG-ID	28	33,482	27,001	70	10	98.89
				MIT-BIH	28	30,752	102,295	70	9	95.83

With manual ECG beat detection, 100% authentication accuracy can be obtained, starting from eight beats and above for the ECG-ID database and from six beats and above for the MIT-BIH database. This could be because the number of the persons to be identified from the ECG-ID database (90 persons) is almost twice the number of the patients to be identified from the MIT-BIH database (48 patients). In contrast, with Hamilton's automatic detection, the authentication accuracy for both ECG-ID and MIT-BIH cannot reach 100%. This can happen because not all the ECG beats detected by Hamilton's method are good beats, so they might require further preprocessing. Nevertheless, manual ECG beat detection (using annotated files) is only feasible in the lab or offline authentication, as the beat detection need to be verified manually by cardiologists. Therefore, the optimum numbers of ECG beat using Hamilton's beat detection is seven and five for ECG-ID and MIT-BIH databases, respectively (the accuracy is highlighted in bold in Table 4).

4.4. Experiment on the Various CNN Models and ECG Beat Detection

As described in Section 3.2 (see Table 1), five profiles were carried out for ECG-ID and MIT-BIH database, i.e., a total of 10 experiments. Table 5 shows the experimental results for various CNN models (various experimental profiles). Of the particular interest is the third profile, which has the following configuration: using manual ECG beat detection (annotated files), using both depthwise separable CNN (DSC) and residual CNN. The third experimental profile produced good results for both the ECG-ID and MIT-BIH database, i.e., 100% accuracy, for eight and six ECG beats, respectively.

Table 5. Experiment results of a various number of ECG beats.

Number of ECG Beats	Beat Detection	ECG-ID Accuracy (%)	MIT-BIH Accuracy (%)
1	Manual	83.33	89.58
	Hamilton	83.33	89.58
2	Manual	37.78	18.75
	Hamilton	38.89	16.67
3	Manual	85.56	87.50
	Hamilton	81.11	87.50
4	Manual	93.33	93.75
	Hamilton	94.44	91.67
5	Manual	95.56	97.92
	Hamilton	96.67	**97.92**
6	Manual	97.78	100
	Hamilton	96.67	95.83
7	Manual	97.78	100
	Hamilton	**98.89**	97.92
8	Manual	100	100
	Hamilton	98.89	97.92
9	Manual	100	100
	Hamilton	97.78	97.92
10	Manual	100	100
	Hamilton	97.78	97.92

Although the third profile produced good authentication results, manual beat detection is only feasible in the lab or in offline authentication, as the beat detection needs to be conducted manually by cardiologists. Note that the first profile produced a good result as well, but it still uses manual beat detection, and the required consecutive beats per patient for the MIT-BIH database was higher (8) compared to the third scenario (6). Comparing the fourth and fifth profiles, it was found that Hamilton's automatic beat detection produced better accuracy compared to hybrid beat detection (training using manual beat detection, while testing using Hamilton's automatic beat detection). Therefore, the fourth experimental profile was found to be the optimal CNN configuration.

From measurement results using the computing platform described in Section 3.1, automatic ECG beat detection in the ECG-ID database using Hamilton's method required around 141,318 µs to detect around 2412 ECG beats or 59 µs per beat on average. In contrast, in the MIT-BIH database, beat detection required 10,827,746 µs to detect around 108,709 beats, or 100 µs per beat on average. Therefore, the maximum detection time required to segment 10 beats is around 1750 µs. This is much faster than acquiring 10 ECG beats from someone's finger.

4.5. Comparison with Other Algorithms

Table 6 shows the benchmarking results with the other 11 algorithms. Of the various algorithms, we highlighted the top three highest accuracies tested with both databases, i.e., Salloum & Kuo [24], Bassiouoni et al. [25], and Wang et al. [9]. Note that, for a fair comparison, we used accuracy results from manual ECG beat detection and segmentation, as used by the other top three algorithms. Nevertheless, for real-time implementation, we could use Hamilton's method for automatic beat detection and segmentation, as it only provides a slight decrease in accuracy performance.

Table 6. Benchmarking of the proposed algorithm with other algorithms using ECG-ID and MIT-BIH databases.

No	Author and Year	Methods	Database	Training	Testing	Accuracy (%)
1	Lugovaya, 2005 [10]	PCA, LDA, MVC	ECG-ID	6 beats	6 beats	96.00
			MIT-BIH	-	-	-
2	Tan et al., 2016 [26]	Wavelet, Random Forest	ECG-ID	9 fiducial points	9 fiducial points	98.79
			MIT-BIH	9 fiducial points	9 fiducial points	99.43
3	Zhang, 2017 [11]	Wavelet, CNN	ECG-ID	-	-	-
			MIT-BIH	single random segment	single random segment	93.50
4	Hussein, 2017 [27]	DCT, feature correlation	ECG-ID	-	-	-
			MIT-BIH	-	single RR segment	97.78
5	Salloum & Kuo, 2017 [24]	LSTM	ECG-ID	9 beats	9 beats	100
			MIT-BIH	18 beats	18 beats	100
6	Bassiouoni et al., 2018 [25]	ANN (MIT-BIH) SVM (ECG-ID)	ECG-ID	3 sets of features	3 sets of features	100
			MIT-BIH	3 sets of features	3 sets of features	99.00
7	Wang et al., 2019 [6]	DWT, S-AE	ECG-ID	QRST features	QRST features	98.87
			MIT-BIH	QRST features	QRST features	96.82
8	Wang et al., 2019 [9]	PCA-Net, SVM	ECG-ID	QRST features	QRST features	97.75
			MIT-BIH	QRST features	QRST features	100
9	Lynn et. al., 2019 [8]	BGRU	ECG-ID	80%	9 beats	98.60
			MIT-BIH	80%	9 beats	98.40
10	Altan, 2019 [28]	LGA, SODP, kNN	ECG-ID	5s ECG segment	5s ECG segment	91.96
			MIT-BIH	5s ECG segment	5s ECG segment	95.12
11	Yifan Chu, 2019 [7]	Multiscale 1D-residual CNN	ECG-ID	500 beats	2 beats	91.96
			MIT-BIH	500 beats	2 beats	95.12
12	Proposed	RDS-CNN	ECG-ID	10 beats	8 beats	100
			MIT-BIH	100 beats	6 beats	100

In Salloum & Kuo [24], ECG beat detection and segmentation was conducted using Tompkins's algorithm [29], in which they manually selected the best-quality eighteen or nine consecutive ECG beats to be fed to the Long Short Term Memory (LSTM) classifier. From each patient of MIT-BIH, they selected 18 ECG beats for training and testing. Each beat has 250 sample points, which is 125 points before and after R_{peak}. For each subject of ECG-ID, they selected nine ECG beats for training and testing.

Each beat has 300 sample points, which is 150 before and after R_{peak}. Our proposed algorithm used a lower number of consecutive ECG beats, i.e., eight for ECG-ID and six for MIT-BIH, and no selection of the quality of ECG signal. Moreover, LSTM is rather more complicated compared to our proposed RDSCNN algorithm, so we can predict that our algorithm will be faster in the authentication process.

In Bassiouoni et al. [25], the features were extracted from fiducial, non-fiducial, and its fusion. There were two different classifiers used, i.e., an artificial neural network (ANN) for MIT-BIH and a support vector machine (SVM) for ECG-ID. Only 30 out of 48 subjects in the MIT-BIH database were tested, while all 90 subjects in the ECG-ID database were tested. In summary, the algorithm in [25] has three stages, including beat segmentation, feature extraction, and different classifiers for different databases. In contrast, our proposed algorithm has only two (faster) stages, including beat segmentation and an RDSCNN classifier, which can be applied for both databases.

Wang et al. [9] used Tompkin's method for ECG beat detection and segmentation starting from 0.24 s and 0.4 s before and after R_{peak}, respectively. The QRST feature was extracted after the beats were decomposed using the discrete wavelet transform (DWT). They used a CNN model that functions like PCA, called PCA-Net, to extract ECG features from the results of DWT decomposition. For classification and authentication, SVM was used. The algorithm is rather complicated (higher computational time) and the accuracy for the ECG-ID database is only 97.75%.

As shown in Table 6, our proposed algorithm using residual depthwise separable convolution neural network outperforms other algorithms, in which it can achieve 100% accuracy for both the ECG-ID and MIT-BIH databases. Furthermore, using Hamilton's method for automatic beat segmentation, it requires around 100 μs (see Section 4.4). The classification process for eight consecutive ECG beats required around 560 μs. Therefore, our proposed algorithms required around 660 μs for the authentication process, i.e., automatic beat segmentation and classification. For a further comprehensive evaluation, other ECG databases could be used as highlighted in [30], including six public domain databases and twelve private databases.

5. Conclusions

In this paper, we have presented a fast and accurate algorithm for ECG authentication using residual depthwise separable convolutional neural networks (RDSCNN). Two prominent databases were used for performance evaluation, i.e., 90 subjects from the ECG-ID database and 48 patients from the the MIT-BIH database. Two ECG beat detection and segmentation methods were evaluated, including manual, using annotated files, and automatic, using Hamilton's method. It was found that the optimum number of consecutive ECG beats for automatic beat detection was seven and five for the ECG-ID and MIT-BIH databases, achieving authentication accuracy of 98.89% and 97.92%, respectively. Furthermore, using manual beat detection with eight (ECG-ID) and six (MIT-BIH) consecutive ECG beats, 100% authentication accuracy can be achieved. Our proposed algorithm is also rather fast, as it requires around 660 μs to conduct authentication, i.e., automatic segmentation and classification. Further research will include optimizing current automatic ECG beat detection and segmentation to be as close as possible to the manual segmentation by cardiologists. Automatic selection of high-quality ECG beats for training and testing could also be conducted, for example by using the integer numbers of 64 ECG features and fractional numbers in the softmax layer. The number of consecutive ECG beats for authentication purposes could be further optimized, down to a single ECG beat.

Author Contributions: Formal analysis, E.I.; Funding acquisition, K.R.; Methodology, E.I.; Supervision, K.R.; Writing—original draft, E.I.; Writing—review & editing, K.R., D.S. and T.S.G. All authors have read and agreed to the published version of the manuscript.

Funding: International Publication Research Grant (PUTI) Universitas Indonesia 2020 based on the Announcement of Directorate Research and Development No. PENG-1/UN2.RST/PPM.00.00/2020.

Conflicts of Interest: The authors declare no conflict of interest.

References

1. Forsen, G.E.; Nelson, M.R.; Staron, R.J., Jr. *Personal Attributes Authentication Techniques*; Pattern Analysis & Recognition Corporation, Rome Air Development Center: St. Utica, NY, USA, 1977.
2. Evans, G.F.; Shirk, A.; Muturi, P.; Soliman, E.Z. Feasibility of using mobile ECG recording technology to detect atrial fibrillation in low-resource settings. *Global Heart* **2017**, *12*, 285–289. [CrossRef] [PubMed]
3. Biel, L.; Pettersson, O.; Philipson, L.; Wide, P. ECG analysis: A new approach in human identification. *IEEE Trans. Instrum. Meas.* **2001**, *50*, 808–812. [CrossRef]
4. Coutinho, D.P.; Silva, H.; Gamboa, H.; Fred, A.; Figueiredo, M. Novel fiducial and non-fiducial approaches to electrocardiogram-based biometric systems. *IET Biom.* **2013**, *2*, 64–75. [CrossRef]
5. Shen, T.-W.; Tompkins, W.; Hu, Y. One-lead ECG for identity verification. In Proceedings of the Second Joint 24th Annual Conference and the Annual Fall Meeting of the Biomedical Engineering Society, Houston, TX, USA, 23–26 October 2002; pp. 62–63.
6. Wang, D.; Si, Y.; Yang, W.; Zhang, G.; Li, J. A Novel Electrocardiogram Biometric Identification Method Based on Temporal-Frequency Autoencoding. *Electronics* **2019**, *8*, 667. [CrossRef]
7. Chu, Y.; Shen, H.; Huang, K. ECG Authentication Method Based on Parallel Multi-Scale One-Dimensional Residual Network With Center and Margin Loss. *IEEE Access* **2019**, *7*, 51598–51607. [CrossRef]
8. Lynn, H.M.; Pan, S.B.; Kim, P. A Deep Bidirectional GRU Network Model for Biometric Electrocardiogram Classification Based on Recurrent Neural Networks. *IEEE Access* **2019**, *7*, 145395–145405. [CrossRef]
9. Wang, D.; Si, Y.; Yang, W.; Zhang, G.; Liu, T. A Novel Heart Rate Robust Method for Short-Term Electrocardiogram Biometric Identification. *Appl. Sci.* **2019**, *9*, 201. [CrossRef]
10. Lugovaya, T.S. Biometric human identification based on ECG. Available online: Availabeonline:https://physionet.org/content/ecgiddb/1.0.0/ (accessed on 14 November 2019).
11. Zhang, Q.; Zhou, D.; Zeng, X. HeartID: A multiresolution convolutional neural network for ECG-based biometric human identification in smart health applications. *IEEE Access* **2017**, *5*, 11805–11816. [CrossRef]
12. Chollet, F. Xception: Deep learning with depthwise separable convolutions. In Proceedings of the IEEE Conference on Computer Vision and Pattern Recognition, Seattle, WA, USA, 14–19 June 2019; pp. 1251–1258.
13. Kaiser, L.; Gomez, A.N.; Chollet, F. Depthwise separable convolutions for neural machine translation. *arXiv* **2017**, arXiv:1706.03059 2017.
14. Ihsanto, E.; Ramli, K.; Sudiana, D.; Gunawan, T.S. An Efficient Algorithm for Cardiac Arrhythmia Classification Using Ensemble of Depthwise Separable Convolutional Neural Networks. *Appl. Sci.* **2020**, *10*, 483. [CrossRef]
15. He, K.; Zhang, X.; Ren, S.; Sun, J. Deep residual learning for image recognition. In Proceedings of the IEEE Conference on Computer Vision and Pattern Recognition, Seattle, WA, USA, 14–19 June 2019; pp. 770–778.
16. Hannun, A.Y.; Rajpurkar, P.; Haghpanahi, M.; Tison, G.H.; Bourn, C.; Turakhia, M.P.; Ng, A.Y. Cardiologist-level arrhythmia detection and classification in ambulatory electrocardiograms using a deep neural network. *Nat. Med.* **2019**, *25*, 65. [CrossRef] [PubMed]
17. Hoang, V.-T.; Jo, K.-H. PydMobileNet: Improved Version of MobileNets with Pyramid Depthwise Separable Convolution. *arXiv* **2018**, arXiv:1811.07083.
18. Hamilton, P.S.; Tompkins, W.J. Quantitative investigation of QRS detection rules using the MIT/BIH arrhythmia database. *IEEE Trans. Biomed. Eng.* **1986**, 1157–1165. [CrossRef] [PubMed]
19. Hamilton, P. Open source ECG analysis. In Proceedings of the Computers in Cardiology, Memphis, TN, USA, 22–25 September 2002; pp. 101–104.
20. Kiranyaz, S.; Ince, T.; Gabbouj, M. Real-time patient-specific ECG classification by 1-D convolutional neural networks. *IEEE Trans. Biomed. Eng.* **2015**, *63*, 664–675. [CrossRef] [PubMed]
21. Moody, G.B.; Mark, R.G. The impact of the MIT-BIH arrhythmia database. *IEEE Eng. Med. Biol. Mag.* **2001**, *20*, 45–50. [CrossRef] [PubMed]
22. Abadi, M.; Barham, P.; Chen, J.; Chen, Z.; Davis, A.; Dean, J.; Devin, M.; Ghemawat, S.; Irving, G.; Isard, M. Tensorflow: A system for large-scale machine learning. In Proceedings of the 12th USENIX Symposium on Operating Systems Design and Implementation, Savannah, GA, USA, 2–4 November 2016; pp. 265–283.
23. Chollet, F. Keras: Deep learning library for theano and tensorflow. Available online: Availabeonline:https://pypi.org/project/Keras/2.2.0/ (accessed on 12 October 2019).

24. Salloum, R.; Kuo, C.-C.J. ECG-based biometrics using recurrent neural networks. In Proceedings of the 2017 IEEE International Conference on Acoustics, Speech and Signal Processing (ICASSP), New Orleans, LA, USA, 5– March 2017; pp. 2062–2066.

25. Bassiouni, M.M.; El-Dahshan, E.-S.A.; Khalefa, W.; Salem, A.M.J.S.; Processing, V. Intelligent hybrid approaches for human ECG signals identification. *Signal Image Video Process.* **2018**, *12*, 941–949. [CrossRef]

26. Tan, R.; Perkowski, M. ECG biometric identification using wavelet analysis coupled with probabilistic random forest. In Proceedings of the 2016 15th IEEE International Conference on Machine Learning and Applications (ICMLA), Anaheim, CA, USA, 18–20 December 2016; pp. 182–187.

27. Hussein, A.F.; AlZubaidi, A.K.; Al-Bayaty, A.; Habash, Q.A. An IoT real-time biometric authentication system based on ECG fiducial extracted features using discrete cosine transform. *arXiv* **2017**, arXiv:08189 2017.

28. Altan, G.; Kutlu, Y.; Yeniad, M. ECG based human identification using Second Order Difference Plots. 2019, 170, 81–93. *Comput. Methods Programs Biomed.* **2019**, *170*, 81–93. [CrossRef] [PubMed]

29. Pan, J.; Tompkins, W.J. A real-time QRS detection algorithm. *IEEE Trans. Biomed. Eng.* **1985**, *32*, 230–236. [CrossRef] [PubMed]

30. Odinaka, I.; Lai, P.-H.; Kaplan, A.D.; O'Sullivan, J.A.; Sirevaag, E.J.; Rohrbaugh, J.W. ECG biometric recognition: A comparative analysis. *IEEE Trans. Inf. Forensics Secur.* **2012**, *7*, 1812–1824.

 *applied sciences*

Article

A Novel P300 Classification Algorithm Based on a Principal Component Analysis-Convolutional Neural Network

Feng Li [1,2,†], Xiaoyu Li [1,2,†], Fei Wang [1,2,3,*], Dengyong Zhang [1,2], Yi Xia [1,2] and Fan He [1,2]

1   School of Computer and Communication Engineering, Changsha University of Science and Technology, Changsha 410114, China; Lif@csust.edu.cn (F.L.); csustecolee@foxmail.com (X.L.); zhdy@csust.edu.cn (D.Z.); csustxy@gmail.com (Y.X.); hf0208@stu.csust.edu.cn (F.H.)
2   Hunan Provincial Key Laboratory of Intelligent Processing of Big Data on Transportation, Changsha University of Science and Technology, Changsha 410114, China
3   School of Software, South China Normal University, Guangzhou 510631, China
*   Correspondence: scutauwf@foxmail.com
†   These authors contributed equally to this work.

Received: 11 January 2020; Accepted: 19 February 2020; Published: 24 February 2020

Abstract: Aiming at enhancing the classification accuracy of P300 Electroencephalogram signals in a non-invasive brain–computer interface system, a novel P300 electroencephalogram signals classification algorithm is proposed which is based on improved convolutional neural network. In the data preprocessing part, the proposed P300 classification algorithm used the Principal Component Analysis algorithm to not only remove the noise and artifacts in the data, but also increase the data processing speed. Furthermore, the proposed P300 classification algorithm employed the parallel convolution method to improve the traditional convolutional neural network framework, which can increase the network depth and improve the network's ability to classify P300 electroencephalogram signals. The proposed algorithm was evaluated by two datasets (the dataset from the competition and the dataset from the laboratory). The results show that, in the dataset I, the proposed P300 classification algorithm could obtain accuracy rates higher than 95%, and achieve one of the best performances in four classification algorithms, while, in the dataset II, the proposed P300 classification algorithm can get accuracy rates higher than 90%, and is superior to the other three algorithms in all ten subjects. These demonstrated the effectiveness of the proposed algorithm. The proposed classification algorithm can be applied in the actual brain–computer interface systems to help people with disability in the daily lives.

Keywords: brain–computer interface (BCI); electroencephalogram (EEG); P300

1. Introduction

Brain–computer interfaces (BCI) can provide a direct communication method between the brain and a computer or other external devices [1–3]. There are several types of electroencephalograms (EEG) signals used in BCI, such as P300 potential [4], steady state visual evoked potential (SSVEP) [5], motor imagery (MI) [6], and so on. Specifically, P300-based BCI is one of the most common BCI systems, as the P300 potential is easy to be stimulated. Compared with other signals, the P300-based BCI system has some advantages: (1) P300 signal is extremely easy to measure and non-invasive; (2) less training time; (3) suitable for most subjects, including those with severe neurological diseases; and (4) users only need to provide a simple control signal [7]. It can implement a variety of different functions, and can even be used in the home of people with disability [7,8].

Farwell and Donchin of the United States introduced the first P300-based character input system in 1988 [9,10], which has been applied until now. The system contained a 6*6 matrix of visual stimulation interface, which was composed of English letters, numbers, and spaces. Before the experiment, the subjects were told that a specified character in the visual stimulator was the target character, and each experiment randomly assigned a character. During the experiment, the subjects were asked to keep an eye on the target character position in the visual stimulator, while any row or column in the visual stimulator flashed randomly. When the target character's row or column was flashing, a positive potential (called P300 ERP) related to the event could be detected in the subject's scalp (about 300 ms after receiving the stimulus); if not, the detected EEG data were non-P300 event-related potentials (N-P300 ERP) [11]. In addition to this standard speller system, there are other paradigms, such as row-column (RC) paradigm [12], single character (SC) paradigm [13], region-based(RB) paradigm [14], and so on. For all these systems, how to identify quickly and accurately is critical to improving the performance of BCI systems.

Due to the collected P300 signals often being high dimensional and feature dependent, some methods were proposed to enhance the feature extraction. PCA (Principal Component Analysis), a principal component analysis method, is widely used in feature extraction and data dimensionality reduction. The principle of PCA is to transform the original signal matrix into a covariance matrix [15] through linear transformation [16], and obtain a new signal matrix by filtering the eigenvalues and eigenvectors of the matrix [17]. The new signal matrix retains some of the most important original signal features in the original signals matrix, and eliminates noise and unimportant features to achieve the purpose of dimensionality reduction. In recent years, many researchers have applied the PCA to reduce the dimensionality of the obtained EEG signals. Salma Tayeb used different dimensionality reduction algorithms to process the EEG signals for the dimensionality reduction part of the data, and found that the use of PCA for dimensionality reduction of P300 signals performed best, compared to independent component analysis (ICA) and linear discriminant analysis (LDA) [18]. Kundu and Sourav used PCA to reduce the dimensionality of P300 signals, and then used SVM to classify the reduced-dimensional signals. PCA reduced the computational burden of weighted classifiers and speeds up the classification speed [19]. Like combined multi-scale filters and PCA to classify EEG signals, the classification accuracy can reach 91.13% [20]. PCA fits in a brain–computer interface, especially the P300 brain–computer interface.

Previous works on P300 classification mainly employed traditional machine learning algorithms, such as support vector machine (SVM), linear discriminant analysis (LDA), and so on. Rakotomamonjy divided P300 EEG signals into several equal parts, and then used SVM to train corresponding classifiers for each part, which improved the accuracy of P300 EEG signals recognition [21]. Chandra S. Throckmorton proposed a Bayesian P300 recognition method to complete classification by determining the maximum regression target probability value. Although the accuracy of classification was improved, it took too much calculation time [22]. With the rapid development of deep learning, many scholars began to use convolutional neural network (CNN) to classify P300 EEG signals [23]. Cecotti realized the use of convolutional neural networks in deep learning to recognize and classify P300 EEG signals. He used convolutional layers to separate the time and space domains of P300 EEG signals. The convolutional neural network is fast but very easy to overfit, which affects the accuracy of recognition [24]. Lawhern Vernon used compact convolutional neural network (CNN) to classify four types of EEG data including P300 EEG signals, and the results shown that the compact CNN has the best classification effect on P300 EEG signals [25]. Sobhani proposed to use deep belief network (DBN) to classify P300 EEG signals which were extracted from each channel, but only a few subjects have good recognition accuracy [26]. Maddula proposed a 3D recurrent convolution neural network (3DRCNN) based on a recurrent convolution neural network (RCNN) to classify P300 EEG signals. It was processed into 3D-EEG signals, and then input them into the RCNN to achieve nice classification [27]. LIU improved the convolutional neural network on the basis of Cecotti's algorithm which was named BN3 algorithm, taking the batch normalization (BN) layer and the dropout (DP) layer to deepen the network layers and overcome the problem of overfitting. The BN3 algorithm achieved good results in classification, but still need to improve the recognition accuracy when the number of experiments are reduced [28].

In this paper, we combined PCA algorithms with new convolution neural network framework to complete the P300 signals classification and recognition (named PCA-CNN). Specifically, the PCA algorithm was used to reduce the dimension of the EEG signal, which not only reduced the calculation time, but also improved the signal-to-noise ratio of the data. The new convolution neural network framework improved single convolution kernel model of traditional convolutional neural network. It contacted multiple convolution kernels to classify P300 EEG signals in the convolution layer, which improved the recognition ability of the convolutional neural network. Two datasets (competition data and self-collected data) were used to verify the effectiveness of the algorithm. The results indicated that the proposed algorithm had a significant effect on the recognition accuracy of P300 EEG signals.

2. Method

2.1. The Dataset

Two sets of experimental data were analyzed in this paper. One was the dataset in the BCI Competition III provided by the Wadsworth Research Center NYS Department of Health [29]. The other was provided by South China University of Technology using a different paradigm. There are two and ten subjects, respectively, in the two datasets. All subjects are healthy persons, who were selected randomly. Specifically, the two subjects (A and B) were chosen from five people in the public BCI Competition III in 2004. The other ten subjects in the dataset II were recruited randomly and participated in the brain–computer interface experiments for the first time, provided by the South China University of Technology.

Dataset I: The graphical user interface (GUI) of the competition was presented in Figure 1a, which is a 6*6 character matrix. When the experiment began, each of the 12 rows and columns flashed randomly. A flashing lasts 100 ms and the interval between two flashing is 75 ms. A subject was asked to focus on the target character, and silently count the flashing repetitions of the row and column containing the target character. Each row or column repeats 15 times when outputting one character. The dataset was consisted of one training (85 characters) and one test (100 characters) sets for each of the two subjects A and B. All EEG signals were collected by a 64-electrode scalp, which were bandpass filtered from 0.1–60 Hz and digitized at 240 Hz. The information details can be found in the BCI competition webpage.

Dataset II: The second dataset was collected in the laboratory from South China University of Technology using a 4*10 paradigm (see Figure 1b). Different from the first dataset, each character flashed separately and randomly. A flashing lasts 100 ms and the interval is 30 ms. A subject was asked to focus on the target character, and silently count the flashing repetitions of the target character. Each character in paradigm repeats 10 times when outputting one character. The dataset was consisted of one training (20 characters) and one test (30 characters) set for each of ten subjects. All EEG signals were collected by a 32-electrode scalp, which were bandpass filtered from 0.1–60 Hz and digitized at 250 Hz.

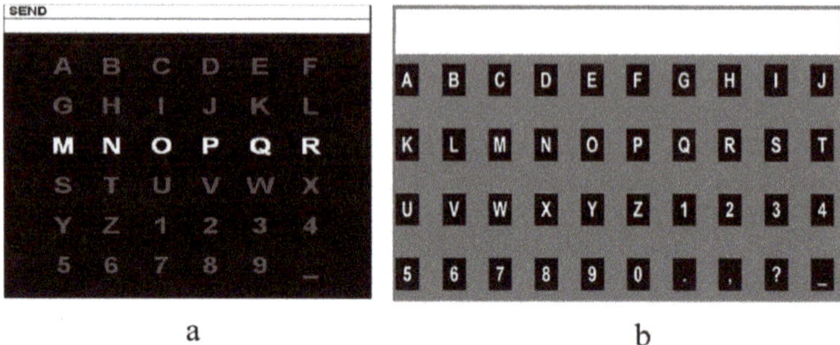

Figure 1. Experimental set-up. (**a**) 6*6 GUI used in the brain–computer interfaces (BCI) competition; (**b**) 4*10 graphical user interface (GUI) used in the laboratory.

2.2. Data Preprocessing

As the raw EEG signals are weak and mixed with non-EEG signals or background noise, the raw data should preprocess first. In order to remove the influence of these interference information [30], we used the 8th-order bandpass Butterworth filter [31] to filter the collected raw data and intercept the potential frequency to 0.1–20 Hz. Then, the number of positive samples should be increased before the next step, in order to prevent the classification problems caused by the imbalance of positive and negative samples. We will increase the number of P300 EEG signals to the number of non-P300 EEG signals; that is, copying P300 EEG signals so that the number of P300 EEG signals is the same as or close to non-P300 signals.

2.3. PCA Algorithm

PCA transforms the original data into a set of linearly independent data vectors in various dimensions through linear transformation, which can be used to extract the main feature components of the data and often be used for dimensionality reduction of high-dimensional data. Therefore, before EEG signals were input into the neural network, we used the PCA algorithm to reduce the signal dimension. After preprocessing of the raw signals, a data matrix X is obtained, of which the abscissa and ordinate are the time and space domains, respectively. We input the matrix X into the PCA algorithm and calculated the covariance matrix (Cov(X)); then, the eigenvalue eigenvectors of the covariance matrix were obtained. We could select a matrix of eigenvectors corresponding to the features with the largest eigenvalues. In this way, the data matrix could be transformed into a new space, and the dimension reduction of the data features could be realized. Through dimensionality reduction using PCA, the size of feature vectors changed from 64*240 and 30*160 to 64*120 and 30*80 (two datasets respectively). The mathematical formula is as follows:

Input the matrix $X(x1, x2, x3, ..., xn)$, and reduce the matrix X to K dimensions, $0 \leq K \leq n$.

Find the covariance matrix of matrix X:

$$\text{Cov}\,X = \frac{1}{n}XX^T \tag{1}$$

Find the eigenvalues and eigenvectors of the covariance matrix CovX:

$$\text{Cov}\,X = \Lambda L \tag{2}$$

where $\Lambda = diag[\lambda 1, \lambda 2, \ldots, \lambda n]$ is the eigenvalue of the X covariance matrix, and L is the eigenvector of the X covariance matrix.

Sort the eigenvalues $\Lambda(set\lambda1 \geq \lambda2 \geq ... \geq \lambda n \geq 0)$ from large to small, and select the largest k of them. Then, use the corresponding k feature vectors as row vectors to form a feature vector matrix P and the data are transformed into a new matrix Y constructed by k feature vectors:

$$Y = PX \tag{3}$$

where Y is the matrix after dimensionality reduction.

2.4. Parallel Convolutional Network

In this paper, we proposed an improved neural network architecture (as shown in Figure 2). The specific network contained 9 layers, the parameters of which were illustrated in Tables 1 and 2. L1 is the data input layer. L2 is the spatial domain convolution layer. L3 to L6 are the parallel convolution layer [32,33] to extract time domain features. L7 is the pooling layer. L8 and L9 are fully connected layer and softmax layer. The computations in each layer (following, as an example, Table 1 Parameters setting) were described in detail as follows:

L1: The input layer loads the pre-processed EEG data into the network and uses a to represent the data tensor transmitted to the neural network.

L2: Convolution layer, which is a spatial filter for all channels of the input signal, can improve the signal-to-noise ratio [34], and remove redundant signals in the spatial domain. The formulas are as follows:

$$a_i^1(j) = \sum_{i=1}^{i=64} a_i^0(j)w_0 + b_0 \tag{4}$$

where $a_i^1(j)$ is output data for L2, i $(1 \leq i \leq 64)$ denotes the Spatial dimension, and j $(1 \leq j \leq 120)$ denotes the time dimension. w_0 denotes the weight, b_0 denotes the deviation (all w_n, and b_n $(0,1,2,...,n)$ denotes the deviation of different values below).

L3 and L4: convolutional layer and dropout layer. This layer is arranged in parallel by three convolutional layers of different convolution sizes. Each convolution and size is the same. Different convolution kernels can be extracted to different values for the same input: information, increasing the complexity of features. After filtering in the time domain after L2 layer spatial filtering, we use 16*5*1, 16*10*1 and 16*15*1 convolution kernels for convolution. After convolution, we can get 16*1*24, 16*1*12 and 16*1*8 feature vectors; these feature vectors are combined into 16*1*44 feature vectors. A dropout layer is added after the convolutional layer to prevent overfitting in the case of too many model parameters [35]. The formula is as follows:

$$a_j^2(s) = \sum_{j=1}^{j=120} a_j^1(s)w_1 + b_1 + a_j^1(s)w_2 + b_2 + a_j^1(s)w_3 + b_3 \tag{5}$$

$$r^1 = \text{Bernoulli}\left(p^1\right) \tag{6}$$

$$a_j^3(s) = \sum_{j=1}^{j=44} a_j^2(s)r^1 w_4 + b_4 \tag{7}$$

where $a_j^2(s)$ is output data for L3, s $(1 \leq s \leq 20)$ denotes the depth of convolution kernels, j denotes the time dimension, and $a_j^3(s)$ is output data after dropout layer. r and p denote the dropout value, $r = 0.5$.

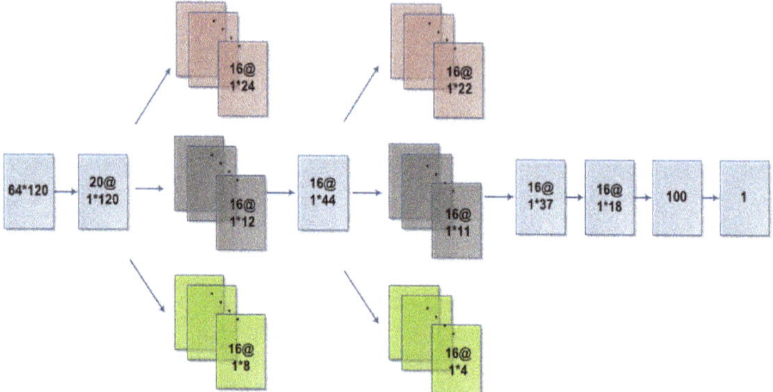

Figure 2. The framework model of the 9-Layer convolution neural network used for classification. The feature map is of dimension Depth @ Height * Weight (such as 16@1*24), layers with colors indicate that different size convolution kernels are used. The parameters in this figure are an example of the parameters in Table 1.

L5 and L6: Structures of these two layers are the same as that of L3 and L4. These layers are convolved with 16*2*1, 16*4*1, and 16*11*1 convolution kernels, which can get 16*1*22, 16*1*11, and 16*1*4 feature vectors, and are combined into 16*1*37 feature vectors. A dropout(DP) layer is added after the convolution layer, and the value is 0.5. The calculation formula is as follows:

$$a_j^4(s) = \sum_{j=1}^{j=44} a_j^3(s)w_5 + b_5 + a_j^3(s)w_6 + b_6 + a_j^3(s)w_7 + b_7 \tag{8}$$

$$r^2 = \text{Bernoulli}\left(p^2\right) \tag{9}$$

$$a_j^5(s) = \sum_{j=1}^{j=37} a_j^4(s)r^2 w_8 + b_8 \tag{10}$$

L7: The pooling layer consists of a pooling filter of size 2, which is used to reduce the parameters of the network.

L8 and L9: Fully connected layer and softmax layer. Data after L7 which are connected with 100 neurons, and then classification by softmax layer. We use rectified linear unit (ReLU) as the activation function [36]. $a^6(k)$ and $a^7(k)$ denote the output data after fully connected(FC) layer. k is the number of feature maps. The calculation formula is as follows:

$$a^6(k) = \sum_{k=1}^{k=288} a^5(k)w_9 + b_9 \tag{11}$$

$$a^7(k) = relu\left(\sum_{k=1}^{k=100} a^6(k)w_{10} + b_{10}\right) \tag{12}$$

When the calculated network output probability is greater than or equal to 0.5, the current input signals are determined to be P300 signals, otherwise not. The judgment is as follows:

$$Q = \begin{cases} 1 & (P \geq 0.5) \\ 0 & (P < 0.5) \end{cases} \tag{13}$$

where Q represents the judgment result and P represents the probability value.

The scintillation of 6 rows and 6 columns in the experiment will be repeated 15 times. This is because the position of the target character can not be accurately determined by single experiment scintillation. The probability values corresponding to multiple row scintillation and multiple column scintillation can be accumulated. The target character can be determined by selecting the rows and columns corresponding to the maximum probability values. The formula is as follows:

$$\begin{cases} X = \arg_i \max \sum_{K=1}^{K=n} p(K,i) & (1 \leq i \leq 6) \\ Y = \arg_i \max \sum_{K=1}^{K=n} p(K,i) & (7 \leq i \leq 12) \end{cases} \tag{14}$$

X denotes the position of column target characters, Y denotes the position of row target characters, n denotes the number of experiments, P denotes the probability value, K denotes the serial number of experiments, and i denotes the row and column numbers.

Dataset II is a random single flicker of 40 characters, from which the location of the maximum probability value can be selected to determine the target character. The formula is as follows:

$$Z = \arg_i \max \sum_{K=1}^{K=n} p(K,i) \quad (1 \leq i \leq 40) \tag{15}$$

Z denotes the position of the target character, n denotes the number of experiments, P denotes the probability value, K denotes the serial number of experiments, and i denotes the position number.

Table 1. The network parameter settings for dataset I.

Number of Layers	Input	Convolution Kernel and Operation	Output	Activation Function
L1	1*64*120	None	1*64*120	None
L2	1*64*120	20*64*1	20*1*120	ReLU
L3	20*1*120 20*1*120 20*1*120	16*1*5 16*1*10 16*1*15	16*1*24 16*1*12 16*1*8	ReLU
L4	16*1*24 16*1*12 16*1*8	Concat and DP	16*1*44	ReLU
L5	16*1*44 16*1*44 16*1*44	16*1*2 16*1*4 16*1*11	16*1*22 16*1*11 16*1*4	ReLU
L6	16*1*22 16*1*11 16*1*4	Concat and DP	16*1*37	ReLU
L7	16*1*37	Maxpool	16*1*18	ReLU
L8	16*1*18	FC	100*1	ReLU
L9	100*1	Softmax	2*1	ReLU

Table 2. The network parameter settings for dataset II.

Number of Layers	Input	Convolution Kernel and Operation	Output	Activation Function
L1	1*30*80	None	1*30*80	None
L2	1*30*80	20*30*1	20*1*80	ReLU
L3	20*1*80 20*1*80 20*1*80	16*1*5 16*1*8 16*1*10	16*1*18 16*1*10 16*1*8	ReLU
L4	16*1*18 16*1*10 16*1*8	Concat and DP	16*1*36	ReLU
L5	16*1*36 16*1*36 16*1*36	16*1*2 16*1*4 16*1*9	16*1*18 16*1*9 16*1*4	ReLU
L6	16*1*18 16*1*9 16*1*4	Concat and DP	16*1*31	ReLU
L7	16*1*31	Maxpool	16*1*15	ReLU
L8	16*1*15	FC	100*1	ReLU
L9	100*1	Softmax	2*1	ReLU

2.5. Evaluation

To measure the performance of the algorithms, we used two indices, the accuracy rate and the information translate rate (ITR) [24,37] to compare the proposed PCA-CNN with other CNN algorithms in the literature. The accuracy can evaluate the effectiveness of the algorithm. The accuracy rate in the article is the number of correct classified characters classified compared with the number of total actual test characters. The calculation formula is as follows:

$$T_{acc} = P_n / S_n \tag{16}$$

where T_{acc} is the accuracy rate of character recognitions, P is the number of correct detected characters, n is the number of repeats, and S is the number of total characters.

ITR can display the recognition speed of the test characters by the classification algorithm in bits per minute. The formula is as follows:

$$\text{ITR} = \frac{60(P \log_2(P) + (1-P) \log_2((1-P)/(N-1)) + \log_2(N))}{T} \tag{17}$$

where N represents the number of classes, P represents the accuracy rate of character recognitions, and T represents the time taken for character recognitions [24,28].

In dataset I, as each flash lasts for 100 ms followed by a pause of 75 ms ($12 * (75 + 100) = 2100$) and a pause of 2.5 s between each character epoch, T can thus be defined as:

$$T = 2.5 + 2.1n \tag{18}$$

where n is the number of repeats, $1 \leq n \leq 15$.

In dataset II, as each flash lasts for 100 ms followed by a pause of 30 ms ($40 * (30 + 100) = 5200$) and a pause of 1.2 s between each character epoch, so T can be defined as:

$$T = 1.2 + 5.2n \tag{19}$$

where n is the number of repeats, $1 \leq n \leq 10$.

3. Experimental Results

In this paper, we used the accuracy rate and ITR to evaluate the P300 signals detection performance on two datasets of different subjects. In dataset I, there are 85 training and 100 test characters for each subject, each of which is repeated 15 times. Tables 3 and 4 present the test accuracy rates of the proposed PCA-CNN and other classification methods in the literature, including CNN algorithms BN3 [28] and CNN-1 [24], and a traditional SVM algorithm [21], on the datasets I of subjects A and B. Bold numbers in the table indicate the best accuracy rate in the $N(1, 2, \ldots, N)$ repeats. For both subjects, the proposed PCA-CNN is one of the best algorithms in 15 repeats, and can obtain the accuracy rate higher than 95%. Furthermore, the PCA-CNN is superior to other three methods from 7 repeats for the subject A in the dataset I, and from 8 repeats for the subject B in the dataset I.

Table 3. The accuracy rate of subject A in dataset I.

Algorithms	The Number of Repeats														
	1	2	3	4	5	6	7	8	9	10	11	12	13	14	15
PCA-CNN	24	37	46	61	71	**75**	**84**	**86**	**90**	90	**92**	**94**	**95**	**97**	**98**
BN3	22	**39**	**58**	**67**	**73**	**75**	79	81	82	86	89	92	94	96	**98**
CNN-1	16	33	47	52	61	65	77	78	86	**90**	91	91	91	93	97
SVM	16	32	52	60	72	71	82	81	82	83	87	88	94	95	97

Table 4. The accuracy rate of subject B in dataset I.

Algorithms	The Number of Repeats														
	1	2	3	4	5	6	7	8	9	10	11	12	13	14	15
PCA-CNN	30	51	57	69	73	74	83	**94**	**95**	**96**	**97**	95	**96**	**96**	**96**
BN3	47	**59**	**70**	**73**	76	**82**	**84**	91	94	95	95	**95**	94	94	95
CNN-1	35	52	59	68	**79**	81	82	89	92	91	91	90	91	92	92
SVM	35	53	62	68	75	80	**84**	86	89	91	92	93	**96**	95	**96**

In dataset II, there are 20 training characters and 30 test characters for each subject, each of which is repeated 10 times. Figure 3 presents the test accuracy rates of the proposed PCA-CNN and other classification methods in the literature, including CNN algorithms BN3 [28] and CNN-1 [24], and the traditional SVM algorithm [21], on datasets II of 10 subjects. The different color line in the figure records the results of all subjects that used different methods on each repeat. For all subjects, the average accuracy rate of PCA-CNN is higher than the other three algorithms in 4th repeats. Furthermore, the average accuracy rate of PCA-CNN is the best algorithm in 10 repeats, and can obtain an average accuracy rate higher than 90%. The experimental results show that the PCA-CNN is superior to others in character recognition.

We record the information translate rates of datasets I and II in different algorithms. As shown in Figures 4 and 5, the ITR value of the PCA-CNN is higher than that of the other three algorithms (BN3 [28], CNN-1 [24], and SVM [21]) in the maximum repeat. In Figure 4, after the 7th repeat, the ITR value of PCA-CNN is higher than the other three algorithms (BN3 [28], CNN-1 [24] and SVM [21]). In Figure 5, after the 4th repeat, the ITR value of PCA-CNN is higher than the other three algorithms (BN3 [28], CNN-1 [24] and SVM [21]). From the overall results of Figures 4 and 5, the PCA-CNN is faster than the CNN-1 [24] and SVM [21]. These indicate that, on the basis of ensuring the characters recognition accuracy rate, the characters' recognition speed of PCA-CNN is still fast, and the PCA-CNN algorithm has application value.

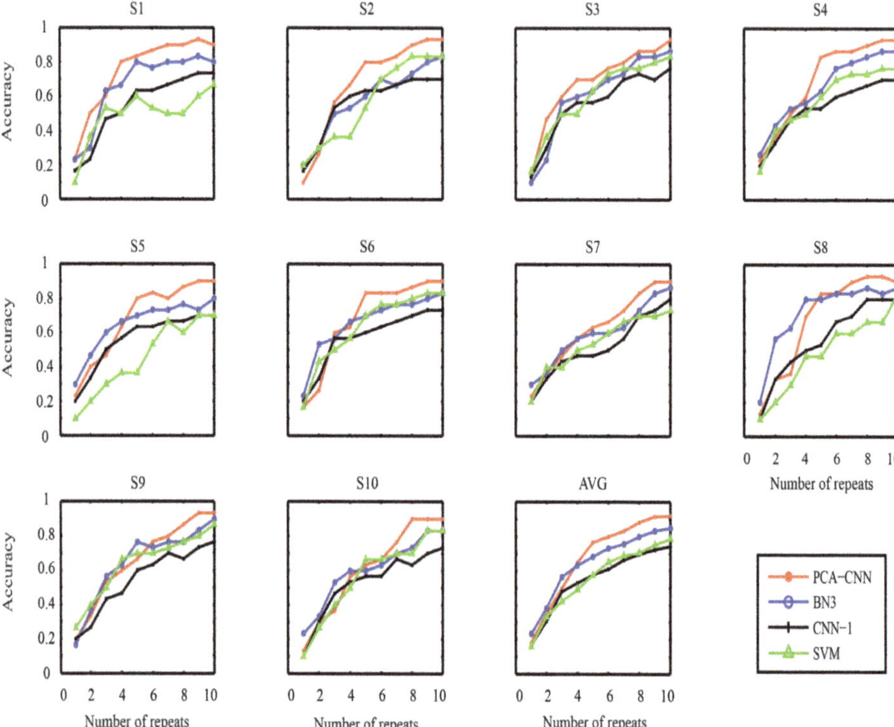

Figure 3. The accuracy rates of all ten subjects in dataset II. The vertical axis corresponds to the accuracy rate, and the horizontal axis corresponds to the number of repeats. The color curves represent respectively the different classification accuracy rates of each subject and the average accuracy rate (AVG) of all 10 subjects (S1, S2, ..., S10) tested by four methods.

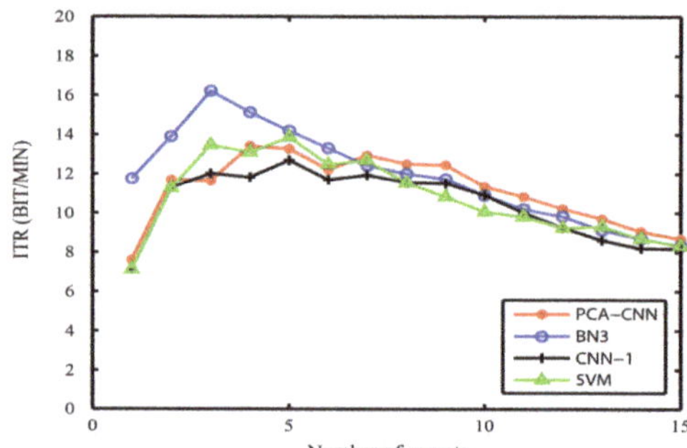

Figure 4. The information translate rate graph for four algorithms in the dataset I calculated by the average information translate rates of subject A and subject B.

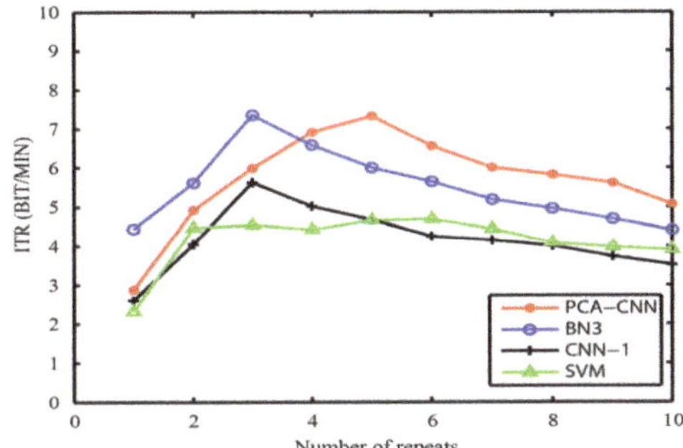

Figure 5. The information translate rate graph for four algorithms in dataset II calculated by the average information translate rates of all ten subjects.

4. Discussion

This paper proposed a PCA-CNN algorithm to improve the classification performance of P300-based BCI system. The PCA was used to reduce the P300 EEG signals dimension, as PCA could not only removed the noise and unimportant features of P300 EEG signals, but also improved the speed of EEG data processing. Furthermore, we used the improved convolutional neural network to classify P300 EEG signals and recognition. The experiment results show that, with the same experimental data and the number of experiments, the classification accuracy of the PCA-CNN algorithm is one of the best among algorithms SVM [21], CNN-1 [24], and BN3 [28].

For dataset I and dataset II, the algorithm produced some identical and a few different results. In both datasets, the PCA-CNN has higher accuracy rates of characters recognition than the other three algorithms (BN3 [28], CNN-1 [24], and SVM [21]) with repeat number increases, and can obtain the highest accuracy rate of character recognition in the last repeat. However, in dataset I, the comparison of the classification accuracy results of the algorithm PCA-CNN and the other three algorithms (BN3 [28], CNN-1 [24] and SVM [21]) is not as obvious as that in the dataset II. This difference may be due to the different number of subjects in dataset I and dataset II. There are only two subjects in dataset I. Thus, too small sample size may lead to insignificant differences in the comparison of classification accuracy results. In both dataset I and dataset II, the ITR value of the PCA-CNN is higher than that of the other three algorithms (BN3 [28], CNN-1 [24] and SVM [21]) in most repeats. The difference is that the comparison of the ITR value results of the algorithm PCA-CNN and the other three algorithms (BN3 [28], CNN-1 [24] and SVM [21]) is not obvious in two datasets. One possible reason is related to the difference of classification accuracy in the two datasets. According to the ITR calculation formula, its value is proportional to the classification accuracy. In dataset I, the classification accuracy rate of the PCA-CNN algorithm is not significantly higher than that of other algorithms in data II, at the maximum repeat.

The PCA algorithm was widely used in various fields of research data analysis, especially suitable for analyzing two-dimensional data matrix [38]. Researchers found that PCA could well express the basic features of the original data with less data [39]. Therefore, P300 EEG signals still retain the integrity of original EEG signals after PCA dimensionality reduction. Although the convolutional neural network can also directly extract features by itself, we find that the classification accuracy rates are higher after using PCA. As shown in Table 5, when the PCA algorithm is not added in the classification process, the accuracy of classification is lower than the accuracy of adding the PCA algorithm.The average recognition accuracy rates (subject A and subject B in the dataset I) of the PCA-CNN algorithm is 97%, while the average accuracy of recognize characters without the PCA algorithm is 94%. The former is 3% higher than the accuracy in the latter category. In Figure 6, the average accuracy rate (ten subjects in the dataset II) of recognized characters related to the CNN algorithm which added the PCA algorithm is 90%, while the average accuracy of recognize characters without the PCA algorithm is 80%. The former is 10% higher than the accuracy in the latter category. These may be caused by the overfitting problem when the convolutional neural network processes a large amount of data, which affects the experimental performance. Adding the PCA algorithm in the convolutional neural network may solve this problem that has been proven by some literature [40,41].

Table 5. The comparison of PCA and NO-PCA character recognition accuracy rate in dataset I.

Methods	Subjects	Accuracy Rate
PCA	A	**98**
	B	**96**
NO-PCA	A	96
	B	93

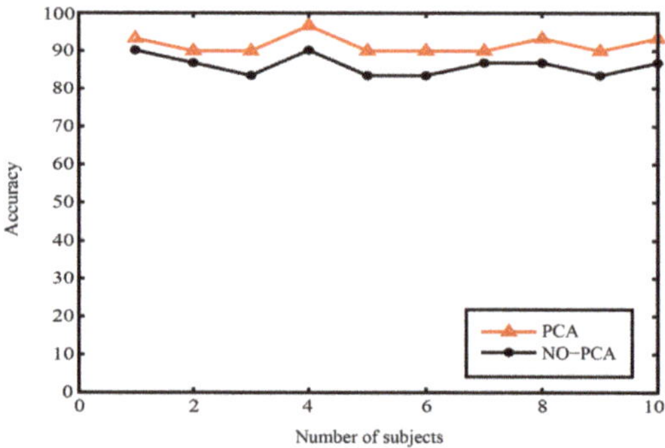

Figure 6. The comparison of PCA and NO-PCA characters recognition accuracy rate in dataset II.

In order to obtain a classifier with better classification ability, we improved the convolutional layer of the traditional convolutional neural network to a parallel convolution layer [42]. The parallel convolution layer adds multiple convolution kernels of different sizes to filter out the P300 EEG signal features. The number of convolution kernels determines the output of the convolution layer, so the convolution layer needs to appropriately increase the number of convolution kernels in order to more fully extract the features of signals [43]. In the previous works on the convolution neural network, such as CNN-1 [24] and BN3 [28], a single convolution kernel layer was used in the convolution part. However, when the number of signals is too large, the effect of a single convolution kernel layer in filtering features will become worse, and it is easy to ignore some features. The parallel convolution layer could increase the data capacity of the network, and may overcome the lack of features caused by improper selection of the convolution kernel size [44].

Based on the traditional convolution neural network, this paper constructed the algorithm PCA-CNN, a new algorithm for the P300 EEG signals classification. Compared with some traditional convolutional neural algorithms, the PCA-CNN algorithm has a higher classification accuracy rate for the P300 EEG signals' recognition. As shown in Tables 3 and 4, and Figure 3, the results show that the PCA-CNN algorithm has good classification performance in datasets I and II. As shown in Figures 4 and 5, the ITR of PCA-CNN is higher than the traditional SVM classification algorithm, which proves the stability of the algorithm performance. The PCA-CNN algorithm can obtain the classification accuracy rate higher than 90% on both datasets I and II; when the classification accuracy rate of P300 EEG signals is higher than 80%, the classification algorithm is effective [45]. The proposed PCA-CNN algorithm can be employed in a brain–computer interface system, and be applied for people with disability in daily lives.

5. Conclusions

Our work focuses on P300 EEG signals preprocessing and the convolution neural network structure designing. In the P300 EEG signals preprocessing part, the PCA is used to retain the data features of the original P300 EEG signals, which reduced the dimension of the original signals and reduced the computational cost of subsequent algorithms. In the convolution neural network structure designing part, this paper used a deep convolutional neural network to implement the classification and recognition of P300 EEG signals. The convolution neural network uses its own powerful feature extraction capabilities to construct a better classifier. The new algorithm changed the single-kernel convolutional layer in the convolution neural network to a multi-kernel convolutional layer, that is, using a multi-kernel convolution filter to extract P300 EEG signals, which improved the classification ability of the network. Compared with some traditional classification algorithms, the PCA-CNN algorithm has a higher character recognition accuracy rate. In the future, our research will consider how to improve the recognition speed of BCI system and implement an online P300 brain–computer interface system based on the deep convolution neural networks.

Author Contributions: Investigation, Y.X.; Methodology, X.L.; Supervision, F.L.; Validation, F.H.; Writing (original draft), X.L.; Writing (review and editing), F.W. and D.Z. All authors have read and agreed to the published version of the manuscript.

Funding: This research was supported by the National Natural Science Foundation of China (Grant No. 61906019), the Natural Science Foundation of Hunan Province, China (Grant No. 2019JJ50649), the Scientific Research Fund of Hunan Provincial Education Department (Grant Nos. 18C0238 and 19B004), the "Double First-class" International Cooperation and Development Scientific Research Project of Changsha University of Science and Technology (No. 2018IC25), and the Young Teacher Growth Plan Project of Changsha University of Science and Technology (No. 2019QJCZ076).

Conflicts of Interest: The authors declare no conflict of interest.

References

1. Wolpaw, J.R.; Birbaumer, N.; McFarland, D.J.; Pfurtscheller, G.; Vaughan, T.M. Brain–computer interfaces for communication and control. *Clin. Neurophysiol.* **2002**, *113*, 767–791. [CrossRef]
2. Schalk, G.; McFarland, D.J.; Hinterberger, T.; Birbaumer, N.; Wolpaw, J.R. BCI2000: A general-purpose Brain-Computer Interface (BCI) system. *IEEE Trans. Biomed. Eng.* **2004**, *51*, 1034–1043. [CrossRef]
3. Weiskopf, N.; Mathiak, K.; Bock, S.W.; Scharnowski, F.; Veit, R.; Grodd, W.; Goebel, R.; Birbaumer, N. Principles of a brain–computer interface (BCI) based on real-time functional magnetic resonance imaging (fMRI). *IEEE Trans. Biomed. Eng.* **2004**, *51*,966–970. [CrossRef] [PubMed]
4. Donchin, E.; Coles, M.G.H. Is the P300 component a manifestation of context updating? *Behav. Brain Sci.* **1988**, *11*, 357–374. [CrossRef]
5. Lin, Z.; Zhang, C.; Wu, W.; Gao, X. Frequency Recognition Based on Canonical Correlation Analysis for SSVEP-Based BCIs. *IEEE Trans. Biomed. Eng.* **2006**, *77*, 2610–2614. [CrossRef] [PubMed]
6. Decety, J. The neurophysiological basis of motor imagery. *Behav. Brain Res.* **1996**, *77*, 45–52. [CrossRef]
7. Fazel-Rezai, R.; Allison, B.Z.; Guger, C.; Sellers, E.W.; Kleih, S.C.; Kübler, A. P300 brain computer interface: Current challenges and emerging trends. *Front. Neuroeng.* **2004**, *5*, 73–77. [CrossRef]
8. Iturrate, I.; Antelis, J.M.; Kubler, A.; Minguez, J. A Noninvasive Brain-Actuated Wheelchair Based on a P300 Neurophysiological Protocol and Automated Navigation. *IEEE Trans. Robot.* **2009**, *25*, 614–627. [CrossRef]
9. Salvaris, M.; Sepulveda, F. Visual modifications on the P300 speller BCI paradigm. *J. Neural Eng.* **2009**, *6*, 046011. [CrossRef]
10. Krusienski, D.J.; Sellers, E.W.; Cabestaing, F.; Bayoudh, S.; McFarland, D.J.; Vaughan, T.M.; Wolpaw, J.R. A comparison of classification techniques for the P300 Speller. *J. Neural Eng.* **2006**, *3*, 299–305. [CrossRef]
11. Ming, D.; An, X.; Xi, Y.; Hu, Y.; Wan, B.; Qi, H.; Cheng, L.; Xue, Z. Time-locked and phase-locked features of P300 event-related potentials (ERPs) for brain–computer interface speller. *Biomed. Signal Process. Control* **2010**, *5*, 243–251. [CrossRef]

12. Townsend, G.; LaPallo, B.K.; Boulay, C.B.; Krusienski, D.J.; Frye, G.E.; Hauser, C.; Schwartz, N.E.; Vaughan, T.M.; Wolpaw, J.R.; Sellers, E.W. A novel P300-based brain–computer interface stimulus presentation paradigm: Moving beyond rows and columns. *Clin. Neurophysiol.* **2010**, *121*, 1109–1120. [CrossRef] [PubMed]

13. Pan, J.; Li, Y.; Gu, Z.; Yu, Z. A comparison study of two P300 speller paradigms for brain–computer interface. *Cogn. Neurodyn.* **2013**, *7*, 523–529. [CrossRef] [PubMed]

14. Fazel-Rezai, R.; Abhari, K. A region-based P300 speller for brain–computer interface. *Can. J. Electr. Comput. Eng.* **2009**, *34*, 81–85. [CrossRef]

15. Matrix, C.; Reynolds, M.R., Jr.; Cho, G.Y. Multivariate Control Charts for Monitoring the Mean Vector and Covariance Matrix. *J. Qual. Technol.* **2006**, *52*, 365. [CrossRef]

16. Tipping, M.E.; Bishop, C.M. Probabilistic Principal Component Analysis. *J. R. Stat. Soc. Ser. B* **1999**, *61*, 611–622. [CrossRef]

17. Fox, R.L.; Kapoor, M.P. Rates of change of eigenvalues and eigenvectors. *AIAA J.* **1968**, *6*, 2426–2429. [CrossRef]

18. Tayeb, S.; Mahmoudi, A.; Regragui, F.; Himmi, M.M. Efficient detection of P300 using Kernel PCA and support vector machine. In Proceedings of the 2014 Second World Conference on Complex Systems (WCCS), Agadir, Morocco, 10–12 November 2014.

19. Kundu, S.; Ari, S. P300 Detection with Brain–Computer Interface Application Using PCA and Ensemble of Weighted SVMs. *IETE J. Res.* **2017**, *38*, 1–9. [CrossRef]

20. Ke, L.; Li, R. Classification of EEG Signals by Multi-Scale Filtering and PCA. In Proceedings of the 2009 IEEE International Conference on Intelligent Computing and Intelligent Systems, Shanghai, China, 20–22 November 2009.

21. Rakotomamonjy, A.; Guigue, V. BCI competition III: Dataset II-ensemble of SVMs for BCI P300 speller. *IEEE Trans. Biomed. Eng.* **2008**, *55*, 1147–1154. [CrossRef]

22. Throckmorton, C.S.; Colwell, K.A.; Ryan, D.B.; Sellers, E.W.; Collins, L.M. Bayesian Approach to Dynamically Controlling Data Collection in P300 Spellers. *IEEE Trans. Neural Syst. Rehabil. Eng.* **2013**, *21*, 508–517. [CrossRef]

23. Xiang, L.; Guo, G.; Yu, J.; Sheng, V.S.; Yang, P. A convolutional neural network-based linguistic steganalysis for synonym substitution steganography. *Math. Biosci. Eng.* **2020**, *17*, 1041–1058. [CrossRef]

24. Cecotti, H.; Graser, A. Convolutional Neural Networks for P300 Detection with Application to Brain-Computer Interfaces. *IEEE Trans. Pattern Anal. Mach. Intell.* **2011**, *33*, 433–445. [CrossRef] [PubMed]

25. Lawhern, V.J.; Solon, A.J.; Waytowich, N.R.; Gordon, S.M.; Hung, C.P.; Lance, B.J. EEGNet: A compact convolutional neural network for EEG-based brain–computer interfaces. *J. Neural Eng.* **2018**, *42*, 117–134. [CrossRef] [PubMed]

26. Sobhani, A. P300 classification using deep belief nets. In Proceedings of the European Symposium on Artificial Neural Networks, Belgium, European, 23–25 April 2014.

27. Maddula, R.; Stivers, J.; Mousavi, M.; Ravindran, S.; de Sa, V. Deep Recurrent Convolutional Neural Networks for Classifying P300 BCI signals. In Proceedings of the 7th Graz Brain-Computer Interface Conference, Graz, Austria, 18–22 September 2017.

28. Liu, M.; Wu, W.; Gu, Z.; Yu, Z.; Qi, F.; Li, Y. Deep Learning Based on Batch Normalization for P300 Signal Detection. *Neurocomputing* **2018**, *275*, 288–297. [CrossRef]

29. Blankertz, B.; Muller, K.R.; Curio, G.; Vaughan, T.M.; Schalk, G.; Wolpaw, J.R.; Schlogl, A.; Neuper, C.; Pfurtscheller, G.; Hinterberger, T.; et al. The BCI Competition 2003: Progress and perspectives in detection and discrimination of EEG single trials. *IEEE Trans. Bio-Med. Eng.* **2004**, *51*, 1044–1051. [CrossRef]

30. Selim, A.E.; Wahed, M.A.; Kadah, Y.M. Machine learning methodologies in P300 speller Brain-Computer Interface systems. In Proceedings of the International Biomedical Engineering Conference, New Cairo, Egypt, 17–19 March 2009.

31. Selesnick, I.W.; Burrus, C.S. Generalized digital Butterworth filter design. *IEEE Trans. Signal Process.* **1996**, *46*, 1688–1694. [CrossRef]

32. Szegedy, C.; Liu, W.; Jia, Y.; Sermanet, P.; Reed, S.; Anguelov, D.; Erhan, D.; Vanhoucke, V.; Rabinovich, A. Going deeper with convolutions. *IEEE Comput. Soc.* **2015**, *131*, 1–9.

33. Li, H.; Ota, K.; Dong, M. Learning IoT in edge: Deep learning for the Internet of Things with edge computing. *IEEE Netw.* **2018**, *32*, 96–101. [CrossRef]

34. Clements, J.M.; Sellers, E.W.; Ryan, D.B.; Caves, K.; Collins, L.M.; Throckmorton, C.S. Applying dynamic data collection to improve dry electrode system performance for a P300-based brain–computer interface. *J. Neural Eng.* **2016**, *13*, 6. [CrossRef]
35. Schmidhuber, J. Deep learning in neural networks: An overview. *Neural Netw.* **2015**, *61*, 85–117. [CrossRef]
36. Hinton, G.; Deng, L.; Yu, D.; Dahl, G.E.; Mohamed, A.R.; Jaitly, N.; Senior, A.; Vanhoucke, V.; Nguyen, P.; Sainath, T.N.; et al. Deep Neural Networks for Acoustic Modeling in Speech Recognition: The Shared Views of Four Research Groups. *IEEE Signal Process. Mag.* **2012**, *29*, 82–97. [CrossRef]
37. Obermaier, B.; Neuper, C.; Guger, C.; Pfurtscheller, G. Information transfer rate in a five-classes brain–computer interface. *IEEE Trans. Neural Syst. Rehabil. Eng.* **2001**, *9*, 283–288. [CrossRef]
38. Bakshi, B.R. Multiscale PCA with application to multivariate statistical process monitoring. *AIChE J.* **1998**, *30*, 44. [CrossRef]
39. Harrou, F.; Nounou, M.N.; Nounou, H.N. Enhanced monitoring using PCA-based GLR fault detection and multiscale filtering. In Proceedings of the 2013 IEEE Symposium on Computational Intelligence in Control and Automation (CICA), Singapore, 16–19 April 2013.
40. Schittenkopf, C.; Deco, G.; Brauer, W. Two strategies to avoid overfitting in feedforward networks. *Neural Netw.* **1997**, *10*, 505–516. [CrossRef]
41. Long, J.; Xueyuan, K.; Haihong, H.; Zhinian, Q.I.N.; Yehong, W.A.N.G. Study on the overfitting of the artificial neural network forecasting model. *J. Meteorol. Res.* **2005**, *19*, 216–225.
42. Zhang, D.; Yang, G.; Li, F.; Wang, J.; Sangaiah, A.K. Detecting seam carved images using uniform local binary patterns. *Multimed. Tools Appl.* **2018**, 1–16. [CrossRef]
43. Krizhevsky, A.; Sutskever, I.; Hinton, G. ImageNet Classification with Deep Convolutional Neural Networks. *Adv. Neural Inf. Process. Syst.* **2012**, *25*, 2–134. [CrossRef]
44. Lee, S.Y.; Jake, K. Aggarwal. Parallel 2D convolution on a mesh connected array processor. *IEEE Trans. Pattern Anal. Mach. Intell.* **1987**, *4*, 590–594. [CrossRef]
45. Donchin, E.; Spencer, K.M.; Wijesinghe, R. The mental prosthesis: Assessing the speed of a P300-based brain–computer interface. *IEEE Trans. Rehabil. Eng.* **2000**, *8*, 174–179. [CrossRef]

Article

A Novel Heart Rate Robust Method for Short-Term Electrocardiogram Biometric Identification

Di Wang [1], Yujuan Si [1,2,*], Weiyi Yang [1], Gong Zhang [1] and Tong Liu [3]

[1] College of Communication Engineering, Jilin University, Changchun 130012, China; wangdi17@mails.jlu.edu.cn (D.W.); yangwy17@mails.jlu.edu.cn (W.Y.); zhanggong18@mails.jlu.edu.cn (G.Z.)

[2] School of Electronic and Information Engineering (SEIE), Zhuhai College, Jilin University, Zhuhai 519041, China

[3] School of Information and Electrical Engineering, Ludong University, Yantai 264025, China; liut@ldu.edu.cn

* Correspondence: siyj@jlu.edu.cn

Received: 8 December 2018; Accepted: 3 January 2019; Published: 8 January 2019

Abstract: In the past decades, the electrocardiogram (ECG) has been investigated as a promising biometric by exploiting the subtle discrepancy of ECG signals between subjects. However, the heart rate (HR) for one subject may vary because of physical activities or strong emotions, leading to the problem of ECG signal variation. This variation will significantly decrease the performance of the identification task. Particularly for short-term ECG signal without many heartbeats, the hardly measured HR makes the identification task even more challenging. This study aims to propose a novel method suitable for short-term ECG signal identification. In particular, an improved HR-free resampling strategy is proposed to minimize the influence of HR variability during heartbeat processing. For feature extraction, the Principal Component Analysis Network (PCANet) is implemented to determine the potential difference between subjects. The proposed method is evaluated using a public ECG-ID database that contains various HR data for some subjects. Experimental results show that the proposed method is robust to HR change and can achieve high subject identification accuracy (94.4%) on ECG signals with only five heartbeats. Thus, the proposed method has the potential for application to systems that use short-term ECG signals for identification (e.g., wearable devices).

Keywords: ECG identification; short-term ECG signals; HR-free resampling strategy; principal component analysis network; ECG-ID

---

1. Introduction

Biometric systems play an important role in security applications and have been deployed around the world in past decades. Currently, common used biometrics in practice include face, fingerprint, iris, etc. However, neither can these biometrics effectively avoid being stolen, nor are they robust enough to falsification. For instance, with finger marks left behind on objects, crackers can recreate fingerprint using latex; Iris images can be captured in a few meters distance and falsified by using contact lenses with copied iris feature printed on [1]; Biometric systems utilizing facial recognition can be easily fooled by high-resolution still photos. In recent years, it has been observed that electrocardiogram (ECG) is a significant signature for individuals even within siblings or twins [2]. Unlike other biometrics, ECG is an inner signal, whose presence automatically ensures the liveness [3,4]. This property makes ECG far more difficult to be stolen or falsified.

The ECG signal is the recording of periodic variation with heart beating. A heartbeat represents one period of the ECG signal, which conveys rich identity information and is an important sign for subject sidentification. However, visual interpretation of beats is difficult because the changes in amplitude and duration are subtle. To deal with this problem, pattern recognition methods are preferred in ECG identification due to their reliable, quick, and objective analysis.

Within the last decade, many methods, based on neural networks [5,6], support vector machine [7], and k-nearest neighborhood [8] have been proposed in literature for the purpose of automatic identification. The literature [9] presented a recurrent neural network (RNN)-based method, which could achieve automatic feature extraction, to improve the identification performance on ECG signals from both the same session and different sessions. Discrete wavelet transform was used to extract wavelet coefficients as the feature vector and KNN was applied as the classifier in literature [10]. A novel automatic ECG identification approach combining back propagation neural network (BP-NN) with Frequency Rank Order Statistics (FROS) was introduced to distinguish different subjects in the literature [11]. The literature [12] utilized neural networks to both identify QRS complex segments and perform user authentication on these segments. All these methods mainly focus on the development of the part of feature extraction and classification. By properly combing the approaches of feature extraction with improved classifiers, these methods have achieved good performance. However, it is noted that most of them are evaluated on signals with stable heart rate (HR). Actually, there are the variation of ECG signals of one person due to HR change. For example, HR increase will shorten the duration of the ventricular depolarization period, leading to T wave shift. In other ECG applications, such as atrial fibrillation (AF) [13], the HR variability can be used to distinguish the AF episode from normal sinus rhythm. However, in ECG biometric, this variation will result in low identification task and make the identification become far more difficult without appropriate processing [14].

To overcome this limitation, a popular method is to normalize QT interval according to correction formulas. For example, Lugovaya et al. [15] scaled the ST-fragment based on Framingham and Bazett's formulas. Francesco et al. [16] preferred to use a different formula based on the suggestion presented by Tawfik in [17]. Besides methods based on QT correction, Kiran et al. [18] also proposed an effective feature extraction depending on the characteristic points, which were P, Q, R, S, and T. By taking less HR related parameters as features, this method was least affected by HR change. However, the performance of these above methods depends on the accurate localization of wave boundaries for QT interval estimation or HR measurement. Though the R and T wave detection techniques have started to provide acceptable results in most cases, detecting P, Q, and S is still challenging [19]. Furthermore, HR measurement of ECG signals requires multiple ECG signals and may not be desirable for systems that use short-term ECG signals for identification. Thus non-fiducial or partial-fiducial identification methods should be investigated.

To solve this problem, Wonki Lee et al. [20] proposed a novel partial-fiducial method, whose idea was that one heartbeat was resampled and mapped into a regular interval by ignoring temporal period information. Taking a pattern matching algorithm based on Euclidean distance as the classifier, they achieved a maximum performance of 98.36% accuracy using finger ECG data. However, according to the research of Mikhail Matveev1 et al. [21], QRS negative area, total area, slope from R to S peak and sum of the absolute QRS velocities values have a strong correlation between ECG recordings acquired 5 years apart. These features will be distorted during the resampling process proposed by [20], hindering a further increase of identification accuracy. Identification performance can be potentially improved by appropriately reserving the original information of the QRS complex.

The aim of this study is to propose an HR robust ECG identification method suitable for short-term ECG signals. Our work contains two main parts: (1) On the basis of the research of reference [20], we propose a QRS-centered resampling strategy for heartbeat processing. The method aims to completely preserve the original temporal and morphological information of the QRS complex while solving the problem of T wave shift. As a result, more potentially helpful information with less HR influence will be provided for the subsequent steps; (2) For feature extraction, Convolution Neural Network (CNN) has proven its effectiveness in medical research such as health informatics and computed tomography image analysis in recent years [22–25], e.g., Acharya et al. [22] conducted a CNN study for automatic arrhythmia detection and recorded accuracy, sensitivity and specificity of 92.50%, 98.09%, and 93.13%, respectively, for two seconds of ECG segments. In our work, Principal Component Analysis Network (PCANet) proposed by Tsung-Han Chan et al. [26], a new kind of CNN that employs PCA to learn multistage filter banks, is adopted as the tool to mine more useful components from the processed heartbeats. By using PCA filters as the convolution kernel, this network is sensitive to the local difference among heartbeats from different subjects. At last, a linear Support Vector Machine (linear-SVM) is used to address the identification for faster training and classification.

The rest of this paper is organized as follows: Section 2 illustrates the proposed methodology; Database and experimental results are shown in Section 3; the results of our approach are discussed in Section 4; and Section 5 concludes the paper.

2. Methods

The whole proposed identification process is mainly composed of five parts: (1) preprocessing; (2) HR variability removal; (3) feature extraction; (4) beat identification; and (5) subject identification. Figure 1 depicts the diagram of the ECG identification methodology proposed in this paper.

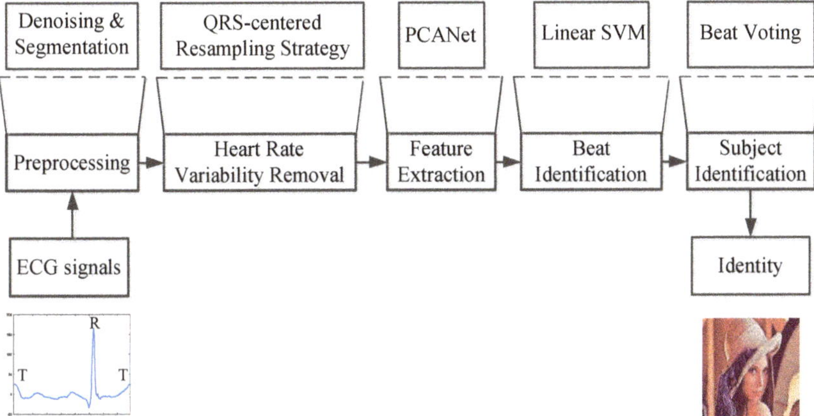

Figure 1. Diagram of electrocardiogram (ECG) identification methodology proposed.

2.1. Preprocessing

Denoising: Multiple factors will interfere with the quality of the ECG signal. The origins of interference are usually various. For example, during the acquisition of ECG signals, power-line interference generates because of the frequency influence of the used power. Interference will display in the form of noise and distort the waveform of ECG signals, leading to the decrease of ECG identification performance. Real raw ECG signals contain three major noise, namely, baseline drift, power-line interference and Electromyogram (EMG) artifact. Generally, the frequency of baseline drift is less than 0.5 Hz [27] and that of power-line interference is 50 Hz (or 60 Hz) [28]. While EMG artifact is a random noise that spreads over the entire frequency range [29]. In this paper, wavelet transform (WT) is employed as the de-nosing method due to its property of sparsity, locality and multi-resolution [30]. The wavelet-based de-noising process is summarized as follows: Raw signals are decomposed to 9 levels by lifting wavelet transform with wavelet db4; Obtained detail coefficients of different levels are thresholded by shrinkage (soft) strategy; Reconstructing the original sequence from the thresholded wavelet detail coefficients leads to removal of noise. Here the shrinkage strategy uses the universal 'VisuShrink' threshold given by [31]:

$$Thr = \sigma\sqrt{2\log(N)} \tag{1}$$

where N is the number of data points and σ represents the estimated noise level, which is obtained according to [32]:

$$\sigma_i = \frac{median(|\omega_i|)}{0.6745} \tag{2}$$

where σ_i is the noise level of the i-th level and $median(x)$ can output the median value of input sequence x. The functions *lwt*, *wthresh*, and *ilwt* in MATLAB were employed as the method for wavelet decomposition, coefficient thresholding and signal reconstruction respectively. Figure 2 shows the comparison between the original signal and the denoised signal.

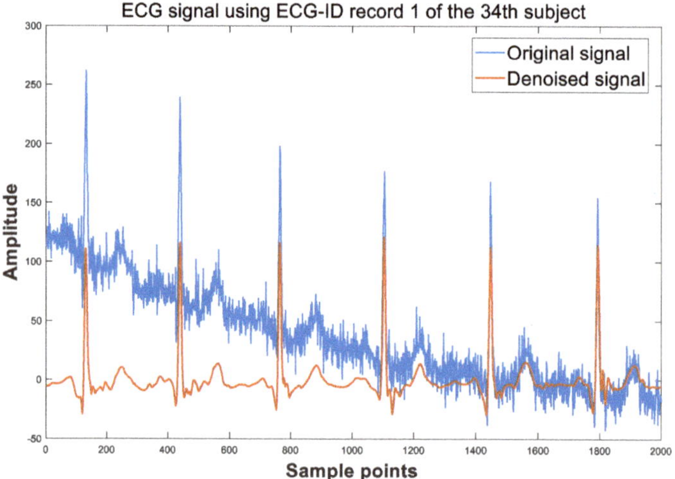

Figure 2. Comparison between the original signal and the de-noising signal.

Segmentation: After denoising, the R and T peak detection task is performed on denoised signals by using the ECGPUWAVE tool box [33]. Then the detected peak points at T are taken as the delimiters for segmentation. Figure 3 shows a beat extracted in the T-T way. It can be seen that the beat starts with its former T peak and ends with its own T peak. The duration is exactly a cardiac cycle.

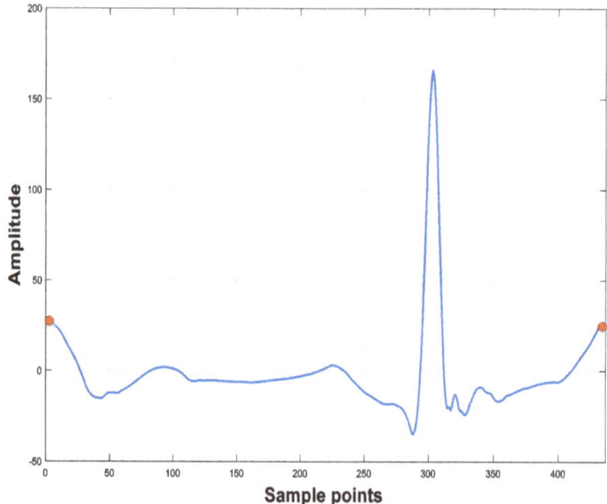

Figure 3. An extracted heart beat taking T as the delimiters.

2.2. QRS-Centered Resampling Strategy

As mentioned above, an ECG signal can be segmented into several heartbeats. Comparison of original and processed heartbeats is shown in Figure 4. Figure 4a shows several beats extracted by taking T peaks as delimiter. It is noted that even beats α, β, and γ, which comes from the same subject, do not have the same ECG waveform pattern. This nonstandard format cannot satisfy the requirement of subsequent pattern matching, thus framing an arbitrary length of beat into a regular interval of the same length is necessary.

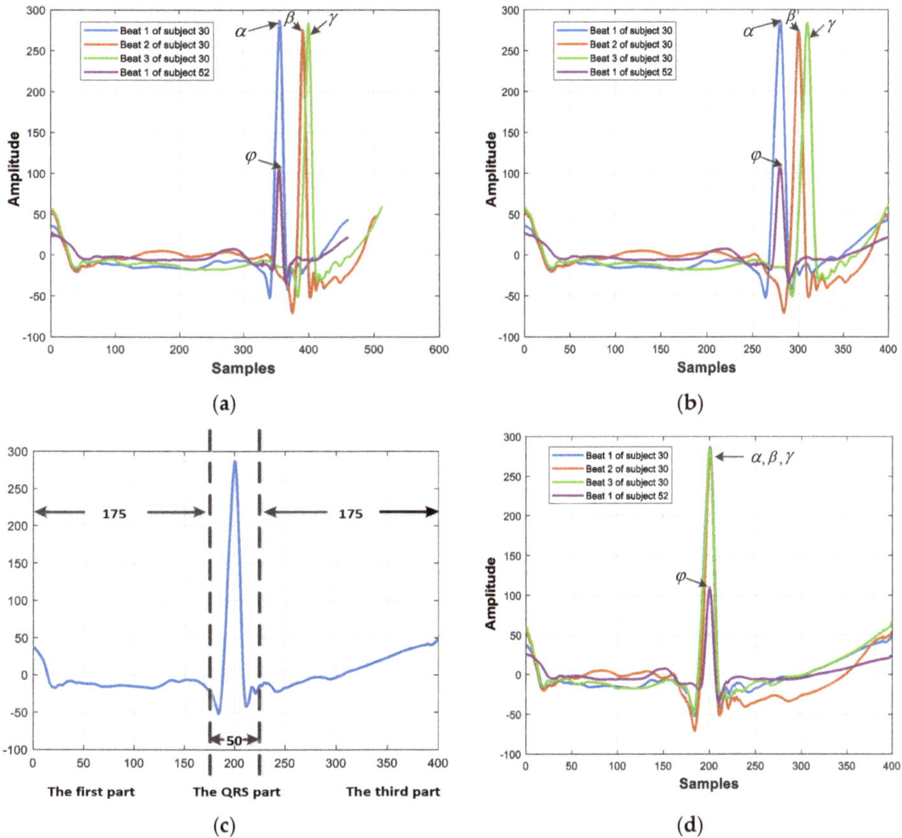

Figure 4. Comparison of heartbeats with different resampling manners. (**a**) Original heartbeats; (**b**) Heartbeats with direct resampling; (**c**) Normalization progress of the QRS-centered resampling; and (**d**) Heartbeats processed by the proposed method.

Generally, resampling-process [34] can provide a sufficient way for format standardization. Heartbeats with direct resampling can be seen in Figure 4b. It is noted that QRS waveforms of beats α, β, and γ are similar to each other, but locate in different positions on the beats. Beat φ has a different QRS waveform from the above three, but its position is similar to α. As a result, it is likely found that the similarity measurements among α, β, and γ are larger than that between α and φ, leading the following decision to an opposite result.

To solve this problem, a QRS-centered resampling strategy is proposed in this section. Based on the research of [5,20,21], our idea is that any temporal and morphological information of the QRS complex may potentially contribute to ECG identification. Meanwhile, only the amplitude is useful in identifying a subject for the rest part. Thus, we remain QRS complex original to preserve all QRS information and resample the rest part to correct T wave shift by ignoring its temporal information. Figure 4c shows the proposed normalization progress. In our strategy, each heartbeat is considered as three parts, namely, the first part, the QRS part and the third part. The first part is from the start-point of beat to the start-point of the QRS part; the QRS part is centered on the detected R point and has unified 50 points; the third part is from the end-point of the QRS part to the end-point of beat. The process of the proposed strategy is summarized as follows: Firstly, considering that the QRS duration of a healthy subject is generally 60~100 ms, a 50-point width window (the digitalized frequency of ECG-ID is 500 Hz and 100 ms corresponds to 50 points on this database) centered on R point is used to determine boundaries of three parts on heartbeats; Then extracted QRS parts are aligned centered on R peaks; Lastly, resampling process is utilized to normalize the first and third part both to 175 points. Extracted beats with our strategy can be seen in Figure 4d. The morphology of beat α, β, and γ, becomes more similar to each other, and shows a significant difference from beat φ at the same time. The function *resample* of MATLAB is employed as the method for resampling.

2.3. PCANet

Principal Component Analysis Network (PCANet) is a simple deep learning method to extract high-level features from the original input. Unlike traditional deep learning networks such as Convolution Neural Network (CNN) or Deep Belief Network (DBN), it does not need complicated iterative process for numerous parameters optimization. Only a few network parameters are necessary for determining a PCANet. Once these parameters are fixed, training the PCANet will be extremely simple and efficient. Figure 5 shows the structure of PCANet for ECG feature extraction. Code implementation of PCANet is available on [35].

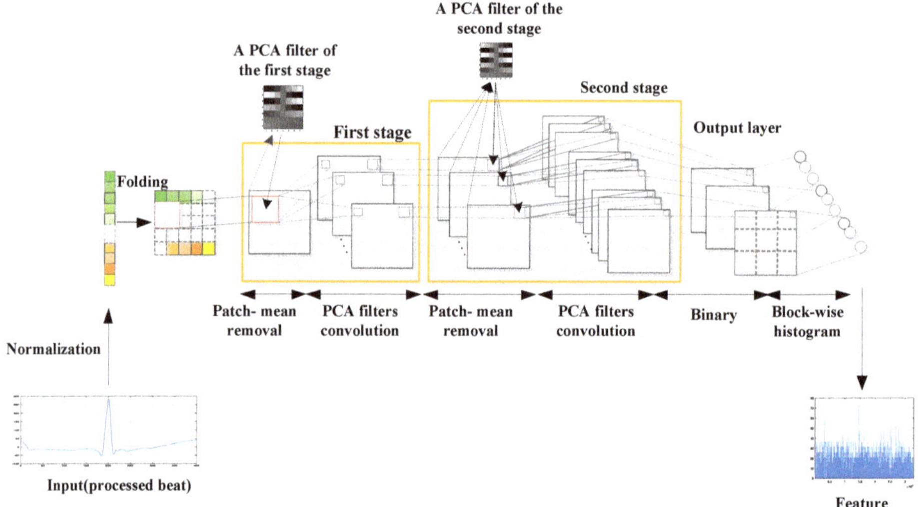

Figure 5. The structure of Principal Component Analysis Network (PCANet) model for ECG feature extraction.

2.3.1. Normalization

Given a heartbeat sample x processed by QRS-centered resampling strategy, it should be normalized before feature extraction. Here, we use min-max normalization to realize this process. The function of *mapminmax* in MATLAB is employed as the method for normalization.

2.3.2. Input Layer

The main function of the input layer is to fold the normalized heartbeat vectors into matrices, which are suitable for the use in the convolution process of PCANet. Assume the normalized heartbeat vector $x^* = [x_1^*, x_2^*, x_3^*, \cdots, x_{mn}^*]$, its specific folding process can be seen in Equation (3):

$$
[x_1^*, x_2^*, x_3^*, \cdots, x_{mn}^*] \rightarrow
\begin{bmatrix}
x_1^*, & x_2^*, & \cdots, & x_n^* \\
x_{n+1}^*, & x_{n+2}^*, & \cdots, & x_{2n}^* \\
\vdots & \vdots & \vdots & \vdots \\
x_{(m-2)n+1}^*, & x_{(m-2)n+2}^*, & \cdots, & x_{(m-1)n}^* \\
x_{(m-1)n+1}^*, & x_{(m-1)n+2}^*, & \cdots, & x_{mn}^*
\end{bmatrix}
\tag{3}
$$

where mn is the number of sampling point in the heartbeat vector. In this paper, since the dimension of nomalized heartbeat is 400, we set both m and n to 20. The function of *reshape* in MATLAB is employed as the method for folding.

2.3.3. The First Stage (PCA)

In this stage, we firstly use a $k_1 \times k_2$ patch to scan one heartbeat matrix with step 1 to collect its all patches. Then we make each patch subtract the mean of all patches and obtain all mean-removal patch matrices of the heartbeat matrix. By constructing the same matrix for N heartbeat matrices and combining them, we get their combination Y.

Then we perform Principal Component Analysis (PCA) on the combination Y. In this way, principal eigenvectors which are ordered based on the decrement of the corresponding eigenvalues can be obtained. By selecting the first L_1 principal eigenvectors and respectively reconstructing them to matrices with size $k_1 \times k_2$, we can get the PCA filter bank of the first convolution layer. Equation (4) shows the specific process:

$$
W_l^1 = mat_{k_1 k_2}(q_l(YY^T)) \in R^{k_1 \times k_2}, l = 1, 2, \cdots, L_1
\tag{4}
$$

where W_l^1 is the l-th PCA filter of the first convolution layer, YY^T is the covariance matrix of Y, $q_l()$ extracts the l-th principal eigenvector of YY^T, and $mat_{k_1,k_2}()$ maps the extracted principal eigenvector to a matrix $W \in R^{k_1 \times k_2}$.

2.3.4. The Second Stage (PCA)

After obtaining PCA filters of the first layer, we can get the filter output by doing convolution on the heartbeat matrix and PCA filter W_l^1. Then the solution process of PCA filter banks in the second stage is just the same as that in the first stage. We firstly scan the first layer output using a patch with size $k_1 \times k_2$ and collect a series of patches. Then the patch mean is subtracted from each patch and the mean-removed patches of the l-th filter output of all heartbeat matrix are combined together to obtain patch matrix. For all of the filter, their patch matrices are concatenated to get combination $\widetilde{Y}$. The PCA filters of the second stage are then obtained as shown in Equation (5):

$$
W_\ell^2 = mat_{k_1,k_2}(q_\ell(\widetilde{Y}\widetilde{Y}^T)) \in R^{k_1 \times k_2}, l = 1, 2, \cdots, L_2
\tag{5}
$$

where W_ℓ^2 is the ℓ-th PCA filter of the second convolution layer, $\widetilde{Y}\widetilde{Y}^T$ is the covariance matrix of $\widetilde{Y}$, $q_\ell()$ extracts the ℓ-th principal eigenvector of $\widetilde{Y}\widetilde{Y}^T$, and $mat_{k_1,k_2}()$ maps the extracted principal eigenvector to a matrix $W \in R^{k_1 \times k_2}$. For each input of the second stage, one will output L_2 matrices of size $m \times n$. The specific computation process is doing convolution on the input and its corresponding filter in the second stage.

2.3.5. Output Stage

After the second stage, obtained outputs are firstly binarized by a Heaviside function. By this function, value in outputs will be 1 for positive entries and 0 otherwise. Then we convert the L_2 outputs back into a single integer-valued matrix by Equation (6):

$$\Gamma_i^l = \sum_{\ell=1}^{L_2} 2^{\ell-1} H(y_i^l * W_\ell^2) \tag{6}$$

where y_i^l means the output of l-th filter for the i-th heartbeat matrix y_i in the first layer. Here weights of the outputs are irrelevant since each integer is treated as a distinct "word". After coding, each integer-valued matrix Γ_i^l is partitioned into B blocks with a set overlapping ratio v and histogram block size η. We compute the histogram of the values in each block and concatenate all B histograms into one vector which is denoted as $Bhist(\Gamma_i^l)$. The feature of input heartbeat vector is lastly defined to be the set of block-wise histograms as shown in Equation (7):

$$f_i = [Bhist(\Gamma_i^1), \cdots, Bhist(\Gamma_i^{L_1})]^T \in R^{(2^{L_2})L_1 B} \tag{7}$$

where f_i is the learned feature representation which can be used in following classification.

The detailed information of network parameter setting can be seen in Table 1.

Table 1. Detailed parameter information of PCANet used for ECG identification.

Steps	Project	Parameter
Input	Heartbeat matrix size	20×20
The first stage	Patch size ($k_1 \times k_2$)	7×7
	The number of filters of the first stage (l)	8
The second stage	Patch size ($k_1 \times k_2$)	7×7
	The number of filters of the second stage (ℓ)	8
Output	Histogram block size (η)	7×7
	Block overlap ratio (v)	0.5

2.4. Classifier

Several classifiers such as K-Nearest Neighbors (KNN), Back propagation neural network (BP-NN), Random Forest (RF), Naive Bayes and Support Vector Machines (SVM) are implemented and compared. The detail of these classifiers is presented below.

2.4.1. K-Nearest Neighbor (KNN)

In pattern recognition, k-nearest neighbor algorithm is a common method used for both classification and regression. It is a type of instance-based learning, or lazy learning. It does not attempt to construct a general model, but simply stores instances of the training data. Classification is computed from a simple majority vote of the nearest neighbors of the instance to be predicted. A query instance is assigned the data class which has the most representatives within the nearest neighbor of the instance. Here, we used Euclidean distance as our distance metric. In our experiment, the best result of classification was achieved for K = 3. KNN can be realized by the function *ClassificationKNN.fit* in MATLAB.

2.4.2. Back Propagation Neural Network (BP-NN)

BP-NN is a multilayer feed forward network trained with error back propagation strategy. It is firstly proposed by Rumelhart [36] and has been extensively used at present. In this paper, we used a typical three-layer BP-NN as classifier to classify the extracted features, and the number of units in the hidden layer was set to 50. The function *patternnet* in MATLAB was employed as the method for BP-NN implementation.

2.4.3. Random Forest (RF)

A random forest model [37] is a collection of k decision trees. Here, cart classification trees, which divide attributes based on the Gini index, are developed with different numbers of inputs to form an RF. The classification is determined by the voting results of all decision trees, and the highest ranked class is selected as the final label of a new instance. In our experiment, the best classification results occurred when k was set to 500. RF model was implemented by using the function *TreeBagger* in MATLAB.

2.4.4. Naive Bayes Classifier

Naive Bayes classifier is one of the simplest machine learning algorithms, being also fast and easy to implement. It is a probabilistic classifier based on applying Bayes' theorem with strong independence assumptions between the features, and has proven to work surprisingly well in practice. In this work, a Naive Bayes classifier using Bernoulli distribution was adopted for heartbeat identification. Code implementation of Naive Bayes classifier is available on [38].

2.4.5. Support Vector Machines (SVM)

Support Vector Machine (SVM) is one of the state-of-the-art classifiers which can split a dataset into two or more categories. By using a function called kernel, support vector machine can transform the input samples into a higher dimensional space and classify them linearly. In this paper, since features extracted by PCANet are high-dimensional and sparse, we adopt a linear kernel support vector machine (linear-SVM) according to [25]. Linear-SVM can be realized by freely available Liblinear toolkit [39]. During the use of Liblinear, parameter C known as error penalty factor, which expresses the tolerance to error, was set to 1 for good performance of classifier.

2.5. Signal Identification

With the features extracted by PCANet, linear-SVM can output the identification result of each beat. To get the label of a whole signal, we make beats of the same signal vote. According to the results, the class with maximum number of votes is selected as the class label of the estimated signal.

3. Results

3.1. Database

We use the challenging ECG-ID database [15], which is available on the PhysioNet, to evaluate our proposed method. The database is chosen because it includes more than two recordings for some of its subjects. All recordings in this database are acquired in lead I and digitalized at 500 Hz over a duration of 20 s. Unlike such databases as Massachusetts Institute of Technology-Biotechnology arrhythmia database (MIT-BIH-AHA), whose signals have stable HR, the HR of signals in the ECG-ID is various. Different emotional or physical conditions and acquisition over a large domain of time make it provide a platform for use of ECG as person identification in real world scenarios. In the experiments, we select the same 12 subjects as reference [18] and each subject has five recordings for training and testing purposes. Table 2 shows the detail information of each recording, including its contained heartbeat number and estimated HR. HR is estimated as follows:

$$HR = (Heartbeat\ Number) / (Signal\ Duration) * 60 \tag{8}$$

here since the signal duration is 20 s, HR is three times as much as contained Heartbeat Number.

Table 2. Heart number & estimated heart rate (HR) of used signals for experiments.

Subject Number	Heartbeat Number & Estimated Heart Rate (beat/min)									
	Record-1		Record-2		Record-3		Record-4		Record-5	
3	25	75	25	75	24	72	23	69	23	69
10	28	84	27	81	28	84	33	99	24	72
24	24	72	25	75	25	75	21	63	18	54
25	24	72	24	72	23	69	19	57	23	69
30	23	69	24	72	21	63	21	63	19	57
32	22	66	22	66	23	69	24	72	23	69
34	31	93	30	90	27	81	29	87	30	90
36	19	57	20	60	23	69	25	75	23	69
52	25	75	26	78	28	84	29	87	31	93
53	26	78	27	81	27	81	30	90	27	81
59	25	75	28	84	33	99	21	63	20	60
72	24	72	25	75	20	60	37	111	34	102

3.2. Experimental Setup

1. Experiment 1

As KNN, BP-NN, RF, Naive Bayes, and Linear-SVM have been widely used, these methods were implemented for evaluating the performance of the extracted features. To convincingly estimate the proposed method, two of the five ECG signals from the subject were combinatorially selected as the training set, and the remaining three were utilized as the testing set. As a result, each experiment was repeated ten (C_5^2) times.

2. Experiment 2

In Experiment 2, the HR robustness was evaluated by comparing the proposed method with other five methods. During identification, all the six methods had the same denoising and fiducial point detection process, and the difference among them was the manner of heartbeat resampling and feature extraction. Six methods are shown in Table 3, where "Y" indicates adoption and "N" indicates none. In heartbeat resampling manner, "QRS-centered" represents the proposed novel resampling strategy, and the meaning of "TT" and "TRT" is explained below.

Table 3. Six Methods with different resampling strategy and feature extraction manner.

Main Operating		TT-CNN	TRT-CNN	QRS-CNN	TT-PCANet	TRT-PCANet	Proposed Method
Heartbeat Resampling Manner	TT	Y	N	N	Y	N	N
	TRT	N	Y	N	N	Y	N
	QRS-centered	N	N	Y	N	N	Y
Feature Extraction	CNN	Y	Y	Y	N	N	N
	PCANet	N	N	N	Y	Y	Y

"TT"-resampling: Segmented heartbeats with different sizes are directly resampled to 400 sample points without alignment of R points. This manner just performs a forced alignment of T wave points, ignoring all temporal period and morphological information of heartbeats.

"TRT"-resampling: R and T points of segmented heartbeats are aligned, and divide each heartbeat into two parts: T-R and R-T. Then both T-R and R-T part are resampled to 200 to obtain the unified size as in Method 1. Compared with Method 1, this method further performs alignment of key fiducial points (R). It is firstly proposed by reference [20] and has proved its effectiveness on heart robustness without requiring HR measurement. All the experiments are repeated 10 times.

Here the architecture of the used traditional CNN is shown in Table 4.

Table 4. Detailed parameter information of traditional Convolution Neural Network (CNN) used for ECG identification.

Layers	Type	Number of Neurons (Output Layer)	Kernel Size	Stride
0–1	Convolution	$16 \times 16 \times 6$	5×5	1
1–2	Max-pooling	$8 \times 8 \times 6$	2×2	2
2–3	Convolution	$4 \times 4 \times 12$	5×5	1
3–4	Max-pooling	$2 \times 2 \times 12$	2×2	2

3. Experiment 3

To evaluate the effectiveness of the proposed method on short-term ECG signals, original signals in test set were segmented into hundreds of fragments with three schemes, in which a single fragment contained one, three and five heartbeats respectively. Based on these obtained short-term ECG segmentation, the subject identification accuracies were achieved and compared.

4. Experiment 4

To further validate the proposed method, the experimental results of reference [18] were compared with ours in Experiment 3. Many performance parameters such as sensitivity, specificity, precision, and F1-score were calculated and compared based on the obtained confusion matrix. For fair comparison we followed the same data distribution as reference [18], in which only Record-1 and Record-2 of each subject were employed as the training set.

5. Experiment 5

Further validation was performed by comparing the results of different methods in reference [9], namely, RNN, Gated Recurrent Unit (GRU), and Long Short-Term Memory (LSTM), with that of our method. Following reference [9], we evaluated the proposed method on two public databases, which were ECG-ID and MIT-BIH Arrhythmia database (MITDB). Here, the used subject number of ECG-ID database increased from 12 to 89. The MITDB is a patient information database, which contains 47 subjects and can be employed to evaluate the proposed method on the level of patient. For each subject of the MITDB, five fragments of 18-heartbeat length, which were recorded at different time, were randomly extracted for training and testing purpose. Thus a five-fold cross validation could be performed for evaluation.

All the above experiments were made in MATLAB 2017a (MATLAB, 2017a, MathWorks, Natick, MA, USA).

3.3. Experimental Results

3.3.1. Experiment 1

Figure 6 gives the comparison of different classifiers fed by features extracted using the proposed method. Related to Figure 6, the features extracted by the QRS-centered resampling strategy and PCANet could produce high heartbeat and subject identification accuracy of 83.14% and 94.72% even with the simplest classifier KNN. Meanwhile, all the other classifiers could yield heartbeat accuracy over 85% and subject identification accuracy over 95%. The obtained results demonstrated that the extracted features could reflect the difference between different subjects, and were effective for ECG identification. In our subsequent experiments, we selected Linear-SVM as the classifier not only because its performance was better compared with other classifiers, but also because it is more suitable for dealing with the extracted high-dimensional features.

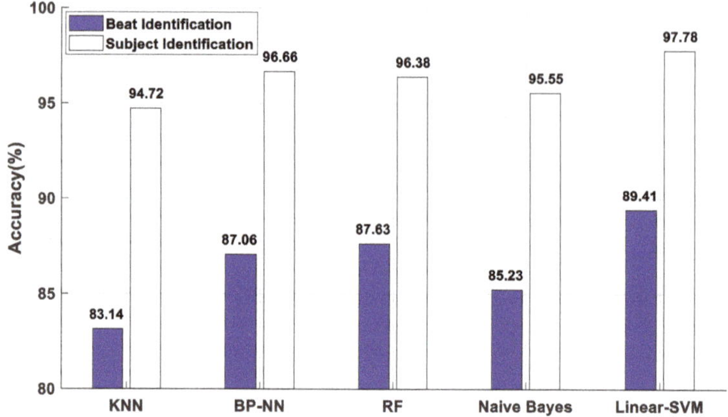

Figure 6. Comparison of different classifiers using the proposed method.

3.3.2. Experiment 2

Figure 7 shows the results of average heartbeat and subject identification accuracy of 10 experiments with six different methods, which are TT-CNN, TRT-CNN, QRS-CNN, TT-PCANet, TRT-PCANet, and the proposed method. Here the subject identification accuracy was obtained after all heartbeats of each signal voted. With TT-CNN and TT-PCANet, the average heartbeat (subject) identification rates were 71.51% (89.44%) and 77.22% (89.72%), respectively. They had the worst performance among the tested methods since it preserved the least information during heartbeat processing. The average accuracies increased to 77.86% (93.88%) and 85.93% (95.27%) with TRT-CNN and TRT-PCANet, in which the R point alignment was additionally taken into account. The proposed method and QRS-CNN showed performance of 83.44% (94.4%) and 89.41% (97.78%), respectively. It was found that when the feature extraction manner was set, methods with the proposed QRS-centered resampling strategy obtained the best heartbeat and subject identification, indicating that the preservation of the temporal period and morphological information of the QRS did benefit for identification accuracy. Also when compared with the CNNs, the PCANets achieved better performance using different heartbeat resampling manner. Furthermore, the CNNs (6c-2s-12c-2s) used in our work needed complicated iterative process for 468 ($5 \times 5 \times 6 + 6 + 5 \times 5 \times 12 + 12 = 468$) parameters optimization, while the number of PCANets parameters is six ($k_1, k_2, l, \ell, \eta, v$) here. Based on the results, we confirmed that our proposed method could achieve good performance under HR variability conditions.

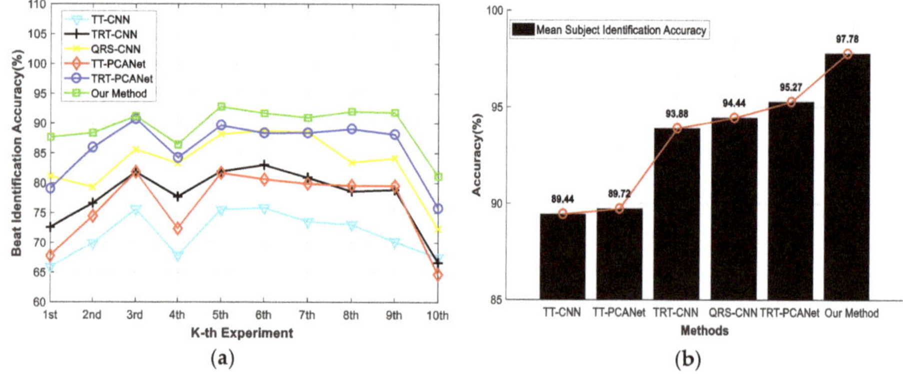

Figure 7. The accuracy comparison of the six identification methods. (**a**) Heartbeat identification comparison; (**b**) Comparison of mean subject identification accuracy of six methods.

3.3.3. Experiment 3

Table 5 shows the variation of average subject identification accuracy with respect to different-length signals during ECG identification using six methods. Compared with the other five methods, higher identification accuracies were achieved by the proposed method under different conditions. The results also showed that the proposed method was able to get high accuracy over 94% even with signals of only five-heartbeat length. Nevertheless, for the rest methods, ECG signals should contain far more heartbeats to make the methods obtain the same accuracy. Generally, the goodness of a heartbeat based method in short-term ECG signal identification is mainly measured by the single heartbeat identification accuracy. Our method gave higher heartbeat identification accuracy of 89.41% on signals than the other five methods, which suggested that our method could provide an efficient way for short-term ECG identification [40,41].

Table 5. The comparison of subject accuracies among short-term ECG signals with different heartbeat length.

Method	ECG Length (in Number of Heartbeats)	Average Subject Identification Accuracy (10 Experiments)
TT-CNN	1	71.51%
TT-CNN	3	77.57%
TT-CNN	5	85.22%
TT-PCANet	1	77.22%
TT-PCANet	3	80.85%
TT-PCANet	5	85.82%
TRT-CNN	1	77.86%
TRT-CNN	3	83.46%
TRT-CNN	5	87.71%
TRT-PCANet	1	85.93%
TRT-PCANet	3	90.28%
TRT-PCANet	5	91.96%
QRS-CNN	1	83.44%
QRS-CNN	3	88.96%
QRS-CNN	5	91.89%
Our Method	1	89.41%
Our Method	3	92.49%
Our Method	5	94.40%

3.3.4. Experiment 4

In this section, we compared the experimental results of [18] with ours to further validate the proposed method. Many performance parameters mentioned in [18] were calculated and compared based on the confusion matrix. The evaluation parameters include Accuracy, Sensitivity, Specificity, Precision and F1-score. The calculation process is as follows:

$$Accuracy = \frac{TP + TN}{TP + FP + TN + FN} \tag{9}$$

$$Sensitivity = Recall = \frac{TP}{TP + FN} \tag{10}$$

$$Specificity = \frac{TN}{TN + FP} \tag{11}$$

$$Precision = \frac{TP}{TP + FP} \tag{12}$$

$$F_1 Score = \frac{2 * Precision * Recall}{Precison + Recall} \tag{13}$$

where *TP, TN, FP,* and *FN*, respectively represent True Positive, True Negative, False Positive, and False Negative. Since subject identification is a multi-class problem, this paper uses the overall statistic of the above indicators to evaluate the performance. The general calculation form of different overall statistics is as follows:

$$Overall\_X = \sum_{i=1}^{N} X_i / N_{person} \tag{14}$$

where N_{person} is the total number of evaluated person and X_i represents the statistic result of the i-th person (X_i can be Accuracy, Recall, Specificity, Precision, or F1-score).

Table 6 shows the confusion matrix for the true classification of ECG beats used for testing the proposed method. The confusion matrix shows the number of beats used for the test with their corresponding True Positive Rate (TPR).

Table 6. Confusion matrix of true classification of ECG beat data of person.

Target Class	Classification Class												Accuracy
	3	10	24	25	30	32	34	36	52	53	59	72	
3	**19**	0	1	0	0	1	0	0	0	0	0	0	90.48%
10	0	**19**	0	0	0	0	0	0	0	0	2	0	90.48%
24	3	1	**16**	1	0	0	0	0	0	0	0	0	76.19%
25	0	0	0	**21**	0	0	0	0	0	0	0	0	100%
30	0	0	0	0	**21**	0	0	0	0	0	0	0	100%
32	0	0	0	0	0	**21**	0	0	0	0	0	0	100%
34	2	0	0	1	0	0	**18**	0	0	0	0	0	85.71%
36	0	2	0	0	0	0	0	**17**	0	0	2	0	80.95%
52	0	0	0	0	0	0	0	0	**21**	0	0	0	100%
53	0	0	0	0	0	0	0	0	0	**21**	0	0	100%
59	0	0	0	0	0	0	0	0	1	0	**17**	3	80.95%
72	0	1	0	0	0	0	0	0	0	0	7	**13**	61.90%

With all heartbeats voting, we can obtain the result of subject identification. Table 7 shows the confusion matrix which includes the number of signals used for testing with their corresponding subject TPR. Based on the confusion matrices of heartbeat and subject identification, the evaluation parameters can be calculated and the results of comparison can be seen in Table 8.

Table 7. Confusion matrix for true identification of person.

Target Class	Classification Class												Accuracy
	3	10	24	25	30	32	34	36	52	53	59	72	
3	**3**	0	0	0	0	0	0	0	0	0	0	0	100%
10	0	**3**	0	0	0	0	0	0	0	0	0	0	100%
24	0	0	**3**	0	0	0	0	0	0	0	0	0	100%
25	0	0	0	**3**	0	0	0	0	0	0	0	0	100%
30	0	0	0	0	**3**	0	0	0	0	0	0	0	100%
32	0	0	0	0	0	**3**	0	0	0	0	0	0	100%
34	0	0	0	0	0	0	**3**	0	0	0	0	0	100%
36	0	0	0	0	0	0	0	**3**	0	0	0	0	100%
52	0	0	0	0	0	0	0	0	**3**	0	0	0	100%
53	0	0	0	0	0	0	0	0	0	**3**	0	0	100%
59	0	0	0	0	0	0	0	0	0	0	**3**	0	100%
72	0	0	0	0	0	0	0	0	0	0	2	**1**	33.33%

In the comparison of our method with seven beats and [18], our method showed improvement on all evaluation indicators. Results showed that the overall heartbeat and subject identification accuracy could reach 88.8889% and 99.0741%, respectively. In terms of specificity, we achieved a prediction success average rate reaching 99.4949% and exceeding 95% for each person. Relative to the results of the previous method, our accuracy, precision, and F1-score were all at a high level.

Table 8. Comparison of our method with seven beats and literature [18].

S. No	Performance Parameter	Value [18]	Value (Our Method)
1	True_Positive Rate (Beat Identification)	80.5556%	88.8889%
2	True_Positive Rate (Subject Identification)	88.8889%	94.4444%
3	Overall_Accuracy	98.1481%	99.0741%
4	Overall_Recall	88.8889%	94.4444%
5	Overall_Specificity	98.9899%	99.4949%
6	Overall_Precision	93.750%	96.6667%
7	Overall_F1 Score	88.5317%	93.7500%

3.3.5. Experiment 5

The goal of this experiment is to compare the results of our method with that of reference [9], which is one of the state-of-the-art methods for ECG identification. Table 9 shows the comparison between reference [9] and the proposed method.

Table 9. Comparison of the proposed method with reference [9] (SI-Subject Identification, HI-Heartbeat Identification).

Methods	ECG-ID: 89		MIT-BIH: 47	
	ECG Length (in Number of Heartbeats)	SI	ECG Length (in Number of Heartbeats)	SI
RNN [9]	18	91.7%	18	93.3%
GRU [9]	18	94.4%	18	95.7%
LSTM [9]	18	100%	18	100%
Proposed method	18	97.75%	18	100%
Proposed method	7	95.25%	7	97.85%
Proposed method	5	92.36%	5	97.80%
Proposed method	3	89.97%	3	96.96%
Proposed method	1	84.50%	1	90.45%

Results on Table 9 show that LSTM network performs better than GRU and traditional RNN in terms of the subject identification accuracy, and that our approach behaves similar to the LSTM network. On the databases, ECG-ID, and the MITDB, subject identification accuracies of 97.75% and 100% were achieved. Besides that, the proposed method could achieve high heartbeat identification accuracy of 84.5% and 90.45% on ECG-ID database and the MITDB, which made it possible to obtain good performance based on short-term signals. For ECG-ID and the MITDB, 95.25% and 97.85% subject identification accuracy were obtained even with signals of only seven-heartbeat length. Especially, the subject identification accuracy of signals of three-heartbeat length could reach 96.96% on the MITDB.

These results further proved the effectiveness of our method for short-term ECG signal identification.

Table 10 shows the state-of-the-art methods proposed for ECG biometric identification. Compared with other methods, the proposed method achieves high-level results in terms of subject identification accuracy on both databases, which suggests that it can serve as an effectively partial-fiducial way for ECG biometric identification.

Table 10. Performance comparison with state-of-the-art works (SI-Subject Identification).

Methods	Year	Feature Extraction (Type)	Decision	ECG Dataset	Performance
Page et al. [12]	2015	QRS complex segments (Fiducial)	NN	ECG-ID: 90	SI: 99.93%
Dar et al. [40]	2015	Haar Transform/GBFS (Non-fiducial)	KNN	ECG-ID: 90 MIT-BIH: 47	SI: 83.2% SI: 95.9%
Dar et al. [10]	2015	Haar Transform and HRV/GBFS (Non-fiducial)	Random Forest	ECG-ID: 90 MIT-BIH: 47	SI: 83.9% SI: 93.1%
Dhou-ha et al. [41]	2016	21 temporal and amplitude features and 10 morphological descriptors (Fiducial)	SVM	MIT-BIH: 44	SI: 98.8%
Tan et al. [42]	2017	Temporal, amplitude, and angle fid. + DWT coefficients (Fiducial)	Random Forests + WDIST KNN	ECG-ID: 89 MIT-BIH: 47	SI: 100% SI: 100%
Yu et al. [43]	2017	PCA (Non-fiducial)	RPROP	ECG-ID: 89	SI: 96.60%
Lynn et al. [44]	2018	Temporal and amplitude features (Fiducial)	BP-NN	ECG-ID: 3 10 20	SI: 98.24% SI: 96.20% SI: 94.00%
Zhao et al. [45]	2018	Generalized S-transformation (Non-fiducial)	CNN	ECG-ID:50	SI: 96.63%
Mahmoud et al. [46]	2018	Mean P-QRS-T fragment + DWT (Fusion of fiducial and non-fiducial)	SVM	ECG-ID: 90 MIT-BIH: 47	SI: 99% SI: 100%
Proposed Method	-	QRS-centered resampling strategy + PCANet (partial-fiducial)	Linear-SVM	ECG-ID: 12 89 MIT-BIH: 47	SI: 97.78% SI: 97.75% SI: 100%

4. Discussion

In this work, a novel HR robust method for short-term ECG biometric identification was developed. Raw ECG signals were filtered with wavelet denoising and segmented into heartbeats by taking the detected T peak points as delimiters. Then the heartbeat was processed by the proposed QRS-centered resampling strategy and standardized to 400 sampling points. The QRS-centered strategy is inspired and based on the prior ECG identification works: Firstly, to our knowledge, all the existing literature about ECG identification has taken QRS complex or its related form as features and QRS complex is very important for identifying a person [9,10,12,18,40–46]. To preserve all potential identity information of the QRS complex, we use a length-fixed window to keep the QRS complex original. Secondly, it is also found that mapping the heartbeat into a regular interval of segments does help to deal with the HR variability problem [5,20]. Thus, we segment the heartbeat into three parts, which are the first, the QRS and the third part respectively, and unify them. In this way, the extracted heartbeat can correct T wave shift without using traditional QT correction formula, and contain sufficient information of QRS such as QRS negative area, amplitude, and so on at the same time. As a result, HR variability is removed and more potential information is provided for the subsequent steps, which is beneficial for the identification accuracy.

After that, PCANet was implemented to learn discrimination among heartbeats from different persons by taking the principal eigenvectors as filter banks. PCANet can be analyzed by comparing it with the Convolutional Neural Network (CNN): Like CNN, the PCANet also has convolution filter bank in each stage; the binary quantization of the PCANet at the output stage performs similar function like the nonlinear layer in CNN; the pooling layer of the PCANet is set to be the block-wise histograms of binary codes. In fact, PCANet can be essentially considered as a CNN model, which has strong capability of feature extraction [26,47,48].

Moreover, the PCANet seems to be more likely to achieve better performance than traditional CNN in ECG biometric identification, especially for ECG data without outlier correction. As we all know, ECG is not a strict periodic signal, and local distortions and variation on heartbeats may appear even without HR variability. Traditional CNN may be influenced by these distortions because its convolutional filter bank is learned in a data-adapting way. Compared with traditional CNN, the filter bank of the PCANet is prefixed by analyzing the main difference between subjects based on the combination of patch matrices of the training data. This learning way gives more holistic observations of the original ECG data, and the learned intra-invariance can essentially capture more identification information [26,47,48]. According to our experimental results, the PCANets did obtain much higher heartbeat identification than the CNNs with different heartbeat resampling manner, which further proved that the PCANet method was robust to distortions.

Compared with other methods in literature, our method has two main advantages. First, the method is HR robust and does not require HR based QT correction. HR variability can be removed under short-term ECG signal condition and avoid the complex operations of accurate Q detection or HR measurement. Based on the experiments, it is found that not only the amplitude, but also the temporal and morphological information of QRS can potentially contribute to identification, which is in accordance with other studies [5,20,21]. Second, the proposed method can achieve high heartbeat identification accuracy, which makes it suitable for systems that use a small quantity of heartbeats to make a decision [49]. Besides that, the naturality of PCANet makes our work easy to be reproduced by other researchers. Because compared with traditional neural networks (CNN), only a few parameters are required to determine a PCANet [25,26,48].

In the future, we will explore the "other class" classification problem. "Other class" refers to the class that classifier has not yet trained and it is totally different from the other trained individuals morphologically. In ECG identification, we can also call it unknown individual. The classification of the unknown individual is an open-set problem and can hardly be solved by some simple methods such as threshold setting [50] or distance matching [51]. So our next stage research is to improve our existing identification architecture and realize "other class" recognition.

5. Conclusions

In this paper, we propose a novel HR robust method for short-term ECG signal identification. In this study, we identified ECG signals by using the QRS-centered resampling strategy and the PCANet. To evaluate the effectiveness of our algorithm, the experiment was performed on ECG fragments with different length and various HRs. Experimental results revealed that this QRS-centered resampling strategy could efficiently remove the influence of HR variability and the PCANet was able to capture important information required for class discrimination from processed heartbeats. Compared with the existing state-of-the-art methods, the proposed approach provides an effectively partial-fiducial way for identification and shows comparative results on both ECG-ID database and the MITDB. Our method is expected to contribute to information security and privacy protection.

Author Contributions: Conceptualization—D.W. and Y.S., Data curation—W.Y. and G.Z., Formal analysis—D.W. and T.L., Writing—Original Draft D.W. and Y.S., Writing—Edit and Review, D.W. and W.Y.

Funding: This work was supported by the Key Scientific and Technological Research Project of Jilin Province under Grant No. 20170414017GH and 20190302035GX; the Natural Science Foundation of Guangdong Province under Grant No. 2016A030313658; the Premier-Discipline Enhancement Scheme Supported by Zhuhai Government under Grant No. 2015YXXK02-2; the Premier Key-Discipline Enhancement Scheme Supported by Guangdong Government Funds under Grant No. 2016GDYSZDXK036; Science Foundation of China under Grant No. 61741311; Natural Science Foundation of China under Grant No. 61702249.

Conflicts of Interest: The authors declare no conflict of interest.

References

1. Komeili, M.; Armanfard, N.; Hatzinakos, D. Liveness Detection and Automatic Template Updating Using Fusion of ECG and Fingerprint. *IEEE Trans. Inf. Forensics Secur.* **2018**, *13*, 1810–1822. [CrossRef]
2. Belgacem, N.; Fournier, R.; Nait-Ali, A.; Bereksi-Reguig, F. A novel biometric authentication approach using ECG and EMG signals. *J. Med. Eng. Technol.* **2015**, *39*, 226–238. [CrossRef] [PubMed]
3. Fang, S.C.; Chan, H.L. QRS detection-free electrocardiogram biometrics in the reconstructed phase space. *Pattern Recognit. Lett.* **2013**, *34*, 595–602. [CrossRef]
4. Karimian, N.; Guo, Z.M.; Tehranipoor, M.; Forte, D. Highly Reliable Key Generation From Electrocardiogram (ECG). *IEEE Trans. Biomed. Eng.* **2017**, *64*, 1400–1411. [CrossRef]

5. Tuerxunwaili; Nor, R.M.; Rahman, A.W.B.A.; Sidek, K.A.; Ibrahim, A.A. Electrocardiogram Identification: Use a Simple Set of Features in QRS Complex to Identify Individuals. In Proceedings of the 12th International Conference on Computing and Information Technology (IC2IT), Khon-Kaen, Thailand, 7–8 July 2016; Springer: Cham, Switzerland, 2016; pp. 139–148. [CrossRef]

6. Ghongade, R.; Ghatol, A. An effective feature set for ECG pattern classification. In Proceedings of the International Conference on Medical Biometrics, Hong Kong, China, 4–5 January 2008; pp. 25–32. [CrossRef]

7. Liu, S.-H.; Cheng, D.-C.; Lin, C.-M. Arrhythmia Identification with Two-Lead Electrocardiograms Using Artificial Neural Networks and Support Vector Machines for a Portable ECG Monitor System. *Sensors* **2013**, *13*, 813–828. [CrossRef] [PubMed]

8. Zhao, Z.; Yang, L.; Chen, D.; Luo, Y. A Human ECG Identification System Based on Ensemble Empirical Mode Decomposition. *Sensors* **2013**, *13*, 6832–6864. [CrossRef] [PubMed]

9. Salloum, R.; Kuo, C.C.J. ECG-based biometrics using recurrent neural networks. In Proceedings of the IEEE International Conference on Acoustics, Speech and Signal Processing (ICASSP), New Orleans, LA, USA, 5–9 March 2017; pp. 2062–2066. [CrossRef]

10. Dar, M.N.; Akram, M.U.; Usman, A.; Khan, S.A. ECG Biometric Identification for General Population Using Multiresolution Analysis of DWT Based Features. In Proceedings of the Second International Conference on Information Security and Cyber Forensics (InfoSec), Cape Town, South Africa, 15–17 November 2015; pp. 5–10. [CrossRef]

11. Tseng, K.-K.; Lee, D.; Hurst, W.; Lin, F.-Y.; Ip, W.H. Frequency Rank Order Statistic with Unknown Neural Network for ECG Identification System. In Proceedings of the 4th International Conference on Enterprise Systems (ES), Melbourne, VIC, Australia, 2–3 November 2016; pp. 160–167. [CrossRef]

12. Page, A.; Kulkarni, A.; Mohsenin, T. Utilizing Deep Neural Nets for an Embedded ECG-based Biometric Authentication System. In Proceedings of the 2015 IEEE Biomedical Circuits and Systems Conference (BioCAS), Atlanta, GA, USA, 22–24 October 2015; pp. 346–349. [CrossRef]

13. Peimankar, A.; Puthusserypady, S. Ensemble Learning for Detection of Short Episodes of Atrial Fibrillation. In Proceedings of the 2018 26th European Signal Processing Conference (EUSIPCO), Roma, Italy, 3–7 September 2018; pp. 66–70. [CrossRef]

14. Poree, F.; Kervio, G.; Carrault, G. ECG biometric analysis in different physiological recording conditions. *Signal Image Video Process.* **2016**, *10*, 267–276. [CrossRef]

15. Nemirko, A.; Lugovaya, T. Biometric human identification based on electrocardiogram. In Proceedings of the XIIIth Russian Conference on Mathematical Methods of Pattern Recognition, Moscow, Russian, 20–26 June 2005; pp. 387–390. [CrossRef]

16. Gargiulo, F.; Fratini, A.; Sansone, M.; Sansone, C. Subject identification via ECG fiducial-based systems: Influence of the type of QT interval correction. *Comput. Meth. Prog. Biomed.* **2015**, *121*, 127–136. [CrossRef]

17. Tawfik, M.M.; Selim, H.; Kamal, T. Human identification using time normalized QT signal and the QRS complex of the ECG. In Proceedings of the 7th International Symposium on Communication Systems, Networks & Digital Signal Processing (CSNDSP 2010), Newcastle upon Tyne, UK, 21–23 July 2010; pp. 755–759.

18. Patro, K.K.; Kumar, P.R. Effective Feature Extraction of ECG for Biometric Application. In Proceedings of the 7th International Conference on Advances in Computing & Communications (ICACC-2017), Cochin, India, 22–24 August 2017; pp. 296–306. [CrossRef]

19. Chen, C.-L.; Chuang, C.-T. A QRS Detection and R Point Recognition Method for Wearable Single-Lead ECG Devices. *Sensors* **2017**, *17*, 1969. [CrossRef]

20. Lee, W.; Kim, S.; Kim, D. Individual Biometric Identification Using Multi-Cycle Electrocardiographic Waveform Patterns. *Sensors* **2018**, *18*, 1005. [CrossRef]

21. Matveev, M.; Christov, I.; Krasteva, V.; Bortolan, G.; Simov, D.; Mudrov, N.; Jekova, I. Assessment of the stability of morphological ECG features and their potential for person verification/identification. In Proceedings of the 21st International Conference on Circuits, Systems, Communications and Computers (CSCC 2017), Crete Island, Greece, 14–17 July 2017. [CrossRef]

22. Acharya, U.R.; Fujita, H.; Lih, O.S.; Hagiwara, Y.; Tan, J.H.; Adam, M. Automated detection of arrhythmias using different intervals of tachycardia ECG segments with convolutional neural network. *Inf. Sci.* **2017**, *405*, 81–90. [CrossRef]

23. Acharya, U.R.; Oh, S.L.; Hagiwara, Y.; Tan, J.H.; Adam, M.; Gertych, A.; Tan, R.S. A deep convolutional neural network model to classify heartbeats. *Comput. Biol. Med.* **2017**, *89*, 389–396. [CrossRef] [PubMed]

24. Andersen, R.S.; Peimankar, A.; Puthusserypady, S. A deep learning approach for real-time detection of atrial fibrillation. *Expert Syst. Appl.* **2019**, *115*, 465–473. [CrossRef]
25. Yang, W.; Si, Y.; Wang, D.; Guo, B. Automatic recognition of arrhythmia based on principal component analysis network and linear support vector machine. *Comput. Biol. Med.* **2018**, *101*, 22–32. [CrossRef] [PubMed]
26. Chan, T.-H.; Jia, K.; Gao, S.; Lu, J.; Zeng, Z.; Ma, Y. PCANet: A Simple Deep Learning Baseline for Image Classification? *IEEE Trans. Image Process.* **2015**, *24*, 5017–5032. [CrossRef] [PubMed]
27. Jané, R.; Laguna, P.; Thakor, N.V.; Caminal, P. Adaptive baseline wander removal in the ECG: Comparative analysis with cubic spline technique. In Proceedings of the Computers in Cardiology, Durham, NC, USA, 11–14 October 1992; pp. 143–146. [CrossRef]
28. Date, A.A.; Ghongade, R.B. Performance of Wavelet Energy Gradient Method for QRS Detection. In Proceedings of the 4th International Conference on Intelligent and Advanced Systems (ICIAS2012), Kuala Lumpur, Malaysia, 12–14 June 2012; pp. 876–881. [CrossRef]
29. Rakshit, M.; Das, S. An efficient ECG denoising methodology using empirical mode decomposition and adaptive switching mean filter. *Biomed. Signal Process. Control* **2018**, *40*, 140–148. [CrossRef]
30. Li, J.; Si, Y.; Lang, L.; Liu, L.; Xu, T. A Spatial Pyramid Pooling-Based Deep Convolutional Neural Network for the Classification of Electrocardiogram Beats. *Appl. Sci.-Basel* **2018**, *8*, 1590. [CrossRef]
31. Donoho, D.L.; Johnstone, J.M. Ideal spatial adaptation by wavelet shrinkage. *Biometrika* **1994**, *81*, 425–455. [CrossRef]
32. Yao, C.; Si, Y. ECG P, T wave complex detection algorithm based on lifting wavelet. *J. Jilin. U Techno Ed.* **2013**, *43*, 177–182. [CrossRef]
33. Goldberger, A.L.; Amaral, L.A.; Glass, L.; Hausdorff, J.M.; Ivanov, P.C.; Mark, R.G.; Mietus, J.E.; Moody, G.B.; Peng, C.-K.; Stanley, H.E. PhysioBank, PhysioToolkit, and PhysioNet: Components of a new research resource for complex physiologic signals. *Circulation* **2000**, *101*, e215–e220. [CrossRef]
34. Wei, J.J.; Chang, C.J.; Chou, N.K.; Jan, G.J. ECG data compression using truncated singular value decomposition. *IEEE Trans. Biomed. Eng.* **2001**, *5*, 290–299. [CrossRef]
35. PCANet Code. Available online: http://mx.nthu.edu.tw/~{}tsunghan (accessed on 5 December 2018).
36. Rumelhart, D.E.; Hinton, G.E.; Williams, R.J. Learning representations by back-propagating errors. *Nature* **1986**, *323*, 533–536. [CrossRef]
37. Breiman, L. Random forests. *Mach. Learn.* **2001**, *45*, 5–32. [CrossRef]
38. Naive Bayes Code. Available online: https://github.com/andreeas26/NaiveBayesClassifier-Matlab (accessed on 28 December 2018).
39. Fan, R.-E.; Chang, K.-W.; Hsieh, C.-J.; Wang, X.-R.; Lin, C.-J. LIBLINEAR: A Library for Large Linear Classification. *J. Mach. Learn. Res.* **2008**, *9*, 1871–1874. [CrossRef]
40. Dar, M.N.; Akram, M.U.; Shaukat, A.; Khan, M.A. Ieee. ECG Based Biometric Identification for Population with Normal and Cardiac Anomalies Using Hybrid HRV and DWT Features. In Proceedings of the 2015 5th International Conference on IT Convergence and Security (ICITCS), Kuala Lumpur, Malaysia, 24–27 August 2015. [CrossRef]
41. Rezgui, D.; Lachiri, Z. ECG Biometric Recognition Using SVM-Based Approach. *IEEJ Trans. Electr. Electron. Eng.* **2016**, *11*, S94–S100. [CrossRef]
42. Tan, R.; Perkowski, M. ECG Biometric Identification Using Wavelet Analysis Coupled with Probabilistic Random Forest. In Proceedings of the 2016 15th IEEE International Conference on Machine Learning and Applications (ICMLA), Anaheim, CA, USA, 18–20 December 2016; pp. 182–187. [CrossRef]
43. Yu, J.; Si, Y.; Liu, X.; Wen, D.; Luo, T.; Lang, L. ECG Identification Based on PCA-RPROP. In Proceedings of the International Conference on Digital Human Modeling and Applications in Health, Safety, Ergonomics and Risk Management, Vancouver, BC, Canada, 9–14 July 2017; pp. 419–432. [CrossRef]
44. Lynn, H.M.; Yeom, S.; Kim, P. ECG-based biometric human identification based on backpropagation neural network. In Proceedings of the 2018 Conference Research in Adaptive and Convergent Systems (RACS 2018), Honolulu, HI, USA, 9–12 October 2018; pp. 6–10. [CrossRef]
45. Zhao, Z.; Zhang, Y.; Deng, Y.; Zhang, X. ECG authentication system design incorporating a convolutional neural network and generalized S-Transformation. *Comput. Biol. Med.* **2018**, *102*, 168–179. [CrossRef]
46. Bassiouni, M.M.; El-Dahshan, E.-S.A.; Khalefa, W.; Salem, A.M. Intelligent hybrid approaches for human ECG signals identification. *Signal Image Video Process.* **2018**, *12*, 941–949. [CrossRef]

Appl. Sci. **2019**, *9*, 201

47. Sun, Z.; Chiong, R.; Hu, Z.-P. An extended dictionary representation approach with deep subspace learning for facial expression recognition. *Neurocomputing* **2018**, *316*, 1–9. [CrossRef]
48. Wu, J.; Qiu, S.; Kong, Y.; Jiang, L.; Chen, Y.; Yang, W.; Senhadji, L.; Shu, H. PCANet: An energy perspective. *Neurocomputing* **2018**, *313*, 271–287. [CrossRef]
49. Chun, S.Y. Single Pulse ECG-based Small Scale User Authentication using Guided Filtering. In Proceedings of the IEEE International Conference on Biometrics (ICB), Halmstad, Sweden, 13–16 June 2016. [CrossRef]
50. Bendale, A.; Boult, T.E. Towards Open Set Deep Networks. In Proceedings of the IEEE Conference on Computer Vision and Pattern Recognition (CVPR), Las Vegas, NV, USA, 27–30 June 2016; pp. 1563–1572. [CrossRef]
51. Mendes Junior, P.R.; de Souza, R.M.; Werneck, R.d.O. Nearest neighbors distance ratio open-set classifier. *Mach. Learn.* **2017**, *106*, 359–386. [CrossRef]

 *applied sciences*

Article

Unsupervised Learning of Total Variability Embedding for Speaker Verification with Random Digit Strings

Woo Hyun Kang and Nam Soo Kim *

Department of Electrical and Computer Engineering and the Institute of New Media and Communications, Seoul National University, Seoul 08826, Korea; whkang@hi.snu.ac.kr
* Correspondence: nkim@snu.ac.kr; Tel.: +82-2-880-1824

Received: 13 March 2019; Accepted: 12 April 2019; Published: 17 April 2019

Abstract: Recently, the increasing demand for voice-based authentication systems has encouraged researchers to investigate methods for verifying users with short randomized pass-phrases with constrained vocabulary. The conventional i-vector framework, which has been proven to be a state-of-the-art utterance-level feature extraction technique for speaker verification, is not considered to be an optimal method for this task since it is known to suffer from severe performance degradation when dealing with short-duration speech utterances. More recent approaches that implement deep-learning techniques for embedding the speaker variability in a non-linear fashion have shown impressive performance in various speaker verification tasks. However, since most of these techniques are trained in a supervised manner, which requires speaker labels for the training data, it is difficult to use them when a scarce amount of labeled data is available for training. In this paper, we propose a novel technique for extracting an i-vector-like feature based on the variational autoencoder (VAE), which is trained in an unsupervised manner to obtain a latent variable representing the variability within a Gaussian mixture model (GMM) distribution. The proposed framework is compared with the conventional i-vector method using the TIDIGITS dataset. Experimental results showed that the proposed method could cope with the performance deterioration caused by the short duration. Furthermore, the performance of the proposed approach improved significantly when applied in conjunction with the conventional i-vector framework.

Keywords: speech embedding; deep learning; speaker recognition

1. Introduction

Speaker verification is the task of verifying the claimed speaker identity in the input speech. The speaker verification process is composed of three steps: acoustic feature extraction, utterance level feature extraction, and scoring. In the first step, spectral parameters—also referred to as acoustic features—are extracted from short speech frames. One of the most popular acoustic features are the Mel-frequency cepstral coefficients (MFCCs), which represent the spectral envelope of the speech within the given time-frame [1]. However, since most classification or verification algorithms operate on fixed dimensional vectors, acoustic features cannot be directly used with such methods due to the variable duration property of speech [1]. Therefore, in the second step, the frame-level acoustic features are aggregated to obtain a single utterance-level feature, which is also known as an embedding vector. The utterance-level feature is a compact representation of the given speech segment or utterance, conveying information on the variability (*variability* refers to the spread of the data distribution) caused by various factors (e.g., speaker identity, recording channel, environmental noise). In the final step, the system compares two utterance-level features and measures the similarity or the likelihood of the given utterances produced by the same speaker.

Many previous studies on utterance-level features focused on efficiently reducing the dimensionality of a Gaussian mixture model (GMM) supervector, which is a concatenation of the mean vectors of each mixture component [2], while preserving the speaker-relevant information via factorization (e.g., eigenvoice adaptation and joint factor analysis) [3,4]. Particularly the i-vector framework [5,6], which projects the variability within the GMM supervector caused by various factors (e.g., channel and speaker) onto a low-dimensional subspace, has become one of the most dominant techniques used in speaker recognition. The i-vector framework is essentially a linear factorization technique which decomposes the variability of the GMM supervector into a total variability matrix and a latent random variable. However, since these linear factorization techniques assume the variability as a linear function of the latent factor, the i-vector framework is not considered to fully capture the whole variability of the given speech utterances.

Recently, various studies [7–13] have been carried out for non-linearly extracting utterance-level features via deep learning. In [7], a deep neural network (DNN) for frame-level speaker classification was trained and the activations of the last hidden layer—namely, the d-vectors—were taken as a non-linear speaker representation. In [8,9], a time delay neural network (TDNN) utterance embedding technique is proposed where the embedding (i.e., the x-vector) is obtained by statistically pooling the frame-level activations of the TDNN. In [10], the x-vector framework was further improved by incorporating long short-term memory (LSTM) layers to obtain the embedding. In [11], gradient reversal layer and adversarial training were employed to extract an embedding vector robust to channel variation. In [12,13], the embedding neural networks were trained to directly optimize the verification performance in an end-to-end fashion. However, since most of the previously proposed deep learning-based feature extraction models are trained in a supervised manner (which requires speaker or phonetic labels for the training data), it is impossible to use them when little to no labeled data are available for training.

At present, due to the increasing demand for voice-based authentication systems, verifying users with a randomized pass-phrase with constrained vocabulary has become an important task [14]. This particular task is called *random digit-string speaker verification*, where the speakers are enrolled and tested with random sequences of digits. The random digit string task highlights one of the most serious causes of feature uncertainty, which is the short duration of the given speech samples [15]. The conventional i-vector is known to suffer from severe performance degradation when short duration speech is applied to the verification process [16–18]. It has been reported that the i-vectors extracted from short-duration speech samples are relatively unstable (as the duration of the speech is reduced, the vector length of the i-vector decreases and its variance increases; this may cause low inter-speaker variation and high intra-speaker variation of the i-vectors) [16–18]. The short duration problem can be critical when it comes to real-life applications, since in most practical systems, the speech recording for enrollment and trial is required to be short [16].

In this paper, we propose a novel approach to speech embedding for speaker recognition. The proposed method employs a variational inference model inspired by the variational autoencoder (VAE) [19,20] to non-linearly capture the total variability of the speech. The VAE has an autoencoder-like architecture which assumes that the data are generated through a neural network driven by a random latent variable (more information on the VAE architecture is covered in Section 3). Analogous to the conventional i-vector adaptation scheme, the proposed model is trained according to the maximum likelihood criterion given the input speech. By using the mean and variance of the latent variable as the utterance-level features, the proposed system is expected to take the uncertainty caused by short-duration utterances into account. In contrast to the conventional deep learning-based feature extraction techniques, which take the acoustic features as input, the proposed approach exploits the resources used in the conventional i-vector scheme (e.g., universal background model and Baum–Welch statistics) and remaps the relationship between the total factor and the total variability subspace through a non-linear process. Furthermore, since the proposed feature extractor is trained in an unsupervised fashion, no phonetic or speaker label is required for training. Detailed descriptions of the proposed algorithm are given in Section 4.

In order to evaluate the performance of the proposed system in the random digits task, we conducted a set of experiments using the TIDIGITS dataset (see Section 5.1 for details on the TIDIGITS dataset). Moreover, we compared the performance of our system with the conventional i-vector framework, which is the state-of-the-art unsupervised embedding technique [21]. Experimental results showed that the proposed method outperformed the standard i-vector framework in terms of equal error rate (EER), classification error, and detection cost function (DCF) measurements. It is also interesting that a dramatic performance improvement was observed when the features extracted from the proposed method and the conventional i-vector were augmented together. This indicates that the newly proposed feature and the conventional i-vector are complementary to each other.

2. I-Vector Framework

Given a universal background model (UBM), which is a GMM representing the utterance-independent distribution of the frame-level features, an utterance-dependent model can be obtained by adapting the parameters of the UBM via a Bayesian adaptation algorithm [22]. The GMM supervector is obtained by concatenating the mean vectors of each mixture component, summarizing the overall pattern of the frame-level feature distribution. However, since the GMM supervector is known to have high dimensionality and contains variability caused by many different factors, various studies have focused on reducing the dimensionality and compensating the irrelevant variability inherent in the GMM supervector.

Among them, the i-vector framework is now widely used to represent the distinctive characteristics of the utterance in the field of speaker and language recognition [23]. Similar to the eigenvoice decomposition [3] or joint factor analysis (JFA) [4] techniques, i-vector extraction can be understood as a factorization process decomposing the GMM supervector as

$$\mathbf{m}(\mathbf{X}) = \mathbf{u} + \mathbf{T}\mathbf{w}(\mathbf{X}), \tag{1}$$

where $\mathbf{m}(\mathbf{X})$, $\mathbf{u}$, $\mathbf{T}$, and $\mathbf{w}(\mathbf{X})$ indicate the ideal GMM supervector dependent on a given speech utterance $\mathbf{X}$, UBM supervector, total variability matrix, and i-vector, respectively. Hence, the i-vector framework aims to find the optimal $\mathbf{w}(\mathbf{X})$ and $\mathbf{T}$ to fit the UBM parameters to a given speech utterance. Given an utterance $\mathbf{X}$, the 0th and the 1st-order Baum–Welch statistics are obtained as

$$n_c(\mathbf{X}) = \sum_{l=1}^{L} \gamma_l(c), \tag{2}$$

$$\tilde{\mathbf{f}}_c(\mathbf{X}) = \sum_{l=1}^{L} \gamma_l(c)(\mathbf{x}_l - \mathbf{u}_c), \tag{3}$$

where for each frame l within $\mathbf{X}$ with L frames, $\gamma_l(c)$ denotes the posterior probability that the lth frame feature $\mathbf{x}_l$ is aligned to the cth Gaussian component of the UBM, $\mathbf{u}_c$ is the mean vector of the cth mixture component of the UBM, and $n_c(\mathbf{X})$ and $\tilde{\mathbf{f}}_c(\mathbf{X})$ are the 0th and the centralized 1st-order Baum–Welch statistics, respectively.

The extraction of the i-vector can be thought of as an adaptation process where the mean of each Gaussian component in the UBM is altered to maximize the likelihood with respect to a given utterance. Let $\boldsymbol{\Sigma}_c$ denote the covariance matrix of the cth mixture component of the UBM and F be the dimensionality of the frame-level features. Then, the log-likelihood given an utterance $\mathbf{X}$ conditioned on $\mathbf{w}(\mathbf{X})$ can be computed as

$$\log P(\mathbf{X}|\mathbf{T}, \mathbf{w}(\mathbf{X})) = \sum_{c=1}^{C} \Big(n_c(\mathbf{X}) \log \frac{1}{(2\pi)^{F/2}|\boldsymbol{\Sigma}_c|^{1/2}} - \frac{1}{2} \sum_{l=1}^{L} \gamma_l(c)(\mathbf{x}_l - \mathbf{m}_c(\mathbf{X}))^t \boldsymbol{\Sigma}_c^{-1}(\mathbf{x}_l - \mathbf{m}_c(\mathbf{X})) \Big), \tag{4}$$

where $\mathbf{m}_c(\mathbf{X})$ is the mean of the cth mixture component of $\mathbf{m}(\mathbf{X})$ and the superscript t indicates matrix transpose. From lemma 1 in [3], the log-likelihood given $\mathbf{X}$ obtained by marginalizing (4) over $\mathbf{w}(\mathbf{X})$ turns out to be

$$\begin{aligned} \log P(\mathbf{X}|\mathbf{T}) &= \log \mathbb{E}_{\mathbf{w}}[P(\mathbf{X}|\mathbf{T}, \mathbf{w})] \\ &= \log \int P(\mathbf{X}|\mathbf{T}, \mathbf{w}) \mathcal{N}(\mathbf{w}|0, \mathbf{I}) d\mathbf{w} \\ &= G(\mathbf{X}) - \frac{1}{2} \log|\xi(\mathbf{X})| \\ &+ \frac{1}{2} \sum_{c=1}^{C} (\tilde{\mathbf{m}}_c(\mathbf{X}) - \mathbf{u}_c)^t \boldsymbol{\Sigma}_c^{-1} \mathbf{q}_c(\mathbf{X}), \end{aligned} \tag{5}$$

where $G(\mathbf{X})$ represents the log-likelihood of the UBM, $\xi^{-1}(\mathbf{X})$ is the covariance matrix of the posterior distribution of the i-vector given utterance $\mathbf{X}$, $\tilde{\mathbf{m}}_c(\mathbf{X})$ is the cth component of the GMM supervector conditioned on $\mathbb{E}[\mathbf{w}(\mathbf{X})]$, and $\mathbf{q}_c(\mathbf{X})$ is the averaged frame of $\mathbf{X}$ centralized by the cth Gaussian component which is defined by:

$$\mathbf{q}_c(\mathbf{X}) = \sum_{l=1}^{L} (\mathbf{x}_l - \mathbf{u}_c). \tag{6}$$

Analogous to the eigenvoice method, the total variability matrix $\mathbf{T}$ is trained to maximize the log-likelihood (5) using the expectation-maximization (EM) algorithm [3], assuming that each utterance is spoken by a separate speaker.

Once the total variability has been obtained, the posterior covariance and mean of the i-vector can be computed as follows [3]:

$$\mathbb{E}[\mathbf{w}(\mathbf{X})\mathbf{w}^t(\mathbf{X})] = (\mathbf{I} + \sum_{c=1}^{C} n_c(\mathbf{X})\mathbf{T}_c^t \boldsymbol{\Sigma}_c^{-1} \mathbf{T}_c)^{-1}, \tag{7}$$

$$\mathbb{E}[\mathbf{w}(\mathbf{X})] = \mathbb{E}[\mathbf{w}(\mathbf{X})\mathbf{w}^t(\mathbf{X})] \sum_{c=1}^{C} \mathbf{T}_c^t \boldsymbol{\Sigma}_c^{-1} \tilde{\mathbf{f}}_c(\mathbf{X}), \tag{8}$$

where $\mathbf{T}_c$ is a partition matrix of $\mathbf{T}$ corresponding to the cth GMM component. Usually, the posterior mean $\mathbb{E}[\mathbf{w}(\mathbf{X})]$ is used as the utterance-level feature of $\mathbf{X}$. Interested readers are encouraged to refer to [5,6] for further details of the i-vector framework.

3. Variational Autoencoder

The VAE is an autoencoder variant aiming to reconstruct the input at the output layer [19]. The main difference between the VAE and an ordinary autoencoder is that the former assumes that the observed data $\mathbf{x}$ is generated from a random latent variable $\mathbf{z}$ which has a specific prior distribution, such as the standard Gaussian. The VAE is composed of two directed networks: encoder and decoder networks. The encoder network outputs the mean and variance of the posterior distribution $p(\mathbf{z}|\mathbf{x})$ given an observation $\mathbf{x}$. Using the latent variable distribution generated by the encoder network, the decoder network tries to reconstruct the input pattern of the VAE at the output layer.

Given a training sample $\mathbf{x}$, the VAE aims to maximize the log-likelihood, which can be written as follows [19]:

$$\log p_\theta(\mathbf{x}) = D_{KL}(q_\phi(\mathbf{z}|\mathbf{x}) \| p_\theta(\mathbf{z}|\mathbf{x})) + \mathscr{L}(\theta, \phi; \mathbf{x}). \tag{9}$$

In (9), ϕ denotes the variational parameters and θ represents the generative parameters [19]. The first term on the right-hand side (RHS) of (9) means the Kullback–Leibler divergence (KL divergence) between the approximated posterior $q_\phi(\mathbf{z}|\mathbf{x})$ and the true posterior $p_\theta(\mathbf{z}|\mathbf{x})$ of the latent variable, which measures the dissimilarity between these two distributions. Since the KL divergence has a non-negative value, the second term on the RHS of (9) becomes the variational lower bound on the log-likelihood, which can be written as:

$$
\begin{aligned}
\log p_\theta(\mathbf{x}) \geq &\, \mathscr{L}(\theta, \phi; \mathbf{x}) \\
= &- D_{KL}(q_\phi(\mathbf{z}|\mathbf{x}) \| p_\theta(\mathbf{z})) \\
&+ \mathbb{E}_{q_\phi(\mathbf{z}|\mathbf{x})}[\log p_\theta(\mathbf{x}|\mathbf{z})],
\end{aligned}
\tag{10}
$$

where $q_\phi(\mathbf{z}|\mathbf{x})$ and $p_\theta(\mathbf{x}|\mathbf{z})$ are respectively specified by the encoder and decoder networks of the VAE.

The encoder and the decoder networks of the VAE can be trained jointly by maximizing the variational lower bound, which is equivalent to minimizing the following objective function [24]:

$$
\begin{aligned}
E_{VAE}(\mathbf{x}) = &\, D_{KL}(q_\phi(\mathbf{z}|\mathbf{x}) \| p_\theta(\mathbf{z})) \\
&- \mathbb{E}_{q_\phi(\mathbf{z}|\mathbf{x})}[\log p_\theta(\mathbf{x}|\mathbf{z})].
\end{aligned}
\tag{11}
$$

The first term on the RHS of (11) is the KL divergence between the prior distribution and the posterior distribution of the latent variable $\mathbf{z}$, which regularizes the encoder parameters [19]. On the other hand, the second term can be interpreted as the reconstruction error between the input and output of the VAE. Thus, the VAE is trained not only to minimize the reconstruction error but also to maximize the similarity between the prior and posterior distributions of the latent variable.

4. Variational Inference Model for Non-Linear Total Variability Embedding

In the proposed algorithm, it is assumed that the ideal GMM supervector corresponding to a speech utterance $\mathbf{X}$ is obtained through a non-linear mapping of a hidden variable onto the total variability space. Based on this assumption, the ideal GMM supervector is generated from a latent variable $\mathbf{z}$ as follows:

$$
\mathbf{m}(\mathbf{X}) = \mathbf{u} + \mathbf{g}(\mathbf{z}(\mathbf{X})),
\tag{12}
$$

where $\mathbf{g}$ is a non-linear function which transforms the hidden variable $\mathbf{z}(\mathbf{X})$ to the adaptation factor representing the variability of the ideal GMM supervector $\mathbf{m}(\mathbf{X})$. In order to find the optimal function $\mathbf{g}$ and the hidden variable $\mathbf{z}(\mathbf{X})$, we apply a VAE model consisting of an encoder and a decoder network as shown in Figure 1. In the proposed VAE architecture, the encoder network outputs an estimate of the hidden variable and the decoder network serves as the non-linear mapping function $\mathbf{g}$.

Analogous to the i-vector adaptation framework, the main task of the proposed VAE architecture is to generate a GMM so as to maximize the likelihood given the Baum–Welch statistics of the utterance. The encoder of the proposed system serves as a non-linear variability factor extraction model. Similar to the i-vector extractor, the encoder network takes the 0th and 1st-order Baum–Welch statistics of a given utterance $\mathbf{X}$ as input and generates the parametric distribution of the latent variable. The latent variable $\mathbf{z}$ is assumed to be a random variable following a Gaussian distribution, and each component of $\mathbf{z}$ is assumed to be uncorrelated with each other. In order to infer the distribution of the latent variable $\mathbf{z}(\mathbf{X})$, it is sufficient for the encoder to generate the mean and the variance of $\mathbf{z}(\mathbf{X})$. The decoder of the proposed system acts as the GMM adaptation model, generating the GMM supervector from the given latent variable according to the maximum likelihood criterion.

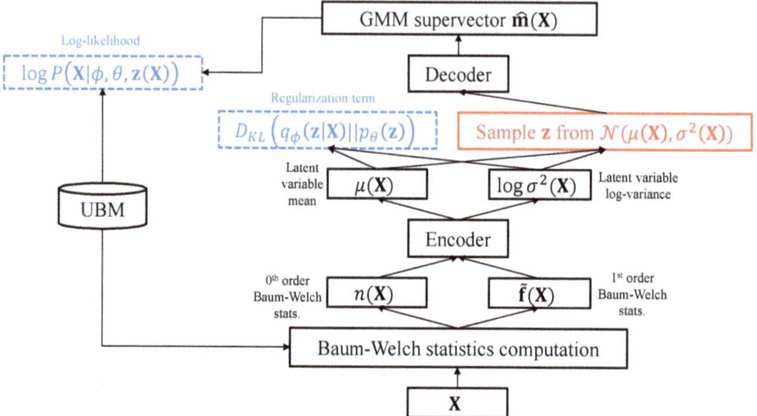

Figure 1. Proposed variational autoencoder (VAE) for non-linear feature extraction. Blue shows the loss terms. Red shows the sampling operations. GMM: Gaussian mixture model; UBM: universal background model.

4.1. Maximum Likelihood Training

Once the GMM supervector $\hat{\mathbf{m}}(\mathbf{X})$ is generated at the output layer of the decoder, the log-likelihood conditioned on the latent variable $\mathbf{z}(\mathbf{X})$ can be defined in a similar manner with (4) as:

$$\log P(\mathbf{X}|\phi,\theta,\mathbf{z}(\mathbf{X})) = \sum_{c=1}^{C} (n_c(\mathbf{X})\log\frac{1}{(2\pi)^{F/2}|\mathbf{\Sigma}_c|^{1/2}}$$

$$-\frac{1}{2}\sum_{l=1}^{L}\gamma_l(c)(\mathbf{x}_l-\hat{\mathbf{m}}_c(\mathbf{X}))^t\mathbf{\Sigma}_c^{-1}(\mathbf{x}_l-\hat{\mathbf{m}}_c(\mathbf{X}))), \tag{13}$$

where $\hat{\mathbf{m}}_c(\mathbf{X})$ denotes the cth component of $\hat{\mathbf{m}}(\mathbf{X})$. Using Jensen's inequality, the lower bound of the marginal log-likelihood can be obtained as follows:

$$\log P(\mathbf{X}|\phi,\theta) = \log\mathbb{E}_{\mathbf{z}}[P(\mathbf{X}|\phi,\theta,\mathbf{z})]$$

$$\geq \mathbb{E}_{\mathbf{z}}[\log P(\mathbf{X}|\phi,\theta,\mathbf{z})]. \tag{14}$$

The marginal log-likelihood can be indirectly maximized by maximizing the expectation of the conditioned log-likelihood (13) with respect to the latent variable $\mathbf{z}$. The reparameterization trick in [19] can be utilized to compute the Monte Carlo estimate of the log-likelihood lower bound as given by

$$\mathbb{E}_{\mathbf{z}}[\log P(\mathbf{X}|\phi,\theta,\mathbf{z})] \simeq \frac{1}{S}\sum_{s=1}^{S}\log P(\mathbf{X}|\phi,\theta,\mathbf{z}_s(\mathbf{X})), \tag{15}$$

where S is the number of samples used for estimation and $\mathbf{z}_s(\mathbf{X})$ is the reparameterized latent variable defined as follows:

$$\mathbf{z}_s(\mathbf{X}) = \mu(\mathbf{X}) + \sigma(\mathbf{X})\epsilon_s. \tag{16}$$

In (16), $\epsilon_s \sim \mathcal{N}(0,\mathbf{I})$ is an auxiliary noise variable, and $\mu(\mathbf{X})$ and $\sigma(\mathbf{X})$ are respectively the mean and standard deviation of the latent variable $\mathbf{z}(\mathbf{X})$ generated from the encoder network. By replacing the reconstruction error term of the VAE objective function (11) with the estimated log-likelihood lower bound, the objective function of the proposed system can be written as:

$$
\begin{aligned}
E_{Prop}(\mathbf{X}) = &D_{KL}(q_\phi(\mathbf{z}|\mathbf{X})\|p_\theta(\mathbf{z})) \\
&- \frac{1}{S}\sum_{s=1}^{S}\log P(\mathbf{X}|\phi,\theta,\mathbf{z}_s(\mathbf{X})).
\end{aligned}
\tag{17}
$$

From (17), it is seen that the proposed VAE is trained not only to maximize the similarity between the prior and posterior distributions of the latent variable, but also to maximize the log-likelihood of the generated GMM by minimizing E_{Prop} via error back-propagation. Moreover, we assume that the prior distribution for $\mathbf{z}$ is $p_\theta(\mathbf{z}) = \mathcal{N}(\mathbf{z}|0,\mathbf{I})$ analogous to the prior for $\mathbf{w}$ in the i-vector framework.

4.2. Non-Linear Feature Extraction and Speaker Verification

The encoder network of the proposed VAE generates the latent variable mean $\mu(\mathbf{X})$ and the log-variance $\log\sigma^2(\mathbf{X})$. Once the VAE has been trained, the encoder network is used as a feature extraction model, as shown in Figure 2. Similar to the conventional i-vector extractor, the encoder network takes the Baum–Welch statistics of the input speech utterance and generates a random variable with a Gaussian distribution, which contains essential information for modeling an utterance-dependent GMM. The mean of the latent variable $\mu(\mathbf{X})$ is exploited as a compact representation of the variability within the GMM distribution dependent on $\mathbf{X}$. Moreover, since the variance of the latent variable $\sigma^2(\mathbf{X})$ represents the variability of the distribution, it is used as a proxy for the uncertainty caused by the short duration of the given speech samples. The features extracted by the proposed VAE can be transformed via feature compensation techniques (e.g., linear discriminant analysis (LDA) [1]) in order to improve the discriminability of the features.

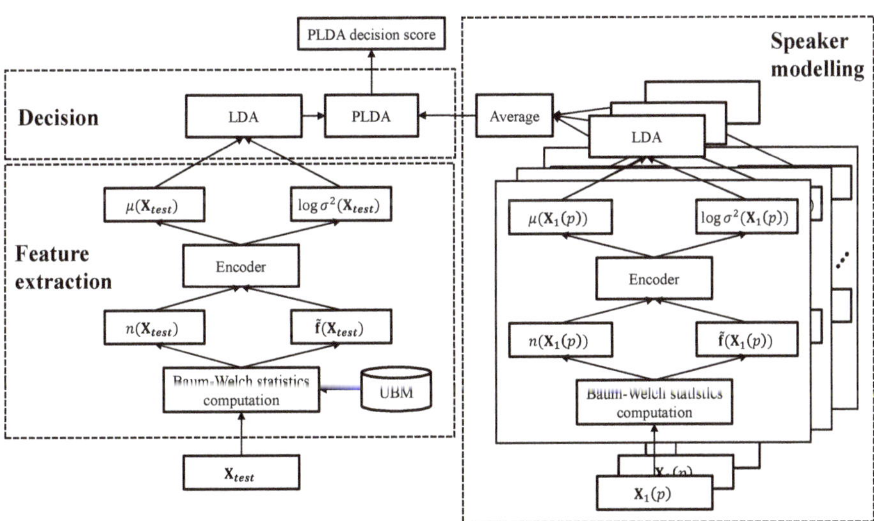

Figure 2. Flow chart of the speaker verification process using the proposed feature extraction scheme.

Given a set of $N(p)$ enrollment speech samples spoken by an arbitrary speaker p

$$
\mathbf{X}(p) = \{\mathbf{X}_1(p),\mathbf{X}_2(p),\cdots,\mathbf{X}_{N(p)}(p)\},
\tag{18}
$$

the speaker model for p is obtained by averaging the features extracted from each speech sample. To determine whether a test utterance $\mathbf{X}_{test}$ is spoken by the speaker p, analogous to the i-vector framework, probabilistic linear discriminant analysis (PLDA) is used to compute the similarity between the feature extracted from $\mathbf{X}_{test}$ and the speaker model of p.

Unlike the conventional i-vector framework, which only uses the mean of the latent variable as feature, the proposed scheme utilizes both the mean and variance of the latent variable to take the uncertainty into account. Providing the speaker decision model (e.g., PLDA) with information about the uncertainty within the input speech, which is represented by the variance of the latent variable, may improve the speaker recognition performance. This is verified in the experiments shown in Section 5.

5. Experiments

5.1. Databases

In order to evaluate the performance of the proposed technique in the random digit speaker verification task, a set of experiments was conducted using the TIDIGITS dataset. The TIDIGITS dataset contains 25,096 clean utterances spoken by 111 male and 114 female adults, and by 50 boys and 51 girls [25]. For each of the 326 speakers in the TIDIGITS dataset, a set of isolated digits and 2–7 digit sequences were spoken. The TIDIGITS dataset was split into two subsets, each containing 12,548 utterances from all 326 speakers, and they were separately used as the enrollment and trial data. In the TIDIGITS experiments, the TIMIT dataset [26] was used for training the UBM, total variability matrix, and the embedding networks.

5.2. Experimental Setup

For the experimented systems, the acoustic feature involves 19-dimensional MFCCs and the log-energy extracted every 10 ms, using a 20 ms Hamming window via the SPro library [27]. Together with the delta (first derivative) and delta-delta (second derivative) of the 19-dimensional MFCCs and the log-energy, the frame-level acoustic feature used in our experiments was given by a 60-dimensional vector.

We trained the UBM containing 32 mixture components in a gender- and age-independent manner, using all the speech utterances in the TIMIT dataset. Training the UBM, total variability matrix, and the i-vector extraction was done by using the MSR Identity Toolbox via MATLAB [28]. The encoders and decoders of the VAEs were configured to have a single hidden layer with 4096 rectified linear unit (ReLU) nodes, and the dimensionality of the latent variables was set to be 200. The implementation of the VAEs was done using Tensorflow [29] and trained using the AdaGrad optimization technique [30]. Additionally, dropout [31] with a fraction of 0.8 and L2 regularization with a weight of 0.01 were applied for training all the VAEs, and the Baum–Welch statistics extracted from the entire TIMIT dataset were used as training data. A total of 100 samples were used for the reparameterization shown in (15).

For all the extracted utterance-level features, linear discriminant analysis (LDA) [1] was applied for feature compensation, and the dimensionality was finally reduced to 200. PLDA [32] was used for speaker verification, and the speaker subspace dimension was set to be 200.

Four performance measures were evaluated in our experiments: classification error (Class. err.), EER, minimum NIST SRE 2008 DCF (DCF08), and minimum NIST SRE 2010 DCF (DCF10). The classification error was measured while performing a speaker identification task where each trial utterance was compared with all the enrolled speakers via PLDA, and the enrolled speaker with the highest score was chosen as the identified speaker. Then, the ratio of the number of wrongly classified trial samples to the total number of trial samples represented the classification error. The EER and minimum DCFs are widely used measures for speaker verification, where the EER indicates the error when the false positive rate (FPR) and the false negative rate (FNR) are the same [1], and the minimum DCFs represent the decision cost obtained with different weights to FPR and FNR. The parameters for measuring DCF08 and DCF10 were chosen according to the weights given by the NIST SRE 2008 [33] and the NIST SRE 2010 [34] protocols, respectively.

5.3. Effect of the Duration on the Latent Variable

In order to investigate the ability of the latent variable to capture the uncertainty caused by short duration, the differential entropies (the differential entropy—or the continuous random variable entropy—measures the average uncertainty of a random variable) of the latent variables were computed. Since the latent variable $\mathbf{z}(\mathbf{X})$ in the proposed VAE has a Gaussian distribution, the differential entropy can be formulated as follows:

$$h(\mathbf{z}(\mathbf{X})) = \frac{1}{2}\log(2\pi e)^K + \frac{1}{2}\log\prod_{k=1}^{K}\sigma_k^2(\mathbf{X}). \tag{19}$$

In (19), K represents the dimensionality of the latent variable and $\sigma_k^2(\mathbf{X})$ is the kth element of $\sigma^2(\mathbf{X})$. From each speech sample in the entire TIDIGITS dataset, 200-dimensional latent variable variance was generated using the encoder network of the proposed framework and used for computing the differential entropy.

In Figure 3, the differential entropies averaged in six different duration groups (i.e., less than 1 s, 1–2 s, 2–3 s, 3–4 s, 4–5 s, and more than 5 s) are shown. As can be seen in the result, the differential entropy computed using the variance of the latent variable gradually decreased as the duration increased. Despite a rather small time difference between the first duration group (i.e., less than 1 s) and the sixth duration group (i.e., more than 5 s), the relative decrement in entropy was 29.91%. This proves that the latent variable variance extracted from the proposed system was capable of indicating the uncertainty caused by the short duration.

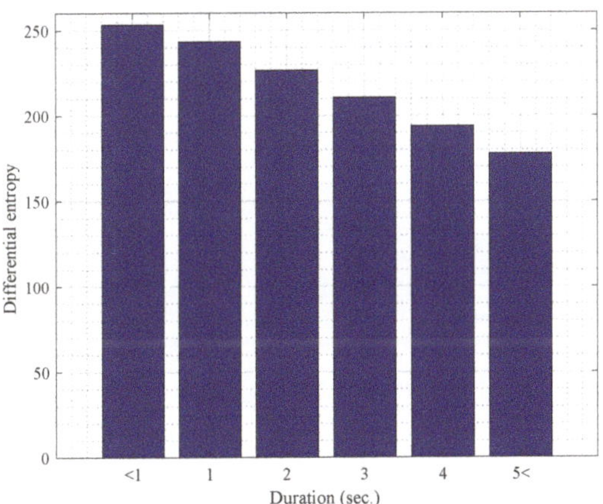

Figure 3. Average differential entropy computed using the latent variable variance extracted from the proposed VAE on different durations.

5.4. Experiments with VAEs

To verify the performance of the proposed VAE trained with the log-likelihood-based reconstruction error function, we conducted a series of speaker recognition experiments on the TIDIGITS dataset. For performance comparison, we also applied various feature extraction approaches. The approaches compared with each other in these experiments were as follows:

- *I-vector*: standard 200-dimensional i-vector;
- *Autoencode*: VAE trained to minimize the cross-entropy between the input Baum–Welch statistics and the reconstructed output Baum–Welch statistics;
- *Classify*: VAE trained to minimize the cross-entropy between the softmax output and the one-hot speaker label;

- *Proposed*: the proposed VAE trained to minimize the negative log-likelihood-based reconstruction error.

Autoencode is a standard VAE for reconstructing the input at the output, and was trained to minimize E_{VAE} (11) given the Baum–Welch statistics as input. On the other hand, *Classify* is a VAE for estimating the speaker label, which was trained to minimize the following loss function:

$$E_{Class}(\mathbf{x}) = D_{KL}(q_\phi(\mathbf{z}|\mathbf{x})\|p_\theta(\mathbf{z})) - \mathbb{E}_{\mathbf{Y}}[\log \hat{\mathbf{Y}}], \tag{20}$$

where $\mathbf{Y}$ denotes the one-hot speaker label of utterance $\mathbf{X}$ and $\hat{\mathbf{Y}}$ is the softmax output of the decoder network. The network structure for *Classify* is depicted in Figure 4. In this experiment, only the mean vectors of the latent variables were used for *Autoencode*, *Classify*, and *Proposed*.

The results shown in Table 1 tell us that the VAEs trained with the conventional criteria (i.e., *Autoencode* and *Classify*) performed poorly compared to the standard i-vector. On the other hand, the proposed VAE with likelihood-based reconstruction error was shown to provide better performance for speaker recognition than the other methods. The feature extracted using the VAE trained with the proposed criterion provided comparable verification performance (i.e., EER, DCF08, DCF10) to the conventional i-vector feature. Moreover, in terms of classification, the proposed framework outperformed the i-vector framework with a relative improvement of 5.8% in classification error. Figure 5 shows the detection error tradeoff (DET) curves obtained from the four tested approaches.

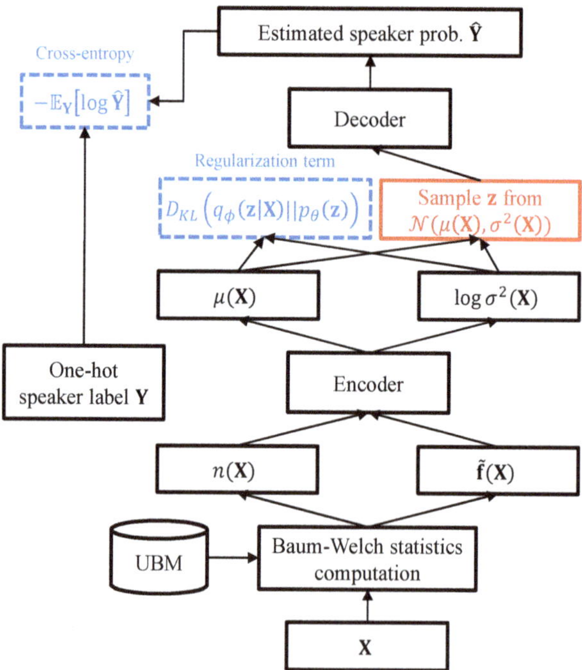

Figure 4. Network structure of the baseline VAE *Classify*.

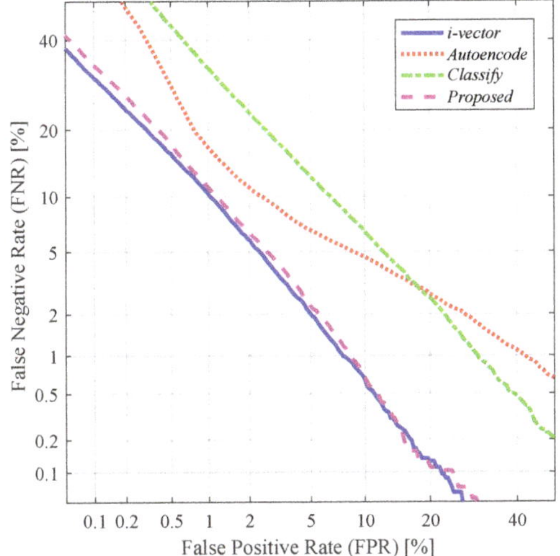

Figure 5. DET curves of the speaker verification experiments using the i-vector and the mean latent variables extracted from VAEs trained for different tasks.

Table 1. Comparison of results between using the i-vector and the mean latent variables extracted from VAEs trained for different tasks. Class. Err.: classification error; DCF08: minimum NIST SRE 2008 DCF; DCF10: minimum NIST SRE 2010 DCF; EER: equal error rate.

	Class. Err. (%)	EER (%)	DCF08	DCF10
i-vector	12.62	**3.36**	**2.00**	**0.07**
Autoencode	24.59	6.06	2.69	0.09
Classify	40.01	8.13	4.21	0.10
Proposed	**11.89**	3.61	2.09	0.07

5.5. Feature-Level Fusion of I-Vector and Latent Variable

In this subsection, we tested the features obtained by augmenting the conventional i-vector with the mean and variance of the latent variable extracted from the proposed VAE. For performance comparison, we applied the following six different feature sets:

- *I-vector(400)*: standard 400-dimensional i-vector;
- *I-vector(600)*: standard 600-dimensional i-vector;
- *LM+LV*: concatenation of the 200-dimensional latent variable mean and the log-variance, resulting in a 400-dimensional vector;
- *I-vector(200)+LM*: concatenation of the 200-dimensional i-vector and the 200-dimensional latent variable mean, resulting in a 400-dimensional vector;
- *I-vector(200)+LV*: concatenation of the 200-dimensional i-vector and the 200-dimensional latent variable log-variance, resulting in a 400-dimensional vector;
- *I-vector(200)+LM+LV*: concatenation of the 200-dimensional i-vector and the 200-dimensional latent variable mean and log-variance, resulting in a 600-dimensional vector.

As seen from Table 2 and Figure 6, the augmentation of the latent variable greatly improved the performance in all the tested cases.

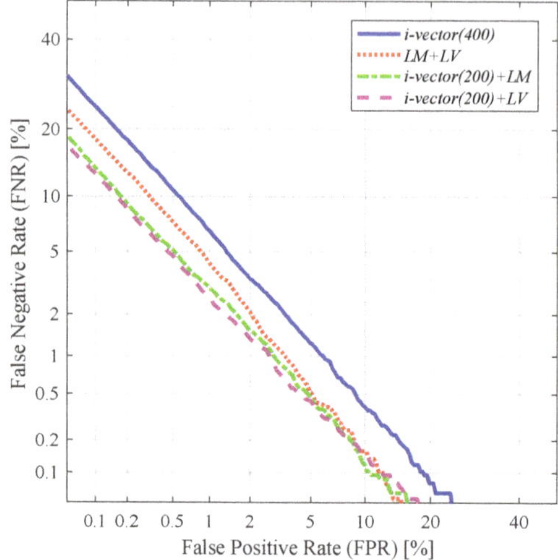

Figure 6. DET curves of the speaker verification experiments using a 400-dimensional i-vector and combinations of two features out of the 200-dimensional i-vector, latent variable mean (LM), and the log-variance of the latent variable (LV).

Table 2. Comparison of results between various feature-level fusions of the conventional i-vector and mean and log-variance of the latent variable extracted from the proposed VAE.

ine	Class. Err. (%)	EER (%)	DCF08	DCF10
i-vector(400)	7.67	2.68	1.54	0.06
LM + LV	6.94	2.03	1.23	0.05
i-vector(200) + LM	5.36	1.78	0.97	0.05
i-vector(200) + LV	**4.99**	**1.65**	**0.94**	**0.04**
i-vector(600)	5.07	2.17	1.29	0.05
i-vector(200) + LM + LV	**2.75**	**0.97**	**0.61**	**0.03**

By using only the mean and log-variance of the latent variable together (i.e., *LM+LV*), a relative improvement of 24.25% was achieved in terms of EER, compared to the conventional i-vector with the same dimension (i.e., *i-vector(400)*). The concatenation of the standard i-vector and the latent variable mean (i.e., *i-vector(200)+LM*) also improved the performance. Especially in terms of EER, *i-vector(200)+LM* achieved a relative improvement of 33.58% compared to *i-vector(400)*. This improvement may be attributed to the non-linear feature extraction process. Since the latent variable mean is trained to encode the various variability within the distributive pattern of the given utterance via a non-linear process, it may contain information that is not obtainable from the linearly extracted i-vector. Thus, by supplementing the information ignored by the i-vector extraction process, a better representation of the speech can be obtained.

The best verification and identification performance out of all the 400-dimensional features (i.e., *i-vector(400)*, *LM+LV*, *i-vector(200)+LM*, and *i-vector(200)+LV*) was obtained when concatenating the standard i-vector and the latent variable log-variance (i.e., *i-vector(200)+LV*). *I-vector(200)+LV* achieved a relative improvement of 38.43% in EER and 34.94% in classification error compared to *i-vector(400)*. This may have been due to the capability of the latent variable variance of capturing the amount of uncertainty, which allows the decision score to take advantage of the duration dependent reliability.

Concatenating the standard i-vector with both the mean and log-variance of the latent variable (i.e., *i-vector(200)+LM+LV*) further improved the speaker recognition performance. Using the *i-vector(200)+LM+LV* achieved a relative improvement of 55.30% in terms of EER, compared to the standard i-vector with the same dimension (i.e., *i-vector(600)*). Figure 7 shows the DET curves obtained when *i-vector(200)+LM+LV* and *i-vector(600)* were applied.

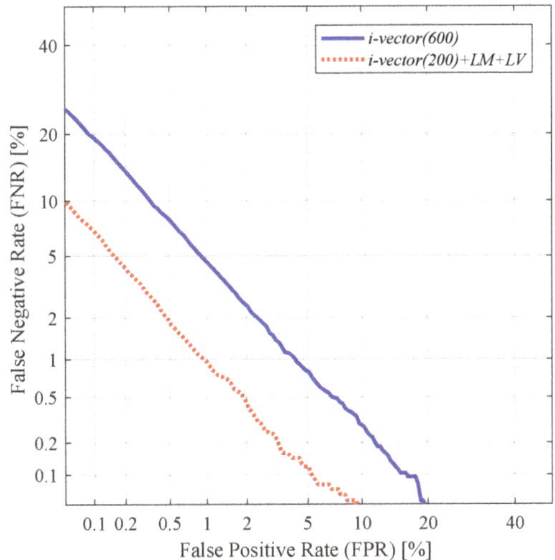

Figure 7. DET curves of the speaker verification experiments using 600-dimensional i-vector and combined feature of 200-dimensional i-vector, latent variable mean, and the log-variance of the latent variable.

5.6. Score-Level Fusion of I-Vector and Latent Variable

In this subsection, we present the experimental results obtained from a speaker recognition task where the decision was made by fusing the PLDA scores of i-vector features and VAE-based features. Given a set of independently computed PLDA scores S_r, $r = 1, \cdots, R$, the fused score S_{fused} was computed by simply adding them as

$$S_{fused} = \sum_{r=1}^{R} S_r. \tag{21}$$

We compared the following scoring schemes:

- *I-vector*: PLDA score obtained by using the standard 200-dimensional i-vector;
- *LM*: PLDA score obtained by using the 200-dimensional latent variable mean;
- *LV*: PLDA score obtained by using the 200-dimensional latent variable log-variance;
- *I-vector+LM*: fusion of the PLDA scores obtained by using the 200-dimensional i-vector and the 200-dimensional latent variable mean;
- *I-vector+LV*: fusion of the PLDA scores obtained by using the 200-dimensional i-vector and the 200-dimensional latent variable log-variance;
- *LM+LV*: fusion of the PLDA scores obtained by using the latent variable mean and log-variance;
- *I-vector+LM+LV*: fusion of the PLDA scores obtained by using the standard 200-dimensional i-vector and the 200-dimensional latent variable mean and log-variance.

Table 3 and Figure 8 give the results obtained through these scoring schemes. As shown in the results, using the latent variable mean and log-variance vectors as standalone features yielded comparable performance to the conventional i-vector method (i.e., *LM* and *LV*). Additionally, fusing the latent-variable-based scores with the score provided by the standard i-vector feature further improved the performance (i.e., *i-vector+LM* and *i-vector+LV*). The best score-level fusion performance was obtained by fusing all the scores obtained by the standard i-vector and the latent variable mean and log-variance vector (i.e., *i-vector+LM+LV*), achieving a relative improvement of 25.89% in terms of EER compared to *i-vector*. However, the performance improvement produced by the score-level fusion methods was relatively smaller than the feature-level fusion methods presented in Table 2. This may have been because the score-level fusion methods compute the scores of the i-vector and the latent variable-based features independently, and as a result the final score cannot be considered an optimal way to utilize their joint information.

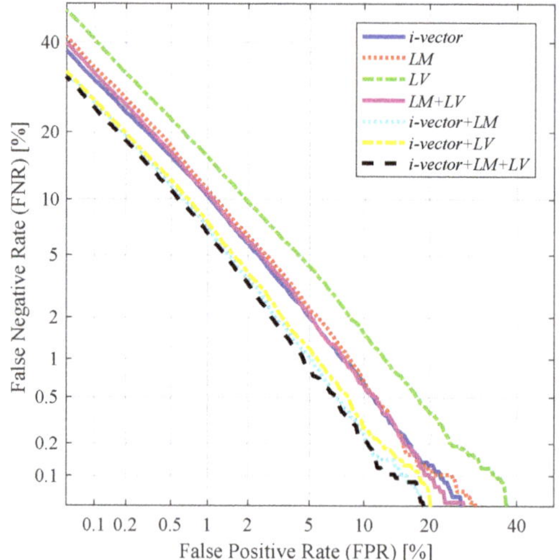

Figure 8. DET curves of the speaker verification experiments using various score-level fusions of the conventional i-vector and the mean and log-variance of the latent variable extracted from the proposed VAE.

Table 3. Comparison of results between various score-level fusions of the conventional i-vector and the mean and log-variance of the latent variable extracted from the proposed VAE.

	Class. Err. (%)	EER (%)	DCF08	DCF10
i-vector	12.62	3.36	2.00	0.07
LM	11.89	3.61	2.09	0.07
LV	17.78	4.65	2.57	0.08
i-vector+LM	7.03	2.63	1.63	**0.06**
ine *i-vector+LV*	7.39	2.76	1.69	**0.06**
LM+LV	10.26	3.50	2.02	0.07
i-vector+LM+LV	**5.75**	**2.49**	**1.57**	**0.06**

6. Conclusions

In this paper, a novel unsupervised deep-learning model-based utterance-level feature extraction for speaker recognition was proposed. In order to capture the variability that has not been fully represented by the linear projection in the traditional i-vector framework, we designed a VAE for GMM adaptation and exploited the latent variable as the non-linear representation of the variability in the given speech. Analogous to the standard VAE, the proposed architecture is composed of an encoder and a decoder network, where the former estimates the distribution of the latent variable given the Baum–Welch statistics of the speech and the latter generates the ideal GMM supervector from the latent variable. Moreover, to take the uncertainty caused by short duration speech utterances into account while extracting the feature, the VAE is trained to generate a GMM supervector in such a way as to maximize the likelihood. The training stage of the proposed VAE uses a likelihood-based error function instead of the conventional reconstruction errors (e.g., cross-entropy).

To investigate the characteristics of the features extracted from the proposed system in a practical scenario, we conducted a set of random-digit sequence experiments using the TIDIGITS dataset. We observed that the variance of the latent variable generated from the proposed network apparently demonstrated the level of uncertainty which gradually decreased as the duration of the speech increased. Additionally, using the mean and variance of the latent variable as features provided comparable performance to the conventional i-vector and further improved the performance when used in conjunction with the i-vector. The best performance was achieved by feature-level fusion of the i-vector and the mean and variance of the latent variable.

In our future study, we will further develop training techniques for the VAE not only to maximize the likelihood but also to amplify the speaker discriminability of the generated latent variable.

Author Contributions: Conceptualization, W.H.K. and N.S.K.; Funding acquisition, N.S.K.; Investigation, W.H.K.; Methodology, W.H.K.; Supervision, N.S.K.; Writing—original draft, W.H.K.; Writing—review and editing, N.S.K.

Funding: This work was supported by Samsung Research Funding Center of Samsung Electronics under Project Number SRFC-IT1701-04.

Conflicts of Interest: The authors declare no conflict of interest.

References

1. Hansen, J.; Hasan, T. Speaker recognition by machines and humans. *IEEE Signal Process. Mag.* **2015**, *32*, 74–99. [CrossRef]
2. Campbell, W.M.; Sturim, D.E.; Reynolds, D.A. Support vector machines using GMM supervectors for speaker verification. *IEEE Signal Process. Lett.* **2006**, *13*, 308–311. [CrossRef]
3. Kenny, P.; Boulianne, G.; Dumouchel, P. Eigenvoice modeling with sparse training data. *IEEE Trans. Audio Speech Lang. Process.* **2005**, *13*, 345–354. [CrossRef]
4. Dehak, N.; Kenny, P.; Dehak, R.; Glembek, O.; Dumouchel, P.; Burget, L.; Hubeika, V.; Castaldo, F. Support vector machines and joint factor analysis for speaker verification. In Proceedings of the 2009 IEEE International Conference on Acoustics, Speech and Signal Processing, Taipei, Taiwan, 19–24 April 2009; pp. 4237–4240.
5. Dehak, N.; Kenny, P.; Dehak, R.; Dumouchel, P.; Ouellet, P. Front-end factor analysis for speaker verification. *IEEE Trans. Audio Speech Lang. Process.* **2011**, *19*, 788–798. [CrossRef]
6. Kenny, P. A small footprint i-vector extractor. In Proceedings of the Odyssey, Singapore, 25–28 June 2012; pp. 1–25.
7. Variani, E.; Lei, X.; McDermott, E.; Moreno, I.L.; Gonzalez-Dominguez, J. Deep neural networks for small footprint text-dependent speaker verification. In Proceedings of the IEEE International Conference on Acoustics, Speech and Signal Processing, Florence, Italy, 4–9 May 2014; pp. 4080–4084.
8. Snyder, D.; Garcia-Romero, D.; Povey, D.; Khudanpur, S. Deep neural network embeddings for text-independent speaker verification. In Proceedings of the INTERSPEECH, Stockholm, Sweden, 20–24 August 2017; pp. 999–1003.
9. Snyder, D.; Garcia-Romero, D.; Sell, G.; Povey, D.; Khudanpur, S. X-vectors: Robust DNN embeddings for speaker recognition. In Proceedings of the IEEE International Conference on Acoustics, Speech and Signal Processing, Prague, Czech Republic, 22–27 May 2018; pp. 5329–5333.
10. Tang, Y.; Ding, G.; Huang, J.; He, X.; Zhou, B. Deep speaker embedding learning with multi-level pooling for text-independent speaker verification. In Proceedings of the IEEE International Conference on Acoustics, Speech, and Signal Processing, Brighton, UK, 12–17 May 2019.
11. Fang, X.; Zou, L.; Li, J.; Sun, L.; Ling, Z.H. Channel adversarial training for cross-channel text-independent speaker recognition. In Proceedings of the IEEE International Conference on Acoustics, Speech and Signal Processing, Brighton, UK, 12–17 May 2019.
12. Chowdhury, F.A.R.R.; Wang, Q.; Moreno, I.L.; Wan, L. Attention-based models for text-dependent speaker verification. In Proceedings of the IEEE International Conference on Acoustics, Speech and Signal Processing, Prague, Czech Republic, 22–27 May 2018; pp. 5359–5363.

13. Wan, L.; Wang, Q.; Papir, A.; Moreno, I.L. Generalized end-to-end loss for speaker verification. In Proceedings of the IEEE International Conference on Acoustics, Speech and Signal Processing, Prague, Czech Republic, 22–27 May 2018; pp. 4879–4883.

14. Yao, S.; Zhou, R.; Zhang, P. Speaker-phonetic i-vector modeling for text-dependent speaker verification with random digit strings. *IEICE Trans. Inf. and Syst.* **2019**, *E102-D*, 346–354. [CrossRef]

15. Saeidi, R.; Alku, P. Accounting for uncertainty of i-vectors in speaker recognition using uncertainty propagation and modified imputation. In Proceedings of the INTERSPEECH, Dresden, Germany, 6–10 September 2015; pp. 3546–3550.

16. Hasan, T.; Saeidi, R.; Hansen, J.H.L.; van Leeuwen, D.A. Duration mismatch compensation for i-vector based speaker recognition systems. In Proceedings of the IEEE International Conference on Acoustics, Speech and Signal Processing, Vancouver, ON, Canada, 26–31 May 2013; pp. 7663–7667.

17. Mandasari, M.I.; Saeidi, R.; McLaren, M.; van Leeuwen, D.A. Quality measure functions for calibration of speaker recognition systems in various duration conditions. *IEEE Trans. Audio Speech Lang. Process.* **2013**, *21*, 2425–2438. [CrossRef]

18. Mandasari, M.I.; McLaren, M.; van Leeuwen, D.A. Evaluation of i-vector speaker recognition systems for forensic application. In Proceedings of the INTERSPEECH, Florence, Italy, 28–31 August 2011; pp. 21–24.

19. Kingma, D.P.; Welling, M. Auto-encoding variational Bayes. *arXiv* **2013**, arxiv:1312.6114.

20. Doersch, C. Tutorial on variational autoencoders. *arXiv* **2016**, arxiv:1606.05908.

21. Park, J.; Kim, D.; Lee, J.; Kum, S.; Nam, J. A Hybrid of deep audio feature and i-vector for artist recognition. In Proceedings of the Joint Workshop on Machine Learning for Music, International Conference on Machine Learning (ICML), Stockholm, Sweden, 14 May 2018.

22. Reynolds, D.; Quatieri, T.F.; Dunn, R.B. Speaker verification using adapted Gaussian mixture models. *Digital Signal Process.* **2000**, *10*, 19–41. [CrossRef]

23. Dehak, N.; Torres-Carrasquillo, P.A.; Reynolds, D.; Dehak, R. Language recognition via ivectors and dimensionality reduction. In Proceedings of the INTERSPEECH, Florence, Italy, 28–31 August 2011; pp. 857–860.

24. Salakhutdinov, R. Learning deep generative models. *Ann. Rev. Stat. Appl.* **2015**, *2*, 361–385. [CrossRef]

25. Leonard, R.G. A database for speaker-independent digit recognition. In Proceedings of the IEEE International Conference on Acoustics, Speech and Signal Processing, San Diego, CA, USA, 19–21 March 1984; pp. 328–331.

26. Lopes, C.; Perdigao, F. Phone recognition on the TIMIT database. *Speech Technol.* **2011**, *1*, 285–302.

27. Gravier, G. SPro: Speech Signal Processing Toolkit. Available online: http://gforge.inria.fr/projects/spro (accessed on 1 March 2019).

28. Sadjadi, S.O.; Slaney, M.; Heck, L. MSR identity toolbox v1.0: A MATLAB toolbox for speaker recognition research. In Proceedings of the Speech and Language Processing Technical Committee Newsletter, Grenoble, France, 21–22 August 2013.

29. Abadi, M.; Agarwal, A.; Barham, P.; Brevdo, E.; Chen, Z.; Citro, C.; Corrado, G.S.; Davis, A.; Dean, J.; Devin, M.; et al. Tensorflow: Large-Scale Machine Learning Heterogenous Systems. Available online: tensorflow.org (accessed on 1 March 2019).

30. Duchi, J.; Hazan, E.; Singer, Y. Adaptive subgradient methods for online learning and stochastic optimization. *J. Mach. Learn. Res.* **2011**, *12*, 2121–2159.

31. Srivastava, N.; Hinton, G.; Krizhevsky, A.; Sutskever, I.; Salakhutdinov, R. Dropout: A simple way to prevent neural networks from overfitting. *J. Mach. Learn. Res.* **2014**, *15*, 1929–1958.

32. Garcia-Romero, D.; Espy-Wilson, C.Y. Analysis of i-vector length normalization in speaker recognition systems. In Proceedings of the INTERSPEECH, Florence, Italy, 28–31 August 2011; pp. 249–252.

33. The NIST Year 2008 Speaker Recognition Evaluation Plan. Available online: http://www.itl.nist.gov/iad/mig//tests/sre/2008/ (accessed on 1 March 2019).

34. The NIST Year 2010 Speaker Recognition Evaluation Plan. Available online: http://www.itl.nist.gov/iad/mig//tests/sre/2010/ (accessed on 1 March 2019).

Article

Attention-Based LSTM Algorithm for Audio Replay Detection in Noisy Environments

Jiakang Li, Xiongwei Zhang *, Meng Sun *, Xia Zou and Changyan Zheng

Lab of Intelligent Information Processing, Army Engineering University, Nanjing 210007, China; jkangli@163.com (J.L.); xiazou@163.com (X.Z.); echoaimaomao@163.com (C.Z.)
* Correspondence: xwzhang9898@163.com (X.Z.); sunmengccjs@gmail.com (M.S.)

Received: 18 March 2019; Accepted: 10 April 2019; Published: 13 April 2019

Featured Application: In this paper, the proposed attention-based long short-term memory (LSTM) algorithm and bagging methods accurately detected the replay spoofing attacks in noisy environments. For example, in voice authentication of bank-accounts or mobile devices, an attacker could be effectively prevented from replaying the voice of a legitimate user by using this algorithm. This algorithm has potential important applications in biometrics and information security, especially for improving the security of the automatic speaker verification systems in noisy environments.

Abstract: Even though audio replay detection has improved in recent years, its performance is known to severely deteriorate with the existence of strong background noises. Given the fact that different frames of an utterance have different impacts on the performance of spoofing detection, this paper introduces attention-based long short-term memory (LSTM) to extract representative frames for spoofing detection in noisy environments. With this attention mechanism, the specific and representative frame-level features will be automatically selected by adjusting their weights in the framework of attention-based LSTM. The experiments, conducted using the ASVspoof 2017 dataset version 2.0, show that the equal error rate (EER) of the proposed approach was about 13% lower than the constant Q cepstral coefficients-Gaussian mixture model (CQCC-GMM) baseline in noisy environments with four different signal-to-noise ratios (SNR). Meanwhile, the proposed algorithm also improved the performance of traditional LSTM on audio replay detection systems in noisy environments. Experiments using bagging with different frame lengths were also conducted to further improve the proposed approach.

Keywords: audio replay attack; noise robustness; attention mechanism; long short-term memory

---

1. Introduction

With the growing popularity of automatic speaker verification (ASV) systems, the attacks on them have posed significant security threats, which requires that the ASV systems have the ability to detect spoofing attacks. Generally, spoofing attacks can be categorized into four types: impersonation, synthesis, conversion, and replay [1]. In order to promote research on anti-spoofing, the automatic speaker verification spoofing and countermeasures (ASVspoof) challenge [2] was first launched in 2015, which focused on discriminating between synthesized or converted voices and those uttered by a human. The second challenge, i.e., the ASVspoof 2017 challenge [3], focused on detecting replay spoofing to discriminate whether a given speech was the voice of an in-person human or the replay of a recorded speech.

Replay attack refers to when an attacker uses a high-fidelity recording device to record the voice of a legitimate authentication system user and then uses the recorded playback through the device on the ASV system, thereby achieving an attack behavior [4]. With the development of electronic technology, the performance of high-fidelity recording and playback equipment has been considerably improved. A replay attack is not difficult to implement and does not require any specialized knowledge, which poses a significant threat to ASV systems [5]. Therefore, the research on replay attack detection has become increasingly important. Villalba et al. [6] used far-field replay speech recordings to investigate the vulnerability of an ASV system and showed that the equal error rate (EER) of the ASV system increased from 1% to nearly 70% when the human voice was recorded directly and replayed to the device. Algre et al. [5] further evaluated the risk of replay attack by using the 2005 and 2006 National Institute of Standards and Technology Speaker Recognition Evaluation dataset (NIST' 05 and NIST' 06) corpus and six kinds of ASV systems. Their results showed that the low-effort replay attack posed a significant risk to all the ASV systems tested. However, the work presented above only reported the results in conditions without background noise, thus it is still not clear how background noise affects the performance of spoofing detection. Therefore, in order to enhance the security of ASV systems, spoofing detection in noisy environment is investigated in this paper.

2. Related Work

Over the years, different methods have been tested to explore spoofing detection. A replay-detection algorithm, based on peakmap audio features, was proposed in [4], which calculated the similarity of the peakmap features between the recorded test utterance and the original enrollment one. If the similarity was above a certain threshold, the utterance was determined to be a replay attack. The replay-detection performance was further improved by adopting a relative similarity score in [7].

Zhang et al. used Mel frequency cepstral coefficients (MFCCs) to detect replay [8]. They believed that the silent segments of a speech would better detect the channel source than the spoken segments. Therefore, voice activation detection (VAD) was utilized to extract the silent segments of a speech and a universal background model (UBM) was created to model the difference between the speaker's testing utterance and their enrollment utterance. However, the use of acoustic signal processing to detect replay attacks is considered to be challenging due to the unpredictable changes in voice recording environments, recording equipment, etc. [3]. For example, artifacts introduced by the acoustic surroundings, such as reverberation, might be confused with the artifacts introduced by playback in some cases. Li et al. [9] tried to use machine learning to detect replay attacks, but only obtained poor performance due to overfitting.

In the ASVspoof 2017 challenge, an official corpus for detecting replay attack and a baseline system, based on the Gaussian mixture model (GMM) with constant Q cepstral coefficient (CQCC) features [10], were provided. The challenge required participants to propose a method that distinguished between genuine speech and a replay recording, where a total of 49 submissions were received from the participants. The average EER of all the submissions was 26.01% [3], where only 20 out of 49 submissions obtained lower EERs than the GMM and CQCC baseline. Among all of the submissions, the best detection performance was reached when using deep convolutional neural networks on spectrograms [11]. This method used an ensemble of three techniques: convolutional neural network (CNN) with recurrent neural network (RNN), light CNN (LCNN), and support vector machine (SVM) i-vector. In addition, Patil et al. [12] proposed variable length Teager energy operator-energy separation algorithm-instantaneous frequency cosine coefficients (VESA-IFCC), and Li et al. [9] utilized a deep neural network (DNN) model with L-Fbanks cepstral coefficient (LFCC) features to identify the genuine and the spoofing speech. All these algorithms achieved better results when compared to the other submissions on the dataset.

The work in this paper is based on the official corpus of the ASVspoof 2017 challenge, i.e., ASVspoof2017 dataset Version 2.0. However, the genuine utterances of the dataset were collected in clean environments (i.e., without any background noises), while the spoof utterances were recorded in environments with various background noises. The genuine utterance setting conflicts with real-world scenarios with complex, noisy environments where the ASV system typically works. In fact, genuine utterances may have background noises, while spoof utterances will contain both the genuine utterances' noises and the noises of the recording environment. Therefore, noisy environments will definitely create challenges in distinguishing between genuine and spoof utterances. Therefore, it is worth exploring the effectiveness of a noise-robust algorithm on spoofing detection, which is the motivation of this paper.

3. Motivation

The previous related work has confirmed that replay attack is still quite a threat to ASV systems, although a couple of replay detection methods have been proposed with success. Towards noise-robust detection, countermeasures and detection algorithms still have room for improvement, especially in complex, noisy environments. Figure 1 shows two pairs of genuine human voices and their corresponding replay speech from the ASVspoof 2017 dataset Version 2.0. The figure tells us the following:

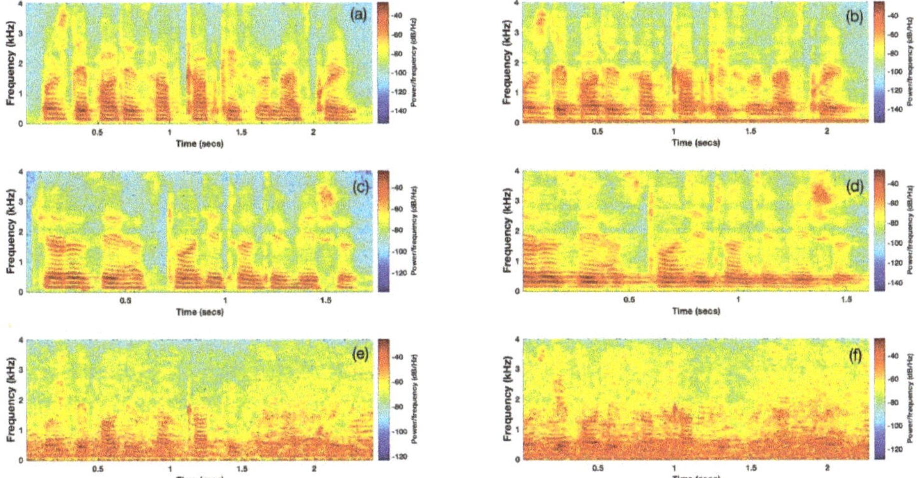

Figure 1. Spectrogram of genuine and replay utterances. (**a**,**c**) are genuine human voices while (**b**,**d**) are replay recordings of (**a**,**c**), respectively. Spectrograms (**e**,**f**) are the noise contaminated versions of (**a**,**b**), respectively, where babble noise with 0 dB signal-to-noise ratio (SNR) was added.

(1) The durations of the utterances are different. Hence, the chosen algorithm should be able to process signals with variable lengths.

(2) The first pair of the genuine (a) and spoof (b) utterances differ greatly in their low frequency regions, while the differences in the second utterance pair, (c) and (d), lie mainly in the high frequency regions. Therefore, distinguishing features can occur over all frequency bands and it is, thus, not appropriate to focus on only high or low frequency bands.

(3) (e) and (f) are the noise-contaminated utterances of (a) and (b), respectively. The spectrograms show that the utterances in noisy environments are more complicated than those of the clean environments. For example, when comparing (a) and (e), it is obvious that some of the distinguishing features of (a) have been dispersed when noise occurs. This phenomenon highlights the difficulties in spoofing detection for noisy utterances, as will be numerically explained in Section 5.3.

In this paper, we aim to improve the spoofing detection algorithm by overcoming the problems presented at the beginning of this section, and propose a robust algorithm for replay attack detection in noisy environments. Our research motivations are summarized as follows:

(1) The sequential nature of speech motivated us to choose long short-term memory (LSTM) to obtain a representation of an utterance. By using LSTM, contextual information can be easily transformed into a vector as input for any back-end classifier.

(2) Considering that different audio frames may contribute differently to the task of replay detection, especially under noisy environments where the frames are contaminated to different degrees by noise, an attention mechanism [13] was used to perform automatic weighting of the frames.

(3) When programming RNNs, for example with TensorFlow, utterance inputs should be the same length, which is normally done by zero-padding the short inputs in the batch to make their length meet a prefixed constant. The lengths of the utterances vary significantly in the ASVspoof 2017 dataset, where the longest utterance has 1423 frames while the shortest one has only 86 frames. There would be too many zeros in the short utterances if simply zero-padding the short ones, which would make the data distribution severely biased toward the padded utterances. In fact, given our research on RNNs, LSTMs, and gated recurrent units (GRUs), the sequence length of the model performs as a hyper-parameter which impacts the model's performance. Therefore, it is also worth investigating the impact of the sequence length on the accuracy of replay detection in the ASV spoofing scenario. Hence, the split/padding method was used to obtain a group of speech segments with uniform length, and is presented in Section 5.2.3. For an input sequence whose length was longer than the prefixed one, splitting was utilized to cut the sequence into smaller sections that met the prefixed length. For an input sequence whose length was shorter than the prefixed one, padding was utilized by repeating the sequence until the prefixed length was reached. In this way, fewer zeros are padded onto the sequences, their segments will have a uniform length, and they are ready to be RNN inputs. The split/padding method is detailed in Section 5.2.3.

(4) Given the split/padding approach described above, a sequence may be divided into several segments, each of which was analyzed to decide on whether it is a genuine or spoofing utterance. Therefore, for the original whole input sequence, it was easy to use the bagging method in ensemble learning to combine the decisions derived on its individual segments. Section 5.2.4 explains the bagging method in detail.

The rest of this paper is organized as follows: Section 4 describes the attention-based LSTM algorithm (AB-LSTM) and the network architecture for replay attack detection, experimental settings and results are presented in Section 5, and the final section, Section 6, concludes the paper.

4. The Proposed Algorithm

In this section, the attention mechanism and the AB-LSTM are introduced. The proposed network architecture for replay attack detection is subsequently presented.

4.1. Attention Mechanism

4.1.1. Traditional Attention

Attention mechanisms have been applied in many tasks and have achieved remarkable results, such as speech recognition [14–16], natural language processing [17], and machine translation [18]. An attention mechanism [13,14] can adaptively learn the relationship of the inputs at several time steps and predict the current time step by using a weight vector. For a sequence with state h_j at each step, a super vector c_t can be computed by,

$$c_t = \sum_{j=1}^{T} \alpha_{tj} h_j,\tag{1}$$

where T is the number of steps of the input sequence and α_{tj} denotes the weight vector, computed at step t for each state h_j. The weight vectors, α_{tj}, are then computed by using an intermediate vector, e_{tj},

$$e_{tj} = \varphi\big(s_{t-1}, h_j\big),\tag{2}$$

where s_{t-1} is the output at step $t-1$ and $\varphi(\cdot)$ is a learned function, which performs as an important scalar determined by h_j and s_{t-1}. Then the value of the weight vectors, α_{tj}, are computed by,

$$\alpha_{tj} = \frac{\exp(e_{tj})}{\sum_{k=1}^{T} \exp(e_{tk})}. \tag{3}$$

This attention mechanism is able to recursively generate weight vectors given the state of the current step. The frames that contribute the most to the final task tend to have higher weights in the entire super vector, while the frames that contribute less to the final task tend to have lower weights. According to Equations (1) to (3), an utterance is encoded to a super vector c, which acts as an input to a back-end genuine/spoof classifier.

4.1.2. Feed-Forward Attention

In the traditional attention mechanism introduced in Section 4.1.1, $\varphi(\cdot)$ in Equation (2) is a learnable function, which depends on the newly generated output, s_{t-1}, and the current state, h_j. In for feed-forward attention, $\varphi(\cdot)$ only depends on the current state h_t, so Equations (2) and (3) become,

$$e_t = \varphi(h_t), \tag{4}$$

and

$$\alpha_t = \frac{\exp(e_t)}{\sum_{k=1}^{T} \exp(e_k)}, \tag{5}$$

respectively, where α_t represent the value of the weight vectors, e_t and e_j represent the intermediate vectors. Thus, the attention mechanism is actually modeled by a super vector, c, of the input sequence by computing an adaptive weighted average of the state sequence h,

$$c = \sum_{t=1}^{T} \alpha_t h_t, \tag{6}$$

The feed-forward attention mechanism is shown in Figure 2. Intuitively, an attention mechanism will put emphasis on the important steps to yield high values in the vector c when modeling sequences.

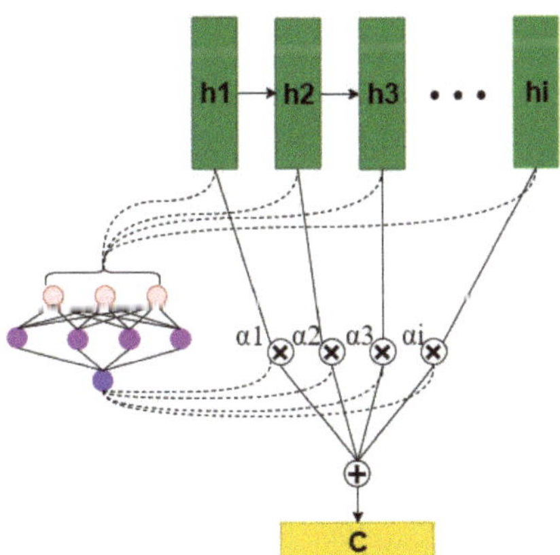

Figure 2. The structure of the "feed-forward" attention mechanism [13,19]. Vectors in the state sequence h_i are fed into the learnable function $\varphi(h_i)$ to produce a weight vector α. The vector c is computed as a weighted average of h_i with the weighting given by the weight vector α.

4.2. Attention-Based LSTM

In this section, for the hidden states of the LSTM output, the feed-forward attention mechanism was utilized to adaptively calculate the weight of every frame. The weights are denoted by $\alpha = \{\alpha_1, \cdots, \alpha_i, \cdots, \alpha_N\}$ for the hidden states, $h = \{h_1, \cdots, h_i, \cdots, h_N\}$, of frames 1, 2, ..., N, where $\{\alpha_i \geq 0, i = 1, \cdots, N\}$. $h_i^{(j)}$ represents the j-th entry of the frame i in the hidden state h. $w^{(j)}$ represents the trainable parameters. Then the weights, computed by the attention mechanism using sigmoid function, can be defined as,

$$u_i = h_i^{(j)} \cdot w^{(j)}, \tag{7}$$

and

$$\alpha_i = \frac{e^{\frac{1}{1+e^{-ui}}}}{\sum_{i=1}^{N} e^{\frac{1}{1+e^{-ui}}} + \varepsilon}, \tag{8}$$

where the weighted, hidden state, $\widetilde{h}_i$, and the super vector, c, are obtained by,

$$\widetilde{h}_i = \alpha_i \cdot h_i^T, \tag{9}$$

and

$$c = \sum_{i=1}^{N} \widetilde{h}_i. \tag{10}$$

4.3. Network Architecture for Replay Attack Detection

The proposed framework is shown in Figure 3. The network architecture is as follows: (1) the frame-level features are extracted from the input utterance, (2) the hidden states of the whole sequence are obtained by LSTM, (3) a batch-normalization layer is utilized to reduce overfitting and to improve the model's performance, (4) attentional weighting of the hidden state is transformed into a super vector, and (5) fully-connected layers followed by a softmax layer are used to make a final decision on whether the voice is genuine or a spoof.

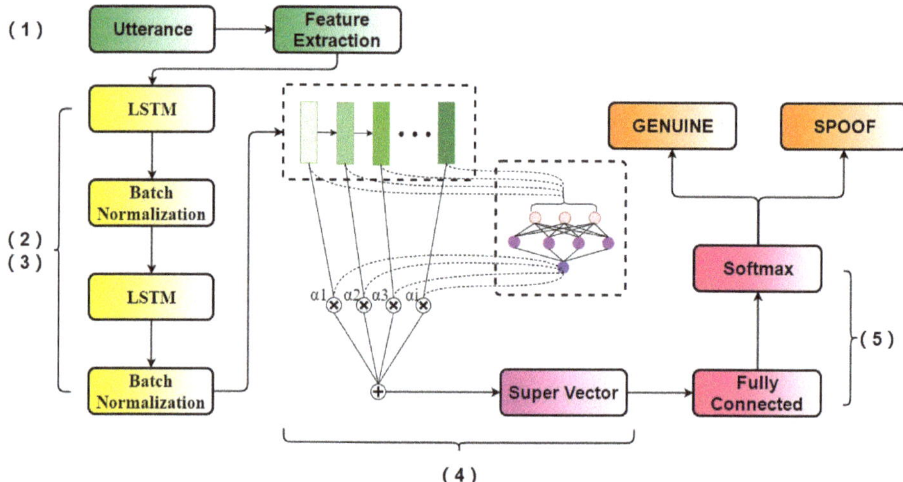

Figure 3. Network architecture for replay attack detection with attention mechanism.

5. Experiments and Results

5.1. ASVspoof 2017 Challenge and Dataset

The ASVspoof 2017 Version 2.0 dataset [3,20] includes three parts: training set, development set, and evaluation set. Each dataset contains both genuine and spoof utterances as shown in Table 1.

Table 1. The detailed information of the ASVspoof 2017 Version 2.0 dataset. The number represents the quantity of utterances in each subset.

Subset	Speakers	Genuine	Spoof
Training	10	1508	1508
Development	8	760	950
Evaluation	24	1298	12008
Total	42	3566	14466

5.2. Experimental Setup

In order to comprehensively evaluate the performance ability of traditional LSTM and AB-LSTM on spoof detection, experiments were conducted in both clean and noisy environments, respectively. Firstly, the CQCC features of each utterance were extracted. Secondly, the split/padding method was used to process the utterance and to obtain segments with the same length. Traditional LSTM and AB-LSTM were then utilized to handle the frame-level features and to decide on genuine or spoof status through bagging methods. Finally, noises with different levels of signal-to-noise ratios (SNRs) were added to the original dataset, and the results on replay detection in noisy environments were obtained by repeating the above steps.

5.2.1. Baseline

The baseline system of the ASVspoof 2017 Challenge used CQCC features [10] with a standard binary GMM classifier for genuine and spoof utterances [3]. For each utterance, the log-likelihood score was obtained from the two GMM classifiers and the final classification decision was based on the log-likelihood ratio. For the experiments in noisy environments, in order to be consistent with the given baseline in the clean background environment [3], we still used CQCC-GMM as the baseline.

5.2.2. Acoustic Features and SNR Settings

In this paper, $D = 90$ dimensional features, with 30 CQCCs, 30 Δ (delta coefficients) and 30 $\Delta\Delta$ (delta-delta coefficients), were utilized. Each frame of an utterance was processed by a 25 ms Hamming window with a 10 ms shift. The number of bins per octave was set to 96. The highest and lowest frequencies were set to 8000 Hz and 16 Hz, respectively. For the experiments in noisy environments, four different kinds of noises were added to the original ASVspoof 2017 Version 2.0 dataset [20]. These four kinds of noises were white, babble, f16, and leopard. The experiments were conducted in four different SNR conditions: −5 dB, 0 dB, 5 dB, and 10 dB.

5.2.3. Split/Padding Method

In the experiments, each utterance had three choices on its prefixed segment length, i.e., 100 frames, 200 frames, and 300 frames. The detailed split/padding methods are shown below. In the padding method (a), for utterances whose length was less than the prefixed value, the front frames were copied and filled to the end of the utterance until reaching the prefixed length. For example, if the specified length was 300 frames, an utterance with 80 frames was extended to 300 frames by repeating its frames, as illustrated in Figure 4.

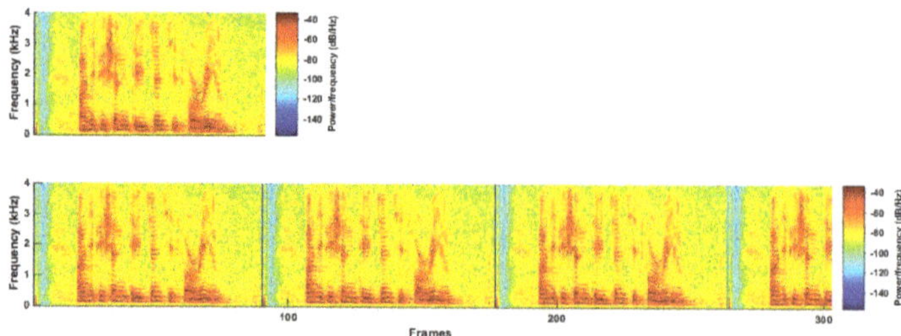

Figure 4. An illustration of the padding method (a).

In the split method (b), for utterances whose length was longer than the prefixed length, the utterance was split. For example, if the prefixed length was 100, an utterance with 250 frames was first extended to 300 frames, according to method (a), and was then be split into three segments, each with 100 frames, as is shown in Figure 5.

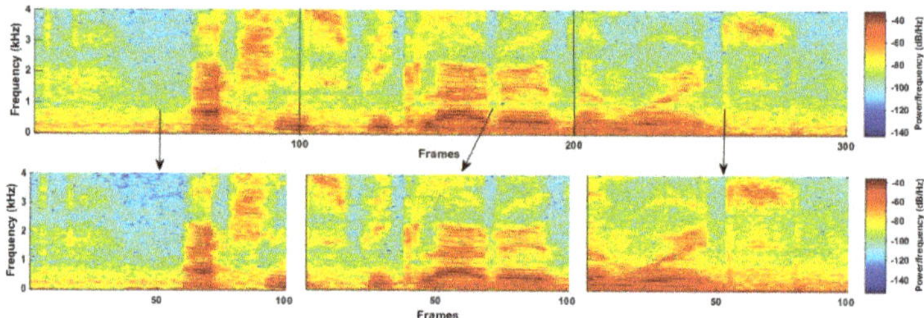

Figure 5. An illustration of the split method (b).

5.2.4. Bagging Methods

By using the split/padding approach presented above, the resulting segments had the same length. A relatively long utterance was, thus, divided into several segments. For each segment, the spoof detection algorithm decided on utterance type. To assess the whole utterance, a bagging approach was used to combine the results of the segments of this utterance. The bagging method was accomplished using two steps (Figure 6). The first step was bagging the decisions derived for each of the segments. That is,

$$\widetilde{F}_a = \frac{\sum_{i=1}^{n} score_i}{n}, \tag{11}$$

where $score_i$ is the classification probabilities of each segment from the utterance, n is the number of segments, and $\widetilde{F}_a$ is the bagging score. The second step was bagging the decisions on the four different kinds of noises, $\widetilde{F}_a$,

$$\widetilde{F}_b = \frac{\sum_{i=1}^{m} \widetilde{F}_{ai}}{m}, \tag{12}$$

where m denotes the number of noises. In the experiments of this paper, m = 4.

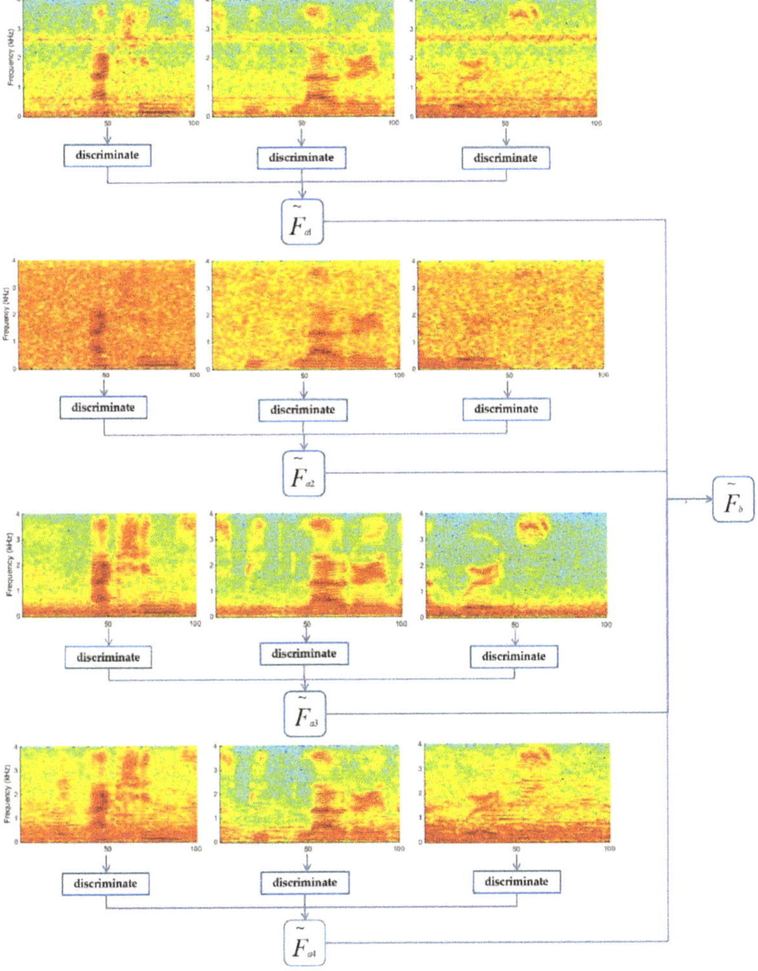

Figure 6. An illustration of the bagging approach used. $\widetilde{F}_{a1}$ denotes the bagging result of the three segments from the same utterance. $\widetilde{F}_{a1}$ to $\widetilde{F}_{a4}$ denote bagging results after adding four different kinds of noises to the original utterance.

5.2.5. Evaluation Metrics

In order to evaluate the performance of the proposed AB-LSTM, contrasting experiments were conducted on AB-LSTM and traditional LSTM, at various SNR levels and for utterances with various lengths. As a classic evaluation metric, EER was used to compare different algorithms. EER is the value where the false alarm rate (FAR) is equal to the false reject rate (FRR). False alarm is defined as the situation where the system incorrectly verifies or identifies a non-target speaker. FAR is the number of false alarms divided by the number of total target trials. False reject is defined as the false rejection of a true target speaker. The FRR is the number of false rejections divided by the total of non-target trials [21]. The CQCC-GMM is adopted from the official MATLAB code in [22]. EER was calculated using the MSR Identity Toolkit Version 1.0 [23].

5.3. Results and Discussion

5.3.1. Experiments in Clean Environments

In this part of the experiment, we used three different frame length segments (100, 200, and 300) as inputs of the network, and calculated the EERs of detection. The architecture of AB-LSTM utilized in the experiments consisted of five recurrent layers with a size of $128 \times 256 \times 256 \times 256 \times 128$. The size of the attention layer was the same as the dimensions of the CQCC features. The last two layers of the model were fully-connected layers with a size of 256×256. The total number of parameters in the network was 1.86 million. As for the traditional LSTM, except for the absence of the attention layer, the remaining parts were the same as those in AB-LSTM. The loss functions of these two methods are binary cross-entropy,

$$\text{loss} = -\sum\nolimits_{i=1}^{n} \hat{y}_i \log y_i + (1 - \hat{y}_i) \log(1 - \hat{y}_i), \tag{13}$$

where n is the number of samples, y_i is the true category, and $\hat{y}_i$ is the predicted classification result.

After processing all the utterances from the three segment length subsets, using the split/padding method, we used the resultant segments from the training set to train the model, and then used the trained model to perform classification on the development and evaluation subsets.

The detailed EERs of traditional LSTM, AB-LSTM, and other baseline methods from the ASVspoof 2017 challenge [3] are presented in Table 2. As can be seen from the experiments under clean background conditions, the AB-LSTM achieved an EER of 16.86% in the evaluation subset, which ranked among the top submissions of the ASVspoof 2017 challenge [3]. As shown in Table 2, AB-LSTM performed better than traditional LSTM, where the EER of AB-LSTM was reduced by 17.0% when compared to traditional LSTM for the cases with 100 frames. For both traditional LSTM and AB-LSTM, the best EER resulted from the settings with 100 frames.

Table 2. Performance of our algorithm and other algorithms used in the ASVspoof 2017 challenge. Results are in terms of the replay/non-replay EER (%).

Methods		*dev* EER	*eval* EER
LCNN$_{FFT}$, SVM$_{i\text{-vect}}$, CNN$_{FFT}$+RNN [11]		3.95	6.73
VESA-IFCC+CFCCIF [12]		0.12	18.33
LCNN [24]		6.47	16.08
Evolving RNN [25]		18.70	18.20
Traditional LSTM	Frame = 100	**13.24**	**20.32**
	Frame = 200	14.18	21.59
	Frame = 300	13.83	20.97
	Average	13.75	20.96
AB-LSTM	Frame = 100	**9.75**	**16.86**
	Frame = 200	10.69	17.69
	Frame = 300	11.94	18.84
	Average	10.79	17.80
CQCC-GMM (baseline)		12.08	29.35

Figure 7 shows the loss curves of traditional LSTM and AB-LSTM methods observed during training. We can see that the decrease in the loss values was faster with the attention mechanism than without it. Although AB-LSTM had an additional attention layer, the convergence speed of the whole model was significantly higher than that of traditional LSTM. This demonstrated that the attention layer accelerated the optimization of the model by making the network concentrate on frames necessary for genuine and spoof utterance detection.

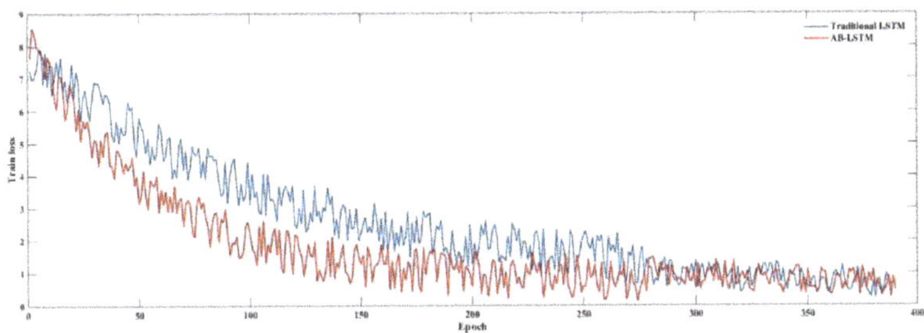

Figure 7. Loss curves of traditional long short-term memory (LSTM) and attention-based LSTM (AB-LSTM).

In order to explore the influence of the network's architecture, another group of experiments were conducted using the 100 frame length segments, combined with the split/padding and bagging methods. All three networks were trained on the training set and evaluated using the evaluation set. Detailed experimental results are shown in Table 3.

Table 3. EER (%) results of the different network's architectures in a clean background environment.

Network's Architecture	Traditional LSTM	AB-LSTM
$128 \times 256 \times 256$	25.74	21.95
$128 \times 256 \times 256 \times 256$	22.61	19.03
$128 \times 256 \times 256 \times 256 \times 128$	**20.32**	**16.86**

5.3.2. Experiments in Noisy Environments

Detailed EERs (%) of the CQCC-GMM baseline, traditional LSTM, and AB-LSTM with different frame lengths under different SNR environments are shown in Tables 4–7. The average EERs of the different frame lengths and the EERs of the bagging methods are also included. The average EERs of different SNR environments are shown in Table 8. Furthermore, Figure 8 illustrates the impact of SNR on EER.

Table 4. EER [a] (%) and EER [b] (%) for different methods when $SNR = -5$ dB.

Methods		EER [a] (%)		EER [b] (%)	
		dev	*eval*	*dev*	*eval*
CQCC-GMM		48.53	47.52	-	-
Traditional LSTM	Frame = 100	**20.58**	**37.78**		
	Frame = 200	20.82	38.09	20.62	38.13
	Frame = 300	23.87	39.86		
	Average	21.76	38.58	-	-
AB-LSTM	Frame = 100	**18.00**	**33.17**		
	Frame = 200	18.42	33.88	18.26	33.31
	Frame = 300	19.71	34.64		
	Average	18.71	33.90	-	-

[a] Bagging EER of segments. [b] Bagging EER of different length of frame.

Table 5. EER [a] (%) and EER [b] (%) for different methods when $SNR = 0$ dB.

Methods		EER [a] (%)		EER [b] (%)	
		dev	*eval*	*dev*	*eval*
CQCC-GMM		39.70	43.51	-	-
Traditional LSTM	Frame = 100	25.04	40.63		
	Frame = 200	**21.37**	**38.11**	22.15	38.46
	Frame = 300	24.74	41.24		
	Average	23.71	39.99	-	-
AB-LSTM	Frame = 100	20.68	33.62		
	Frame = 200	**19.20**	**33.17**	19.12	33.09
	Frame = 300	18.35	32.94		
	Average	19.41	33.24	-	-

Table 6. EER [a] (%) and EER [b] (%) for different methods when $SNR = 5$ dB.

Methods		EER [a] (%)		EER [b] (%)	
		dev	*eval*	*dev*	*eval*
CQCC-GMM		45.99	48.95	-	-
Traditional LSTM	Frame = 100	**19.00**	**36.43**		
	Frame = 200	22.18	39.12	20.77	37.56
	Frame = 300	24.66	40.05		
	Average	21.95	38.53	-	-
AB-LSTM	Frame = 100	**18.43**	**33.27**		
	Frame = 200	19.62	34.06	18.83	33.29
	Frame = 300	19.74	34.85		
	Average	19.26	34.06	-	-

Table 7. EER [a] (%) and EER [b] (%) for different methods when $SNR = 10$ dB.

Methods		EER [a] (%)		EER [b] (%)	
		dev	*eval*	*dev*	*eval*
CQCC-GMM		46.71	46.36	-	-
Traditional LSTM	Frame = 100	**22.38**	**37.91**		
	Frame = 200	23.13	38.14	22.63	38.08
	Frame = 300	23.58	38.33		
	Average	23.03	38.13	-	-
AB-LSTM	Frame = 100	**17.36**	**31.62**		
	Frame = 200	17.57	32.21	17.46	32.11
	Frame = 300	18.09	33.45		
	Average	17.67	32.43	-	-

Table 8. Average EER [a] (%) and EER [b] (%) of all four SNR levels.

Methods		EER [a] (%)		EER [b] (%)	
		dev	*eval*	*dev*	*eval*
CQCC-GMM		45.23	46.59	-	-
Traditional LSTM	Frame = 100	**21.75**	**38.19**		
	Frame = 200	21.88	38.37	21.54	38.06
	Frame = 300	24.21	39.87		
	Average	22.61	38.81	-	-
AB-LSTM	Frame = 100	**18.62**	**32.92**		
	Frame = 200	18.70	33.33	18.42	32.95
	Frame = 300	18.97	33.97		
	Average	18.76	33.41	-	-

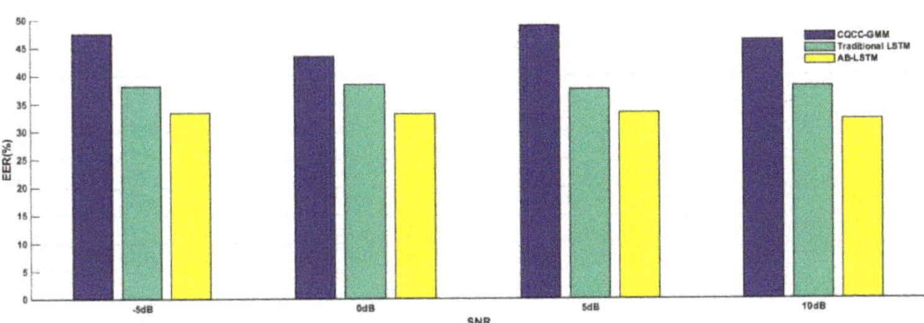

Figure 8. The average EER values of CQCC-GMM, traditional LSTM, and AB-LSTMs at different SNR levels.

Compared with the baseline's EER of 30.60% for the evaluation set in the clean environment [3], the average EER of the four kinds of SNR in noisy environments increased to 45%. This implies that the noise in complex sound environments seriously affected the performance of the baseline system. As for traditional LSTM and the proposed AB-LSTM algorithm, the results of the experiments showed that the AB-LSTM algorithm performed better than the CQCC-GMM baseline for any frame length at any SNR level. It was also seen that the AB-LSTM algorithm achieved better EERs than the traditional LSTM for all parameter settings.

The experiments conducted without the attention mechanism had an average EER of 38.06%, an 8.53% reduction with respect to the baseline system's EER of 46.59%. AB-LSTM obtained an average EER of 32.95% on the evaluation set, which was 13.64% and 5.11% lower than the EER for the baseline system and the traditional LSTM, respectively. The improvement in the EER using the AB-LSTM algorithm, compared to the baseline and traditional LSTM, might be attributed to the attention mechanism, which made the model focus adaptively on the distinguishing utterance information. These results demonstrated the effectiveness of the attention mechanism in replay detection.

The experimental results also demonstrated that 100 was a good choice for the prefixed frame length. The detection methods that used the 100 frame segments performed better than with 200 or 300 frames in all the three SNR environments and for the AB-LSTM algorithm. The experiments also demonstrated that bagging using different lengths of inputs performed better than those using simple averaging. In addition, with the decrease of SNR, the EERs of discrimination also decreased, which indicated that the intensity of the noise affected the accuracy of the replay detection.

As can be seen from Tables 4–7 and Figure 8, EERs did not show a significant decreasing trend with the increase in SNR. The reason for this phenomenon may be that the additional noises caused more serious interference with the original speech signal. However, we found that EERs obtained by AB-LSTM in this study were significantly better than the CQCC-GMM baseline method and traditional

LSTM in noisy environments. The algorithm proposed in this paper deserves further study in replay attack detection.

6. Conclusions

In audio replay detection, it is necessary to pick frames that have important impacts on detection. By introducing an attention mechanism to spoof detection using LSTM, we proposed an algorithm to adaptively highlight the important frames. Moreover, two kinds of bagging methods were proposed and tested, from which accurate and stable performance was observed for spoof detection on noisy speech. Extensive experiments, using the ASVspoof 2017 Version 2.0 dataset contaminated by various noises, demonstrated the effectiveness of the proposed detection methods by achieving a 13.18% reduction in EER when compared to the baseline.

Author Contributions: Conceptualization, J.L., M.S. and X.Z. (Xia Zou); Methodology, J.L., M.S. and C.Z.; investigation, X.Z. (Xiongwei Zhang) and M.S.; data curation, J.L.; writing—original draft preparation, J.L.; writing—review and editing, X.Z. (Xiongwei Zhang) and M.S.

Funding: This research was funded by [the Natural Science Foundation of Jiangsu Province for Excellent Young Scholars] grant number [BK20180080].

Conflicts of Interest: The authors declare no conflict of interest.

References

1. Wu, Z.; Evans, N.; Kinnunen, T.; Yamagishi, J.; Alegre, F.; Li, H. Spoofing and countermeasures for speaker verification: A survey. *Speech Commun.* **2015**, *66*, 130–153. [CrossRef]
2. Wu, Z.; Kinnunen, T.; Evans, N.; Yamagishi, J.; Hanilci, C.; Sahidullah, M.; Sizov, A. ASVspoof 2015: The First Automatic Speaker Verification Spoofing and Countermeasures Challenge. *IEEE J. Sel. Top. Signal Process.* **2017**, *11*, 588–604. [CrossRef]
3. Kinnunen, T.; Sahidullah, M.; Delgado, H.; Todisco, M.; Evans, N.; Yamagishi, J.; Lee, K.A. The ASVspoof 2017 challenge: Assessing the limits of replay spoofing attack detection. In Proceedings of the Annual Conference of the International Speech Communication Association, Stockholm, Sweden, 20–24 August 2017.
4. Shang, W.; Stevenson, M. A playback attack detector for speaker verification systems. In Proceedings of the IEEE International Symposium on Communication, St Julians, Malta, 12–14 March 2008.
5. Alegre, F.; Janicki, A.; Evans, N. Re-assessing the threat of replay spoofing attacks against automatic speaker verification. In Proceedings of the International Conference of the Biometrics Special Interest Group (BIOSIG) IEEE, Darmstadt, Germany, 10–12 September 2014.
6. Villalba, J.; Lleida, E. Speaker Verification Performance Degradation Against Spoofing and Tampering Attacks. In Proceedings of the FALA Workshop, Vigo, Spain, 10–12 November 2010.
7. Shang, W.; Stevenson, M. Score normalization in playback attack detection. In Proceedings of the IEEE International Conference on Acoustics, Speech & Signal Processing, Dallas, TX, USA, 14–19 March 2010.
8. Zhang, L.; Cao, J.; Xu, M.; Zheng, F. Prevention of impostors entering speaker recognition systems. *J. Tsinghua Un. Sci. Technol.* **2008**, *48*, 699–703.
9. Li, L.; Chen, Y.; Wang, D.; Zheng, T.F. A Study on Replay Attack and Anti-Spoofing for Automatic Speaker Verification. *arXiv* **2017**, arXiv:1706.02101.
10. Todisco, M.; Delgado, H.; Evans, N. Constant Q Cepstral Coefficients: A Spoofing Countermeasure for Automatic Speaker Verification. *Comput. Speech Lang.* **2017**, *45*, 516–535. [CrossRef]
11. Galina, L.; Sergey, N.; Egor, M.; Alexander, K.; Oleg, K.; Vadim, S. Audio replay attack detection with deep learning frameworks. In Proceedings of the Annual Conference of the International Speech Communication Association, Stockholm, Sweden, 20–24 August 2017.
12. Patil, H.A.; Kamble, M.R.; Patel, T.B.; Soni, M. Novel variable length teager energy separation based instantaneous frequency features for replay detection. In Proceedings of the Annual Conference of the International Speech Communication Association, Stockholm, Sweden, 20–24 August 2017.
13. Raffel, C.; Ellis, D.P.W. Feed-forward networks with attention can solve some long-term memory problems. In Proceedings of the International Conference on Learning Representations (ICLR), San Juan, Puerto Rico, 12–19 November 2016.

14. Chan, W.; Jaitly, N.; Le, Q.V.; Vinyals, O. Listen, attend and spell: A neural network for large vocabulary conversational speech recognition. In Proceedings of the IEEE International Conference on Acoustics Speech & Signal Processing (ICASSP), Shanghai, China, 20–25 March 2016.

15. Chorowski, J.; Bahdanau, D.; Serdyuk, D.; Cho, K.; Bengio, Y. Attention-Based Models for Speech Recognition. *Comput. Sci.* **2015**, *10*, 429–439.

16. Zhang, S.; Chen, Z.; Zhao, Y.; Li, J.; Gong, Y. End-to-end attention based text-dependent speaker verification. In Proceedings of the IEEE Spoken Language Technology Workshop (SLT), San Diego, CA, USA, 13–16 December 2016.

17. Hu, D. An Introductory Survey on Attention Mechanism in NLP Problems. *arXiv* **2018**, arXiv:1811.05544.

18. Bahdanau, D.; Cho, K.; Bengio, Y. Neural machine translation by jointly learning to align and translate. In Proceedings of the International Conference on Learning Representations (ICLR), San Diego, CA, USA, 7–9 May 2015.

19. Cho, K. Introduction to Neural Machine Translation with GPUs (part 3). Available online: https://devblogs.nvidia.com/introduction-neural-machine-translation-gpus-part-3/ (accessed on 26 July 2015).

20. Delgado, H.; Todisco, M.; Sahidullah, M.; Evans, N.; Kinnunen, T.; Lee, K.A.; Yamagishi, J. ASVspoof 2017 version 2.0: Meta-data analysis and baseline enhancements. In Proceedings of the Odyssey-the Speaker & Language Recognition Workshop, Les Sables d'Olonne, France, 26–29 June 2018.

21. Steven, D. Robust Speaker Verification System with Anti-spoofing Detection and DNN Feature Enhancement Modules. Master's Dissertation, Nanyang Technological University, Singapore, 2015.

22. ASVspoof 2017: Automatic Speaker Verification Spoofing and Countermeasures Challenge. Available online: http://www.asvspoof.org/data2017/baseline_CM.zip (accessed on 25 March 2017).

23. Sadjadi, S.O.; Slaney, M.; Heck, L. MSR Identity Toolbox: A MATLAB Toolbox for Speaker Recognition Research. *Speech Lang. Tech. Comm. Newsl.* **2013**, *1*, 1–32.

24. Lai, C.I.; Abad, A.; Richmond, K.; Yamagishi, J.; Dehak, N.; King, S. Attentive filtering networks for audio replay attack detection. *arXiv* **2018**, arXiv:1810.13048v1.

25. Valenti, G.; Delgado, H.; Todisco, M.; Evans, N.; Laurent, P. *An End-to-End Spoofing Countermeasure for Automatic Speaker Verification Using Evolving Recurrent Neural Networks*; Odyssey-the Speaker & Language Recognition Workshop: Les Sables d'Olonne, France, 2018.

Article

Periocular Recognition in the Wild: Implementation of RGB-OCLBCP Dual-Stream CNN

Leslie Ching Ow Tiong [1,†], Yunli Lee [2,†] and Andrew Beng Jin Teoh [3,*,†]

[1] Computational Science Research Center, Korea Institute of Science and Technology (KIST), Building L0243 14 gil, 5 Hwarangro, Seongbukgu, Seoul 02792, Korea
[2] School of Science and Technology, Sunway University, 5 Jalan Universiti, Bandar Sunway, Petaling Jaya 47500, Selangor, Malaysia
[3] School of Electrical and Electronic Engineering, Yonsei University, 50 Yonsei-ro, Sinchon-dong, Seodaemun-gu, Seoul 03722, Korea
* Correspondence: bjteoh@yonsei.ac.kr
† These authors contributed equally to this work.

Received: 2 June 2019; Accepted: 1 July 2019; Published: 3 July 2019

Featured Application: The proposed periocular biometric network can apply to any application that requires identity management, such as homeland security, border controls, access control, criminal investigation, etc.

Abstract: Periocular recognition remains challenging for deployments in the unconstrained environments. Therefore, this paper proposes an RGB-OCLBCP dual-stream convolutional neural network, which accepts an RGB ocular image and a colour-based texture descriptor, namely Orthogonal Combination-Local Binary Coded Pattern (OCLBCP) for periocular recognition in the wild. The proposed network aggregates the RGB image and the OCLBCP descriptor by using two distinct late-fusion layers. We demonstrate that the proposed network benefits from the RGB image and thee OCLBCP descriptor can gain better recognition performance. A new database, namely an Ethnic-ocular database of periocular in the wild, is introduced and shared for benchmarking. In addition, three publicly accessible databases, namely AR, CASIA-iris distance and UBIPr, have been used to evaluate the proposed network. When compared against several competing networks on these databases, the proposed network achieved better performances in both recognition and verification tasks.

Keywords: periocular recognition in the wild; convolutional neural network; colour-based local binary coded pattern

---

1. Introduction

Biometric systems have been widely deployed since the late 1990s worldwide for identity management, banking, homeland security, etc. [1]. Among different biometric systems, face recognition enjoys flexibility, availability, and user-friendly [2]. However, biometrics experts and the police departments of the United States have agreed that the face recognition technology remains challenging after the "Boston Marathon bombings" in 2013 [3]. For instance, the appearances of subjects such as cosmetic products, plastic surgery or wearing masks may cause the failure of identifying the suspects. To hinder the complexity of the facial region, periocular recognition is gaining attention these days attributed to its promising recognition performance [4].

What does periocular refer to? According to the definition in [5], periocular defines the region around the eyes, which includes the eyelids, eyelashes, and eyebrows (see Figure 1). The periocular region demonstrates more tolerance of variability in expression and occlusion, such as crime scene where perpetrators intentionally mask part of their faces. This creates more capability of matching partial faces [6,7]. In addition, due to the rapid growth of camera use in social networks, surveillance, and smartphones, this arguably increases the interest of periocular recognition [8,9]. For all these reasons, periocular recognition has become an area of intense study in the biometrics and computer vision communities.

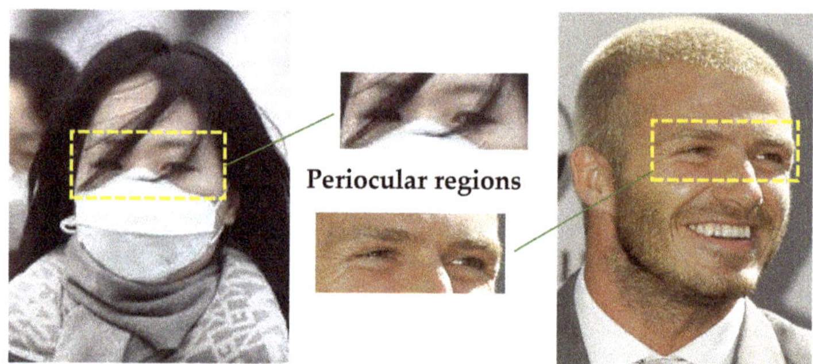

Figure 1. Samples of periocular regions. We demonstrate sample images of the periocular region that including eyebrows. The images are collected from The Korea Times [10] and Kitchen Decor [11].

In this paper, we address the challenges of periocular recognition in the unconstrained or "in-the-wild" environments that remain not well-addressed by the current works [12,13]. This challenge is associated with the issue of dissimilarities in periocular images due to the placement of sensors, pose alignments, illumination levels, occlusions, etc. Thus, we study this problem by means of a fusion approach with dual-stream Convolutional Neural Network (CNN), which accepts RGB ocular image and a novel colour-based texture descriptor, known as Orthogonal Combination-Local Binary Coded Pattern (OCLBCP). We have also developed and shared a new database, namely Ethnic-Ocular database, by collecting the periocular region images in the wild to validate the proposed network.

1.1. Related Works

The early study on periocular biometrics presented in [5] shows promising results in human recognition. The authors adopted several handcrafted descriptors such as Histogram of Oriented Gradients (HOG), Local Binary Patterns (LBP) and Scale-Invariant Feature Transform (SIFT) as periocular feature representation, followed by the score fusion for classification. Fernandez et al. [14] and Cao et al. [15] also introduced a similar approach, which convolves periocular features extracted from HOG or LBP feature matrix with Gabor filters and followed by score fusion. There are several research articles focused on combinations of texture descriptors with fusion algorithm for periocular representation and recognition [16–20]. All these approaches are mainly focused on amalgamation of various handcrafted texture descriptors and followed by learning machines for decent performances in periocular recognition. However, these approaches are less robust to "in the wild" variations such as resolutions, levels of illumination, poses, and occlusions due to inadequacy and inflexibility of handcrafted texture descriptors in representing periocular features. Therefore, the periocular recognition in the wild remains a challenge.

In recent years, CNNs have gained escalating attention in image classification [21,22]. CNNs can be used to extract image texture features from different layers while handcrafted texture descriptor are only limited to low-level features, which is equivalent to the first convolutional (*conv*) layer features of CNNs. Apart from *conv* layers at different level, the features can be extracted from max pooling (*maxpool*) and fully-connected (*fc*) layers of CNNs. Several researchers have employed CNNs for periocular recognition. For instance, Gangwar et al. [23] proposed two CNNs (for left and right oculars), namely DeepIrisNet, which extracts comprehensive information to boost recognition performance. Other studies, e.g., by Proença et al. [24] and Zhao et al. [25], have demonstrated enhanced CNN frameworks for periocular recognition where the prior knowledge is exploited to discard unnecessary information. Proença et al. [24] suggested removal/separation of the iris and sclera from the periocular regions, while Zhao et al. [25] identified the critical regions (only included eyebrow and eye region) that can extract more discriminative information to improve periocular recognition. However, these networks were found to underperform when there are misalignments of periocular images, images missing the eyebrows and images missing ocular.

The relevant works that deal with non-ideal ocular are those by Zhang et al. [26] and Soleymani et al. [27]. Zhang et al. [26] fused iris and periocular modalities through a weighted concatenation. The network achieved significant results when compared to other CNNs. Similarly, Soleymani et al. [27] invented a new multimodal CNN, namely multi-fusion CNN, where the iris, face and fingerprint features are fused at *fc* layer. A fusion layer is designed to fuse different levels of *fc* layers as multi-feature representations with the sole RGB image. However, these works leveraged several biometrics where all of them may not always be available such as occluded face with mouth covered or iris from a distance. Furthermore, the use of multiple biometrics modality may jeopardise the usability of the system such as fingerprint and iris need cooperation from the users.

In the previous work of CNN that consumes face texture descriptor, Levi et al. [28] demonstrated the use of colour-based LBP descriptor as input to CNN rather than raw RGB face image for emotional recognition. The authors showed that colour-based texture descriptor is useful to train their network in the wild environment. This work motivates us to investigate and analyse the impact of colour-based texture descriptor within CNN for periocular recognition in the wild.

1.2. Motivation and Contributions

In the early days of periocular recognition, the problems were mostly concerned about what was the best way to handle periocular in the presence of illuminations, pose alignments, and occlusions [5,6]. Many periocular databases were built using carefully controlled images for each of these issues. UBIPr [12], CASIA-iris database [29], and MICHE database [30] are the most comprehensive efforts in this direction and created in a well-controlled environment.

Presently, the challenges of periocular recognition concern about images that having large variations due to in the wild environments, such as ageing, appearances, cameras location, level of illuminations, occlusions, pose alignments, and others [18,31]. In addition, many existing databases [12,13,29,30] and research communities [18,23,27] still yet to prepare for periocular recognition in the wild challenge. Especially, the appearances of periocular with cosmetic products, and plastic surgery can affect the recognition performance negatively.

This paper offers a solution for periocular recognition in the wild by investigating the fusion of RGB periocular images and a novel texture descriptor, i.e., OCLBCP, by means of a dual-stream CNN. OCLBCP exploits the colour information in the periocular texture to better represent the periocular features for recognition in the wild. The two networks share the parameters and a late fusion takes place at the last *conv* layer before *fc* layer.

For validation of the proposed network, a new database is introduced, namely Ethnic-ocular, by collecting the periocular region images in the wild setup. The databased includes five ethnic groups: *African*, *Asian*, *Latin American*, *Middle Eastern*, and *White*. The database is created in such a way that each ethnic group has a unique shape of periocular and skin texture of periocular regions [32]. Therefore, the database avoids unbalanced selection, as there are differences in the configuration of oculars among different ethnicities.

Hence, the contributions of this paper are as follows:

- To study complementarity between CNN and input features, we investigate and analyse the combination of RGB image and a novel texture descriptor, namely OCLBCP for periocular recognition in the wild.
- Two distinct late-fusion layers are introduced in the proposed CNN. The role of the late-fusion layers is to aggregate the RGB image and OCLBCP descriptor. Hence, the proposed

two-stream CNN is beneficial from these new features of the late-fusion layers to deliver better accuracy performance.

- A new periocular in the wild database, namely Ethnic-ocular, is created and shared in [33]. The images were collected across highly uncontrolled subject–camera distances, appearances, resolutions, locations, levels of illumination, and so on. The database includes training and testing schemes for performance analysis and evaluation.

The paper is organised as follows: Section 2 describes the structure of the proposed colour-based Orthogonal Combination-Local Binary Coded Pattern (OCLBCP) texture descriptor. The proposed network with fusion algorithm is presented in Section 3 and the detailed database information is presented in Section 4. Section 5 discusses the experimental results and analysis. A conclusion is summarised in Section 6.

2. Colour-Based Orthogonal Combination—Local Binary Coded Pattern

This section introduces a new colour-based texture descriptor known as Orthogonal Combination—Local Binary Coded Pattern (OCLBCP). OCLBCP is devised based on the notion of an orthogonal combination of Local Binary Pattern (LBP) [34] and Local Ternary Pattern (LTP) [35]. The OCLBCP descriptor yields a more vibrant texture representation since it is less sensitive to the image noise and levels of illuminations.

Let $\mathbf{I}_p \in \mathbb{R}^{x \times y}$ be the periocular grayscale image, where x and y are the width and height of $\mathbf{I}_p$, respectively. The apparent changes in the images are related to illuminations and poses, thus we deploy the pre-processing method used in [36] to reduce the noise from $\mathbf{I}_p$. First, we transform the $\mathbf{I}_p$ into Fourier domain as Z. Furthermore, we apply the Butterworth filter (B) to Z by reducing the illumination noise and enhancing the reflectance [37]. After that, we apply an inverse Fourier transform to obtain the filtered image $\mathbf{I}_p'$.

To construct the OCLBCP descriptor, $\mathbf{I}_p'$ has to be proposed first according to the LBP [34] and LTP [35] transformation. LBP summarises the local structure in an image by comparing each pixel with its neighbourhood [34]. This descriptor works by thresholding a neighbourhood matrix using the grey level of the central pixel in the binary code. LTP is an extension of the LBP with three-valued codes [35]. The descriptor works by comparing each pixel with its neighbouring pixels. Then, they are combined after thresholding into a ternary pattern. The ternary pattern is split into two binary patterns and called positive and negative matrices.

In this paper, the LBP consists of the 3×3 neighbourhood matrix, and the LTP consists of the positive and negative matrices. To do so, $\mathbf{I}_p'$ is partitioned into sub-matrix with size 3×3 and the neighbourhood values of sub-matrix is binarised according to the centre value of the sub-matrix, which serves as a reference value for thresholding. After that, the descriptor combines the sub-matrix of LBP and LTP into four orthogonal groups: D_1, D_2, D_3, and D_4 (see Figure 2). The orthogonal groups serve to achieve illumination invariance and uncover better texture information by removing outlying disturbances. Specifically, to obtain D_1, the bits from the yellow boxes in the LBP and the bits from green boxes in LTP positive in Figure 2 are combined. The same processes are repeated for D_2, D_3, and D_4.

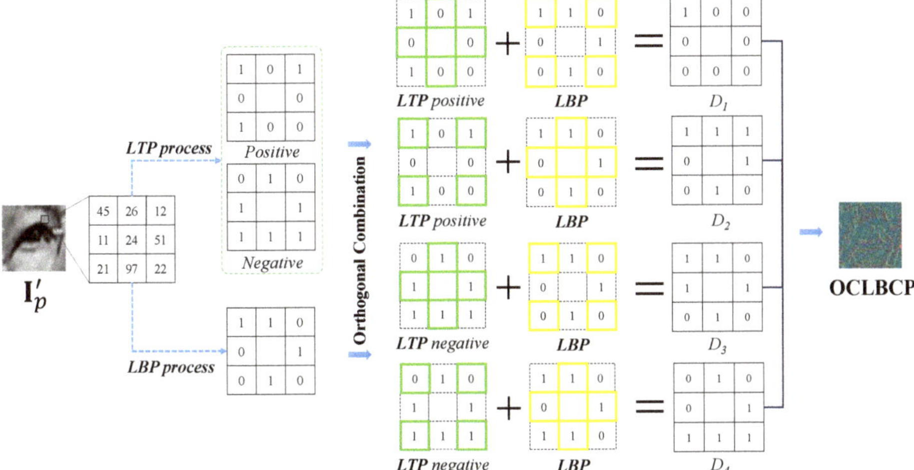

Figure 2. Illustration of Orthogonal Combination–Local Binary Coded Pattern (OCLBCP).

Suppose θ is the OCLBCP descriptor, we first convert the binary codes D_k into a decimal number D_{ck}, $k = 1, 2, 3$, and 4, and then choose the largest value from all the orthogonal groups. Specifically, the θ is formed by combining the groups as follows:

$$\theta(i, j) = \max \left[D_{c1}(i, j), D_{c2}(i, j), D_{c3}(i, j), D_{c4}(i, j) \right], \tag{1}$$

where i and j are the indices of θ.

To map $\theta(i, j)$ into a colour-based texture descriptor, we create a distance pattern matrix Δ to represent the similarity of the image intensity patterns across all possible pixel values based on [28]:

$$\Delta := \begin{bmatrix} \delta_{1,1} & \delta_{1,2} & \delta_{1,3} & \cdots & \delta_{1,c} \\ \delta_{2,1} & \delta_{2,2} & \delta_{2,3} & \cdots & \delta_{2,c} \\ \vdots & \vdots & \vdots & \ddots & \vdots \\ \delta_{r,1} & \delta_{r,2} & \delta_{r,3} & \cdots & \delta_{r,c} \end{bmatrix}, \tag{2}$$

where r and c are defined as the indices of δ. $\delta_{r,c}$ is calculated by Earth Mover's Distance. After that, teh Multi-Dimensional Scaling (MDS) algorithm is adopted to seek the mapping of Δ to the low-dimensional metric space (colour pattern matrix $\mathcal{M}$) [38]:

$$\mathcal{M} = \left[MDS(\theta) + \| \min(MDS(\theta)) \| \right] * \left[\frac{255}{\| \max(MDS(\theta)) \|} \right], \tag{3}$$

$$MDS(\cdot) = \sqrt{\frac{\sum_r \sum_c f(\delta_{r,c})}{\varrho}}, \tag{4}$$

where ϱ is scale factor and $f(\delta_{r,c})$ is a monotonic transformation function of $\delta_{r,c}$. In this paper, we set ϱ to three due to RGB channels in the colour image. Note that $\mathcal{M}$ is a three-colour channels matrix that outputs from $MDS(\cdot)$, which contains R, G, and B pixel values. Finally, we map $\theta(i, j)$ with $\mathcal{M}$ to generate colour-based texture descriptor OCLBCP. The mapping process uses the given pixel values of $\theta(i, j)$ to match the pixel values from the R channel of $\mathcal{M}$. After that, $\theta(i, j)$ is converted with the RGB values from $\mathcal{M}$. Algorithm 1 summarises the process of generating OCLBCP.

Algorithm 1 Creating colour-based texture description OCLBCP.

Input: $I_p \in \mathbb{R}^{x \times y}$
Output: OCLBCP

1: Perform preprocessing to I_p and obtain the filtered image I_p'
2: Construct LBP and LTP process on I_p
3: Perform Equation (1) with the LBP, LTP positive and LTP negative matrices to obtain θ
4: Construct distance pattern matrix Δ using Equation (2)
5: Generate the colour-based pattern matrix $\mathcal{M}$ with δ by using Equations (3) and (4)
6: Map θ with $\mathcal{M}$ to generate OCLBCP

---

3. RGB-OCLBCP of Dual-Stream CNN

We propose a dual-stream CNN that conceives the periocular RGB image and OCLBCP descriptor as the first and second stream to the network. Note that the dual-stream CNN was originally proposed by Feichtenhofer et al. [39] for action detection and recognition. The two input streams refer to temporal and structural streams. In our work, the network accepts and processes periocular colour image and texture descriptor, and then feature fusion layers are devised to extract better feature representation for ocular recognition.

As shown in Figure 3, the architecture of the proposed network consists of 16 convolutional (*conv*) layers and 8 max-pooling (*maxpool*) layers. The *conv* layers are designed to learn the correspondence between the RGB image and OCLBCP descriptor and to discriminate between themselves with the shared weights. Table 1 tabulates the architecture of the proposed network.

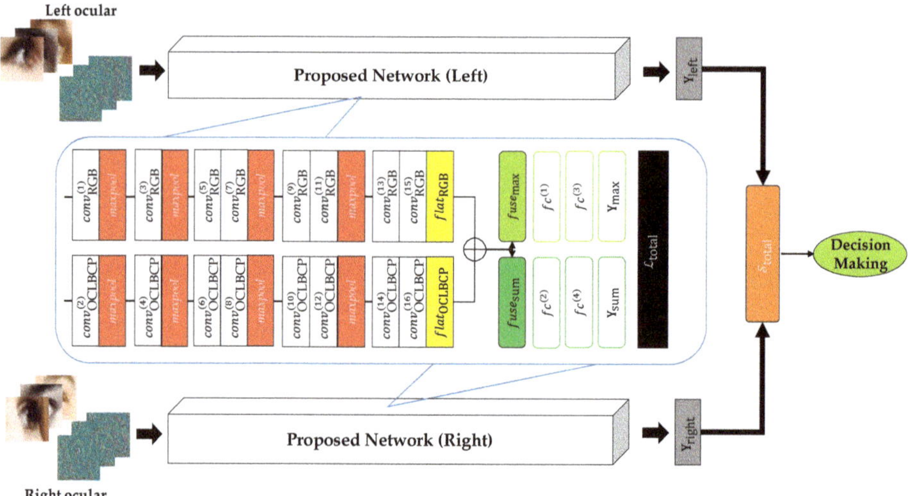

Figure 3. The architecture of the proposed network.

Table 1. Configurations of each layer for the proposed network.

Network Layers	Configurations
$conv_{\text{RGB}}^{(1)}$, $conv_{\text{OCLBCP}}^{(2)}$	$f^{\,1}$: 64@80 × 80; $k^{\,2}$: 2 × 2; *maxpool*: 2 × 2
$conv_{\text{RGB}}^{(3)}$, $conv_{\text{OCLBCP}}^{(4)}$	f: 128@40 × 40; k: 2 × 2; *maxpool*: 2 × 2
$conv_{\text{RGB}}^{(5)}$, $conv_{\text{OCLBCP}}^{(6)}$	f: 256@20 × 20; k: 2 × 2
$conv_{\text{RGB}}^{(7)}$, $conv_{\text{OCLBCP}}^{(8)}$	f: 256@20 × 20; k: 2 × 2; *maxpool*: 2 × 2
$conv_{\text{RGB}}^{(9)}$, $conv_{\text{OCLBCP}}^{(10)}$	f: 512@10 × 10; k: 2 × 2
$conv_{\text{RGB}}^{(11)}$, $conv_{\text{OCLBCP}}^{(12)}$	f: 512@10 × 10; k: 2 × 2; *maxpool*: 2 × 2
$conv_{\text{RGB}}^{(13)}$, $conv_{\text{OCLBCP}}^{(14)}$	f: 512@10 × 10; k: 2 × 2
$conv_{\text{RGB}}^{(15)}$, $conv_{\text{OCLBCP}}^{(16)}$	f: 512@10 × 10; k: 2 × 2
$flat_{\text{RGB}}$, $flat_{\text{OCLBCP}}$	1 × 1 × 12,800
$fuse_{\text{max}}$, $fuse_{\text{sum}}$	1 × 1 × 4096
$fc^{(1)}$, $fc^{(2)}$	1 × 1 × 4096
$fc^{(3)}$, $fc^{(4)}$	1 × 1 × 4096
$\mathbf{Y}_{\text{max}}$, $\mathbf{Y}_{\text{sum}}$	1 × 1 × C

^{1} f refers to the size of the feature map in *conv* layers. ^{2} k is defined as the filter size.

3.1. Fusion Layers

Two fusion layers, namely $fuse_{\text{max}}$ and $fuse_{\text{sum}}$, are designed to aggregate the information from the RGB image and OCLBCP descriptor, as shown in Figure 3. The $fuse_{\text{max}}$ layer takes the largest activation from the $flat_{\text{RGB}}$ and $flat_{\text{OCLBCP}}$ layers with m nodes, where both of them are flattened to $conv_{\text{RGB}}^{(15)}$ and $conv_{\text{OCLBCP}}^{(16)}$, respectively. The $fuse_{\text{max}}$ can be represented as:

$$fuse_{\text{max}}(i) = \max[flat_{\text{RGB}}(i), flat_{\text{OCLBCP}}(i)], \quad i = 1, \cdots, m. \tag{5}$$

On the other hand, $fuse_{\text{sum}}$ takes a sum of activations of $flat_{\text{RGB}}$ and $flat_{\text{OCLBCP}}$. The layer is defined as follows:

$$fuse_{\text{sum}}(i) = flat_{\text{RGB}}(i) + flat_{\text{OCLBCP}}(i), \quad i = 1, \cdots, m. \tag{6}$$

3.2. Total Loss for Training

For training, we define a total loss function, $\mathcal{L}_{\text{total}}$, which is composed of a summation of softmax cross entropy $\mathcal{L}$ of logit vector and their respective encoded label:

$$\mathcal{L}_{\text{total}} = \mathcal{L}\left(\mathbf{V}_{\text{max}}\right) + \mathcal{L}\left(\mathbf{V}_{\text{sum}}\right), \tag{7}$$

$$\mathcal{L}(\mathbf{V}) = -\sum_{n}^{N}\sum_{c}^{C} L_{nc}\log[\text{softmax}(\mathbf{V})_{nc}], \tag{8}$$

$$\text{softmax}(\mathbf{V})_{nc} = \frac{\exp^{\mathbf{V}_{nc}}}{\sum_{c}^{C}\exp^{\mathbf{V}_{nc}}}, \tag{9}$$

where $\mathbf{V} \in \{\mathbf{V}_{\text{max}}, \mathbf{V}_{\text{sum}}\}$. $\mathbf{V}_{\text{max}}$ and $\mathbf{V}_{\text{sum}}$ are defined as the features of $fuse_{\text{max}}$ and $fuse_{\text{sum}}$ layers in the training samples $\mathbf{V}$, respectively. L, N, and C denote class labels, the number of training samples in $\mathbf{V}$, and the number of classes, respectively. Note that a periocular region contains left and right oculars; we therefore train each side with separate networks (Figure 3).

3.3. Score Fusion Layer for Recognition

To recognise an unknown identity, a score fusion layer $\mathcal{S}_{\text{total}}$ is devised to merge the distance scores from the softmax vectors for decision-making. Let $\mathbf{Y}_{\text{max}} = \text{softmax}(\mathbf{V}_{\text{max}}) \in \mathbb{R}^{C}$ and $\mathbf{Y}_{\text{sum}} = \text{softmax}(\mathbf{V}_{\text{sum}}) \in \mathbb{R}^{C}$ be the softmax vectors of $fc^{(3)}$ and $fc^{(4)}$, respectively. Since we train the proposed network for left and right ocular, we thus differentiate the softmax vector $\mathbf{Y}$ to $\mathbf{Y}_{\text{left}}$ and $\mathbf{Y}_{\text{right}}$. Note each individual $\mathbf{Y}$ to $\mathbf{Y}_{\text{left}}$ and $\mathbf{Y}_{\text{right}}$ is still the sum of its corresponding $\mathbf{Y} = \mathbf{Y}_{\text{max}} + \mathbf{Y}_{\text{sum}}$.

We evaluated the proposed system in two common biometric working modes: recognition and verification. For the former, the testing data are divided into a gallery set and a probe set. Each subject in the gallery set is composed of his/her left and right softmax vectors as $\mathbf{Y}_j^G = \{\mathbf{Y}_{j,\text{left}}^G, \mathbf{Y}_{j,\text{right}}^G\}$, where $j = 1, \cdots, C$; the probe set is defined as $\mathbf{Y}^P = \{\mathbf{Y}_{\text{left}}^P, \mathbf{Y}_{\text{right}}^P\}$. The score fusion layer is computed with the sum rule as follows:

$$\mathcal{S}_{\text{total}}(\mathbf{Y}^P, \mathbf{Y}_j^G) = \text{s}(\mathbf{Y}_{\text{left}}^P, \mathbf{Y}_{j,\text{left}}^G) + \text{s}(\mathbf{Y}_{\text{right}}^P, \mathbf{Y}_{j,\text{right}}^G), \tag{10}$$

where $\text{s}(\mathbf{Y}_*^P, \mathbf{Y}_{j,*}^G) = 1 - \cos(\mathbf{Y}_*^P, \mathbf{Y}_{j,*}^G)$ is defined as cosine similarity distance and $* \in \{\text{left}, \text{right}\}$. To identify $\mathbf{Y}^P$, ϕ is decided as follows:

$$\phi = \max_j [\mathcal{S}_{\text{total}}(\mathbf{Y}^P, \mathbf{Y}_j^G)] \tag{11}$$

Verification protocol refers to verifying a person's identity that is claimed as a genuine or an impostor. Let $\mathbf{Y}^R = \{\mathbf{Y}_{\text{left}}^R, \mathbf{Y}_{\text{right}}^R\}$ as the reference set (template) and $\mathbf{Y}^A = \{\mathbf{Y}_{\text{left}}^A, \mathbf{Y}_{\text{right}}^A\}$ as the query set, to decide the $\mathbf{Y}^A$ is a genuine or an impostor, ζ is decided by using Equation (12) as follows:

$$\zeta(\mathbf{Y}^A, \mathbf{Y}^R) = \begin{cases} 1, & \mathcal{S}_{\text{total}}(\mathbf{Y}^A, \mathbf{Y}^R) \leq \tau \\ 0, & \mathcal{S}_{\text{total}}(\mathbf{Y}^A, \mathbf{Y}^R) > \tau \end{cases}, \tag{12}$$

where τ is training dataset dependence threshold value.

4. Database

A large-scale collection of periocular in the wild images from different ethnic groups was created, namely Ethic-ocular database. This database is built for periocular recognition, which contains left and right oculars that were extracted from 85,394 images downloaded from the web. All images were collected in the wild, with uncontrolled subject–camera distances, poses, appearances with and without make-up, and levels of illumination.

We propose this new database to support balanced selection in the configuration of oculars among different ethnicities, and also to stimulate research for periocular recognition in the wild that all periocular images are taken in common and everyday settings. Figure 4 demonstrates several samples of images.

4.1. Collection Setup

To create our database, we selected subject names randomly from BBC News [40], CNN News [41], Naver News [42], and FaceScrub database [43]. The subjects were randomly selected based on different ethnicities. They mostly are celebrities, politicians, athletes, etc.

From the search result, the top 300 images for each subject were downloaded using Python scripts. After that, the images were manually verified to ensure that the subjects correctly labelled the images. We firstly extracted facial regions in these images by using the face detector from Matlab [44] for periocular region extraction. Then, the coordinates of facial feature points were fixed based on the face detector bounding box for image alignment. Then, the images of subjects were labelled manually. After that, we implemented the technique from [45], which allowed us to crop images into left and right oculars. The database contains 85,394 images (including left and right oculars images) of 1034 subjects. Note that the views of these images are between $-45°$ and $45°$.

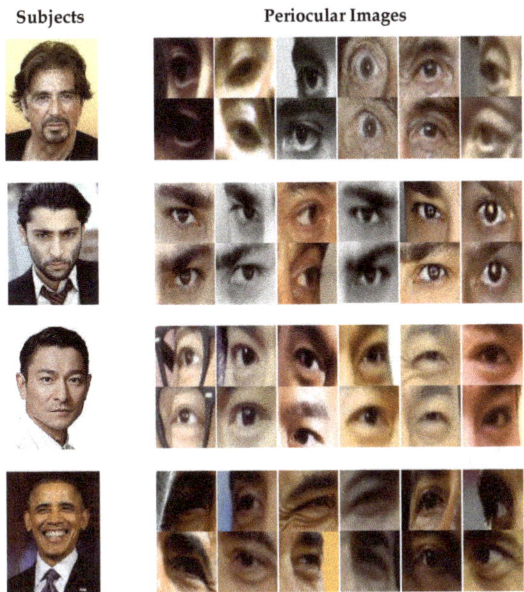

Figure 4. Samples of periocular images in the wild.

4.2. Training Protocol

For the training protocol, 623 subjects were randomly selected. Note that no subjects for training overlapped with the subjects for benchmarking. To develop or train our own models, we designed the protocol by dividing the images for each subject with the ratio of training, testing, and validations as 70:15:15.

4.3. Benchmark Protocol

We selected the remaining 411 subjects as benchmarks. In the benchmarking scheme, we created recognition and verification tasks. For recognition task, images about a specific set of individuals to be recognised (gallery set) were gathered and a new image (the probe set) was presented; the task was to decide which of the gallery identities was represented by the probe set. In the experiments, we divided the images per subject with the ratio of the gallery set to probe set as 50:50. This division process was repeated three times.

For verification task, the task was to analyse two sets of periocular images and decide whether they represent the same person or two different people. In the experiments, we randomly selected 1200 pairs as "same" labels and 1200 pairs as "not same". This selection process was repeated three times.

5. Experiments

We conducted several experiments to evaluate the performance comparisons of recognition and verification between our network and other benchmark networks. All configurations of the networks are described in Section 5.1 and the experimental results are presented in Section 5.2.

5.1. Experimental Setup

5.1.1. Configuration of Proposed Network

The proposed network was implemented using the open source deep learning toolkit TensorFlow [46]. About the configurations, we applied an annealed learning rate and it was started from 1.0×10^{-3}. The rate was subsequently reduced by 10^{-1} for every 10 epochs. The minimum learning rate was defined as 1.0×10^{-5}. We applied an Adam optimiser in this network, where the weight decay and momentum were set to 1.0×10^{-4} and 0.9, respectively.

In our experiments, the batch size was set to 64 and the training was carried out across 200 epochs. The training was done by using our database and following the protocols mentioned in Section 4.2 and it was performed by an NVidia Titan Xp GPU.

5.1.2. Configuration of Benchmark Networks

We selected several deep networks to evaluate the performance of periocular recognition: AlexNet [21], DeepIrisNet-A [23], DeepIrisNet-B [23], FaceNet [47], LCNN29 [48], Multi-fusion CNN [27], and VGG16 [49]. Inspired by the work of Gangwar et al. [23], Soleymani et al. [27], Schroff et al. [47], Wu et al. [48], and Hernandez et al. [50], these networks have been proven to be successful in very large recognition tasks. In the experiments, we utilised the pre-trained models that were provided by the authors to fine-tune and improve the networks themselves by training the left and right oculars, respectively. In the cases of DeepIrisNet-A, DeepIrisNet-B, and Multi-fusion CNN, the networks are not publicly available. Therefore, we did our best effort to implement these networks from scratch by following Gangwar et al. [23] and Soleymani et al. [27], respectively.

5.2. Experimental Results

We present the experimental results on the tasks of periocular recognition and verifications by conducting the databases on periocular recognition in the wild and controlled environments. For the recognition, we evaluated the performance by using Cumulative Matching Characteristic (CMC) curve with 95% confidence interval (CI). For the verification, we evaluated the performance using Receiver Operating Characteristic (ROC) curve with Equal Error Rate (EER) and Area under the ROC curve (AUC).

5.2.1. Performance Analysis on Proposed Network

This section analyses the robustness and performance of our network and other networks using Ethnic-ocular database, which reports the experimental results in Table 2.

Table 2. Performance analysis on the proposed network and other networks with Rank-1 and Rank-5 recognition accuracies. The highest accuracy is highlighted in bold.

Networks	Accuracy (%)		*t.w.* [2]	*flops* [3]
	Rank-1	*Rank-5*		
CNN [1] with RGB image	80.79 ± 1.43	90.42 ± 1.29	131.1 M	2.22 GFLOPS
CNN with OCLBCP	66.65 ± 2.22	89.73 ± 1.91	131.1 M	2.22 GFLOPS
Dual-stream CNN (using unshared weights)	82.09 ± 1.59	92.11 ± 1.32	250.8 M	1.90 GFLOPS
Proposed network	$\mathbf{85.03 \pm 1.88}$	$\mathbf{94.23 \pm 1.26}$	**126.1 M**	**0.90 GFLOPS**

[1] CNN is defined as single-stream CNN; [2] *t.w.*, total weight number; [3] *flops*, floating points operation.

Table 2 shows the proposed network achieved the highest Rank-1 and Rank-5 recognition accuracies with $85.03 \pm 1.88\%$ and $94.23 \pm 1.26\%$, respectively. As compared to CNN, this network using the RGB image only achieved the Rank-1 and Rank-5 accuracies of $80.79 \pm 1.43\%$ and $90.42 \pm 1.29\%$, respectively. In addition, CNN using the OCLBCP can only achieved $66.65 \pm 2.22\%$ and $89.73 \pm 1.91\%$ for Rank-1 and Rank-5 accuracies, respectively. These results indicate that our network provides more complementary information than CNN. This leads to the proposed late-fusion layers that significantly correlate the RGB image and OCLBCP for achieving better recognition performance.

Furthermore, we also evaluated the dual-stream CNN without using shared weights. However, this network only achieved $82.09 \pm 1.59\%$ and $92.11 \pm 1.32\%$ at Rank-1 and Rank-5 accuracies (see Table 2), respectively. The experimental results prove that the proposed network performed well with at least 2.9% improvement as compared to dual-stream CNN without using shared weight. As can be observed, the shared *conv* layers and the fusion layers were utilised in the network to aggregate the RGB image and OCLBCP. Thus, the proposed network successfully transformed new knowledge representations to perform better recognition in the wild.

In Table 2, we also notice the space complexity (total weight number) and time complexity (flops) of the proposed network are significantly smaller than its single network and dual-stream unshared weights networks counterparts while still outperforming them.

5.2.2. Performance Evaluation on Recognition and Verification Tasks

We used Ethnic-ocular, as well as three public databases, the AR [51], CASIA-iris distance [29], and UBIPr [12], to evaluate the performances of the proposed network and other benchmark networks. All the experimental results are outlined in the following sections.

Evaluation on AR Database

The AR database is designed under a constrained environment, which consists of 117 subjects with varying neutrals, expressions, illuminations, and occlusion conditions, who were captured across two sessions. We opted for this database as it provides a good baseline to evaluate the robustness and performance in constrained environments, such as different levels of illuminations and expressions in an indoor environment. Extraction for the periocular regions was done by using the method in [45].

The experimental protocol for recognition was as follows: ten images for each subject were used as gallery sets from Session 1 and another ten per subject as probe sets from Session 2. On the other hand, the verification protocol was designed by randomly selecting 250 reference-query pairs as "same' and another 250 pairs as "not same".

Table 3 presents the performance comparisons on recognition. As can be seen in the table, our network achieved the highest Rank-1 and Rank-5 recognition accuracies with 96.32% and 98.80%, respectively. Likewise, DeepIrisNet-A had the best performance on Rank-1 and Rank-5 among the other benchmark approaches, which only achieved accuracies of 95.24% and 98.38%, respectively. Figure 5a illustrates that the proposed network outperformed other approaches with respect to all the benchmarks from Rank-1 to Rank-10 recognition.

For the verification task, we report the experimental results in Table 4. The proposed network also achieved the best EER and AUC with 5.13% and 0.9880, respectively. DeepIrisNet-A, Multi-fusion CNN, and VGG16 achieved the second-best performances among the other benchmark approaches with 7.69% for EER. Figure 5b illustrates the ROC curve and shows that the proposed network (red solid line with diamond) outperformed the benchmark approaches.

Table 3. Performance evaluation of the recognition task on the AR database, CASIA-iris distance database, UBIPr database, and Ethnic-ocular database. The highest accuracy is written in bold.

Networks	AR		CASIA-Iris		UBIPr		Ethnic-Ocular	
	Rank-1	*Rank-5*	*Rank-1*	*Rank-5*	*Rank-1*	*Rank-5*	*Rank-1*	*Rank-5*
AlexNet	93.59	96.75	95.00 ± 1.8	96.98 ± 2.5	84.88 ± 2.5	96.01 ± 1.8	64.72 ± 3.3	82.98 ± 2.5
DeepIristNet-A	95.24	98.38	95.95 ± 2.1	98.15 ± 0.6	90.30 ± 1.2	97.41 ± 1.1	79.54 ± 3.1	90.43 ± 2.4
DeepIristNet-B	94.44	97.18	95.79 ± 2.6	97.75 ± 0.6	90.20 ± 1.7	97.43 ± 0.5	81.13 ± 3.1	92.37 ± 1.2
FaceNet	94.19	97.75	96.09 ± 2.1	98.10 ± 0.4	90.24 ± 1.4	97.36 ± 0.4	78.71 ± 3.7	92.19 ± 1.6
LCNN29	94.27	97.52	96.01 ± 2.0	97.85 ± 0.9	90.28 ± 1.7	97.18 ± 0.7	79.35 ± 2.6	92.17 ± 1.8
Multi-fusion CNN	96.07	98.71	95.81 ± 1.9	97.67 ± 1.0	90.75 ± 1.0	97.44 ± 0.3	81.79 ± 3.5	93.03 ± 1.3
VGG16	94.20	97.61	95.88 ± 0.1	97.99 ± 0.5	90.24 ± 1.4	97.09 ± 1.1	76.43 ± 2.2	91.29 ± 1.5
Proposed Network	**96.32**	**98.80**	**96.62 ± 1.3**	**98.45 ± 0.4**	**91.28 ± 1.2**	**98.59 ± 0.4**	**85.03 ± 1.9**	**94.23 ± 1.3**

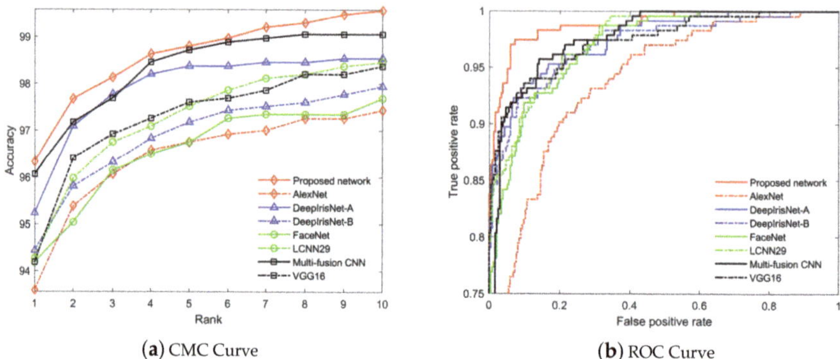

(a) CMC Curve   (b) ROC Curve

Figure 5. Performances of recognition and verification tasks on AR database. The figures are best viewed in colour.

Table 4. Performance evaluation of the verification task on the AR database, CASIA-iris distance database, UBIPr database, and Ethnic-ocular database. The highest accuracy is written in bold.

Networks	AR		CASIA-Iris		UBIPr		Ethnic-Ocular	
	EER (%)	*AUC*	*EER (%)*	*AUC*	*EER (%)*	*AUC*	*EER (%)*	*AUC*
AlexNet	14.53	0.9363	8.06 ± 5.3	0.9533	7.11 ± 2.9	0.9805	16.47 ± 1.6	0.9139
DeepIristNet-A	7.69	0.9751	7.51 ± 1.1	0.9674	5.07 ± 2.2	0.9877	8.79 ± 1.7	0.9689
DeepIristNet-B	8.12	0.9741	5.87 ± 1.5	0.9756	4.29 ± 0.9	0.9890	8.77 ± 1.1	0.9693
FaceNet	9.40	0.9692	6.10 ± 2.2	0.9738	5.46 ± 1.5	0.9870	11.67 ± 1.2	0.9489
LCNN29	9.39	0.9737	6.34 ± 1.6	0.9719	6.34 ± 2.1	0.9849	10.95 ± 1.6	0.9536
Multi-fusion CNN	7.69	0.9756	8.69 ± 1.1	0.9594	4.09 ± 2.1	0.9913	8.63 ± 1.3	0.9681
VGG16	7.69	0.9747	7.42 ± 1.7	0.9681	4.38 ± 1.3	0.9892	9.43 ± 2.5	0.9553
Proposed Network	**5.13**	**0.9882**	$\mathbf{4.35 \pm 0.5}$	**0.9860**	$\mathbf{3.41 \pm 1.8}$	**0.9938**	$\mathbf{6.63 \pm 1.5}$	**0.9818**

Evaluation on CASIA-Iris Distance Database

To evaluate whether our approach performs well on another standard database, we also tested its performance in a more subjective experiment with CASIA-iris distance database. This database consists of 142 subjects under a long-range subject–camera distance and indoor environment. The images were captured by a high-resolution camera so both dual-eye iris and periocular are included in the image region of interest. The further details of the database can be found in [29].

The experimental protocol for recognition was designed with the ratio of the gallery set to probe set as 50:50 and the division process was repeated three times. The experimental protocol for the verification was designed by randomly selecting 250 reference-query pairs as "same" and another 250 pairs as "not same". This selection process was repeated three times.

According to Table 3, the proposed network achieved the highest average accuracies for Rank-1 and Rank-5 recognitions with $96.62 \pm 1.3\%$ and $98.45 \pm 0.4\%$, respectively. Besides, FaceNet achieved the second-best performance with $96.09 \pm 2.1\%$ and $98.10 \pm 0.4\%$ for Rank-1 and Rank-5 recognition accuracies, respectively. We also present in Figure 6a the Rank-1 to Rank-10 recognition results. As can be seen, our network achieved the best results among the benchmark networks.

For the verification, the proposed network achieved the lowest EER accuracy as $4.35 \pm 0.5\%$ and AUC as 0.9860. Interestingly, DeepIrisNet-B attained second lowest performance with $5.87 \pm 1.5\%$ for EER and 0.9756 as AUC. Figure 6b illustrates the ROC curve, which demonstrates that our network obtained the best performance of AUC and the lowest EER. Both recognition and verification results indicate that the proposed network is capable of learning the features of the RGB image and OCLBCP decently for improving the performance of recognition and verification tasks.

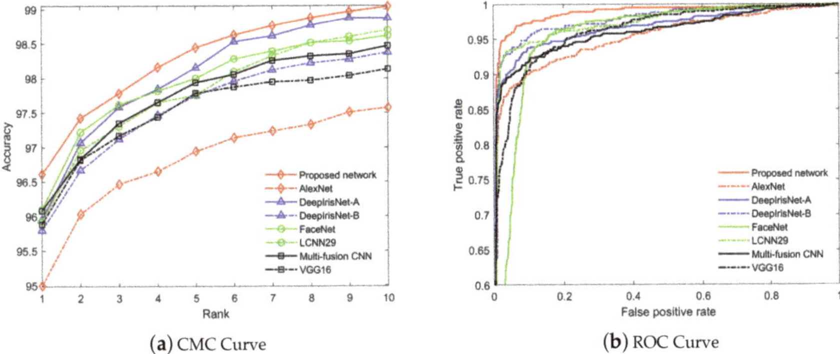

(a) CMC Curve   (b) ROC Curve

Figure 6. Performances of recognition and verification tasks on CASIA-iris distance database. The figures are the best to view in colour.

Evaluation on UBIPr Database

We also conducted another more challenge experiment with the UBIPr database to verify the robustness of the proposed network. This database consists of 342 subjects with varying subject–camera distances, levels of illumination, and poses [12]. This experiment evaluated the performance of all the networks with varying poses and subject–camera distances. Six images from each subject were randomly divided as a gallery set; the remaining images were used as a probe set. The division process was repeated three times. For the verification, we randomly selected 600 reference-query pairs as "same" and another 600 pairs as "not same". This selection process was also repeated three times.

Table 3 presents that our network achieved the highest average Rank-1 and Rank-5 recognition accuracies with 91.28 ± 1.2% and 98.59 ± 0.4%, respectively. The second best was achieved by multi-fusion CNN with 90.75 ± 1.0% and 97.44 ± 0.3% as Rank-1 and Rank-5 accuracies, respectively. Besides, Figure 7a also illustrates the CMC curve and shows that our network achieved the best performance of recognition for all ranks.

For the verification, Table 4 reveals that our network achieved the lowest EER with 3.41 ± 1.8% and AUC was 0.9938. This is concrete evidence to demonstrate that the proposed network can verify the unconstrained periocular robustly. Figure 7b shows that our network outperformed most of the benchmark networks and achieved the highest recall rate against all other approaches.

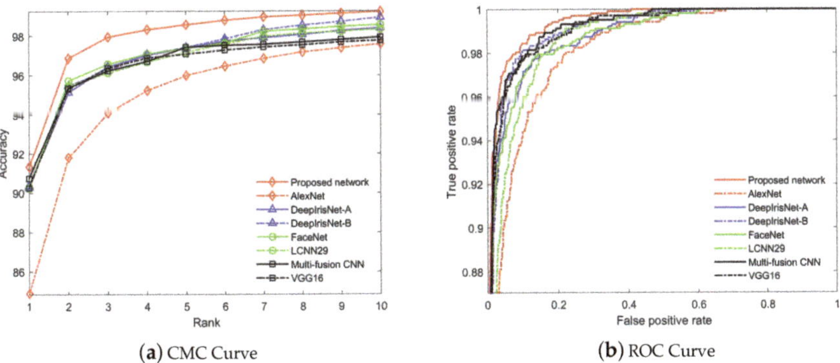

(a) CMC Curve   (b) ROC Curve

Figure 7. Performances of recognition and verification tasks on UBIPr distance database. The figures are the best to view in colour.

Evaluation on Ethnic-Ocular Database

We present the experimental results in Table 3 by following the recognition protocol mentioned in Section 4.3. To evaluate the performance of the proposed approach, we compared our results with seven benchmark approaches (see Table 3). For the results of recognition, our network achieved 84.79 ± 1.9% and 94.23 ± 1.3% as Rank-1 and Rank-5 accuracies, respectively. Figure 8a illustrates the CMC curve of the proposed network, showing that the proposed method outperformed other benchmark methods from Rank-1 to Rank-10 recognition accuracies. The results indicate that the late-fusion layers are capable of correlating the RGB image and OCLBCP descriptor.

Table 4 also shows that the proposed network achieved the lowest EER accuracy with 6.63 ± 1.5% for verification. Figure 8b illustrates the ROC curve, showing that our network outperformed all benchmark networks. The results prove that our approach can learn new features from the late-fusion layers in order to transfer knowledge between the networks to perform better performance of recognition.

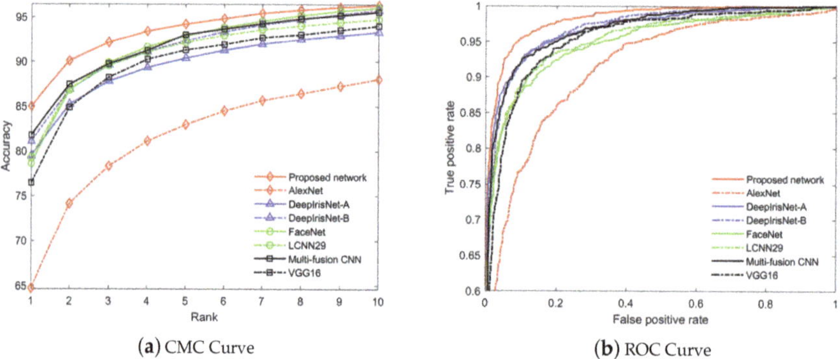

(**a**) CMC Curve　　　　(**b**) ROC Curve

Figure 8. Performances of recognition and verification tasks on Ethnic-ocular database. The figures are the best to view in colour.

5.2.3. Discussion

Through the experimental analysis and results, we observed that having access to the RGB image and OCLBCP descriptor can exploit the discriminatory features as inputs for a better periocular recognition. In addition, the proposed network utilises the colour-based texture information, which contributes to a more robust feature representation for the challenges in recognition and verification in the wild. This is because handcrafted texture descriptor can offer latent and complement information for complex data learning.

By evaluating across constrained environments, our results score higher accuracies consistently. Periocular recognition and verification in the wild bring more challenges as compared to the constrained environment. The experimental results prove that our network is able to perform better recognition due to its ability to learn new features from the proposed late-fusion layers. The effectiveness of fusion layers in the network supports our assumption firmly that multi-feature learning can work much better than just using RGB image in periocular recognition.

6. Conclusions

This paper proposed a dual-stream CNN, which accepts RGB ocular image and OCLBCP for periocular recognition in the wild. By aggregating the RGB image and OCLBCP features into two distinct late-fusion layers, these features offer robust and better recognition performance. We collected and shared a new Ethnic-ocular database, which consists of a large collection of periocular images in the wild based on different ethnic groups. Through extensive experiments by comparing against several competing networks on new Ethnic-ocular database and publicly available databases, the proposed network achieved better performance in both recognition and verification tasks.

In the near future, we plan to investigate different kinds of fusion stages and fusion layers in CNNs, which could improve the performance of multi-feature learning. Periocular recognition is futile for subjects with "wearing sunglasses". As a remedy, we shall incorporate the Generative Adversarial Model, which is useful to recover the periocular area in the face image.

Author Contributions: The authors have equivalent contributions.

Funding: This work was supported by the National Research Foundation of Korea (NRF) grant funded by the Korea government (MSIP) (NO. NRF-2019R1A2C1003306).

Conflicts of Interest: The authors declare no conflict of interest.

References

1. Jain, A.K.; Nandakumar, K.; Ross, A. 50 years of biometric research: Accomplishments, challenges, and opportunities. *Pattern Recog. Lett.* **2016**, *79*, 80–105. [CrossRef]
2. Klare, B.F.; Klein, B.; Taborsky, E.; Blanton, A.; Cheney, J.; Allen, K.; Grother, P.; Mah, A.; Jain, A.K. Pushing the frontiers of unconstrained face detection and recognition: IARPA Janus benchmark A. In Proceedings of the IEEE Conference on Computer Vision and Pattern Recognition (CVPR), Boston, MA, USA, 7–12 June 2015; pp. 1931–1939.
3. Klontz, J.C.; Jain, A.K. A case study of automated face recognition: The Boston Marathon bombings suspects. *Computer* **2013**, *46*, 91–94. [CrossRef]
4. Barroso, E.; Santos, G.; Cardoso, L.; Padole, C.; Proença, H. Periocular recognition: How much facial expressions affect performance? *Pattern Anal. Appl.* **2016**, *19*, 517–530. [CrossRef]
5. Park, U.; Jillela, R.R.; Ross, A.; Jain, A.K. Periocular biometrics in the visible spectrum: A feasibility study. In Proceedings of the International Conferences on Biometrics: Theory, Applications and Systems (BTAS), Washington, DC, USA, 28–30 September 2009; pp. 1–6.
6. Bharadwaj, S.; Bhatt, H.S.; Vatsa, M.; Singh, R. Periocular biometrics: When iris recognition fails. In Proceedings of the International Conferences on Biometrics: Theory, Applications and Systems (BTAS), Washington, DC, USA, 27–29 September 2010; pp. 1–6.
7. Park, U.; Jillela, R.R.; Ross, A.; Jain, A.K. Periocular biometrics in the visible spectrum. *IEEE Trans. Inf. Forensics Secur.* **2011**, *6*, 96–106. [CrossRef]
8. Raja, K.B.; Raghavendra, R.; Stokkenes, M.; Busch, C. Smartphone authentication system using periocular biometrics. In Proceedings of the International Conferences on Biometrics Special Interest Group (BIOSIG), Darmstadt, Germany, 10–12 September 2014; pp. 1–8.
9. Mokhayeri, F.; Granger, E.; Bilodeau, G. Synthetic face generation under various operational conditions in video surveillance. In Proceedings of the International Conferences on Image Processing (ICIP), Quebec City, QC, Canada, 27–30 September 2015; pp. 4052–4056.
10. The Korea Times. Available online: https://www.koreatimes.co.kr/www/nation/2019/01/371_262460.html (accessed on 12 February 2019).
11. Kitchen Decor. Available online: https://kitchendecor.club/files/now-beckham-hairstyle-david.html (accessed on 12 February 2019).
12. Padole, C.N.; Proença, H. Periocular recognition: Analysis of performance degradation factors. In Proceedings of the International Conferences on Biometrics (ICB), New Delhi, India, 29 March–1 April 2012; pp. 439–445.
13. Raja, K.B.; Raghavendra, R.; Stokkenes, M.; Busch, C. Collaborative representation of deep sparse filtered features for robust verification of smartphone periocular images. In Proceedings of the International Conferences on Image Processing (ICIP), Phoenix, AZ, USA, 25–28 September 2016; pp. 330–334.
14. Alonso-Fernandez, F.; Bigun J. Periocular recognition using retinotopic sampling and Gabor decomposition. In Proceedings of the European International Conferences on Vision (ECCV), Firenze, Italy, 7–13 October 2012; pp. 309–318.
15. Cao, Z.; Schmid, N.A. Fusion of operators for heterogeneous periocular recognition at varying ranges. *Pattern Recognit. Lett.* **2016**, *82*, 170–180. [CrossRef]

16. Mahalingam, G.; Ricanek, K. LBP-based periocular recognition on challenging face datasets. *EURASIP J. Image Video Process.* **2013**, *36*, 1–13. [CrossRef]
17. Tan, C.-W.; Kumar, A. Towards online iris and periocular recognition under relaxed imaging constraints. *IEEE Trans. Image Process.* **2013**, *22*, 3751–3765.
18. Nigam, I.; Vatsa, M.; Singh, R. Ocular biometrics: A survey of modalities and fusion approaches. *Inf. Fusion* **2015**, *26*, 1–35. [CrossRef]
19. Raghavendra, R.; Busch, C. Learning deeply coupled autoencoders for smartphone based robust periocular verification. In Proceedings of the International Conferences on Image Processing (ICIP), Phoenix, AZ, USA, 25–28 September 2016; pp. 325–329.
20. Cho, S.R.; Nam, G.P.; Shin, K.Y.; Nguyen D.T.; Pham, T.D.; Lee, E.C.; Park, K.R. Periocular-based biometrics robust to eye rotation based on polar coordinates. *Multimed. Tools Appl.* **2017**, *76*, 11177–11197. [CrossRef]
21. Krizhevsky, A.; Sutskever, I.; Geoffrey, H. Imagenet classification with deep convolutional neural networks. In Proceedings of the International Conferences on Neural Information Processing Systems (NIPS), Lake Tahoe, NV, USA, 3–6 December 2012; pp. 1097–1105.
22. Anwer, R.M.; Khan, F.S.; van de Weijer, J.; Molinier, M.; Laaksonen, J. Binary patterns encoded convolutional neural networks for texture recognition and remote sensing scene classification. *ISPRS J. Photogramm. Remote Sens.* **2018**, *138*, 74–85. [CrossRef]
23. Gangwar, A.; Joshi, A. DeepIrisNet: Deep iris representation with applications in iris recognition and cross-sensor iris recognition. In Proceedings of the International Conferences on Image Processing (ICIP), Phoenix, AZ, USA, 25–28 September 2016; pp. 2301–2305.
24. Proença, H.; Neves, J.C. Deep-PRWIS: Periocular recognition without the iris and sclera using deep learning frameworks. *IEEE Trans. Inf. Forensics Secur.* **2018**, *13*, 888–896. [CrossRef]
25. Zhao, Z.; Kumar, A. Improving periocular recognition by explicit attention to critical regions in deep neural network. *IEEE Trans. Inf. Forensics Secur.* **2018**, *13*, 2937–2952. [CrossRef]
26. Zhang, Q.; Li, H.; Sun, Z.; Tan, T. Deep feature fusion for iris and periocular biometrics on mobile devices. *IEEE Trans. Inf. Forensics Secur.* **2018**, *13*, 2897–2912. [CrossRef]
27. Soleymani, S.; Dabouei, A.; Kazemi, H.; Dawson, J.; Nasrabadi, N.M. Multi-level feature abstraction from convolutional neural networks for multimodal biometric identification. In Proceedings of the International Conferences on Pattern Recognition (ICPR), Beijing, China, 20–24 August 2018; pp. 3469–3476.
28. Levi, G.; Hassner, T. Emotion recognition in the wild via convolutional neural networks and mapped binary patterns. In Proceedings of the International Conferences on Multimodal Interaction, Seattle, WA, USA, 9–13 November 2015; pp. 503–510.
29. CASIA-Iris Distance Database. Available online: http://www.cbsr.ia.ac.cn/china/Iris%20Databases%20CH.asp (accessed on 12 December 2018).
30. Marsico, M.D.; Nappi, M.; Riccio, D.; Wechsler, H. Mobile iris challenge evaluation (MICHE)-I, biometric iris dataset and protocols. *Pattern Recognit. Lett.* **2015**, *57*, 17–23. [CrossRef]
31. Alonso-Fernandez, F.; Raja, K.B.; Raghavendra, R.; Busch, C.; Bigun, J.; Vera-Rodriguez, R.; Fierrez, J. Cross-sensor periocular biometrics: A comparative benchmark including smartphone authentication. *arXiv* **2019**, arXiv:1902.08123.
32. Rhee, S.C.; Woo, K.S.; Kwon, B. Biometric study of eyelid shape and dimensions of different races with references to beauty. *Aesthetic Plast. Surg.* **2012**, *36*, 1236–1245. [CrossRef] [PubMed]
33. Ethnic-Ocular Database. Available online: https://www.dropbox.com/sh/vgg709to25o01or/AAB4-20q0nXYmgDPTYdBejg0a?dl=0 (accessed on 29 January 2019).
34. Ojala, T.; Pietikäinen, M.; Mäenpää, T. Multiresolution gray-scale and rotation invariant texture classification with local binary patterns. *IEEE Trans. Pattern Anal. Mach. Intell.* **2002**, *24*, 971–987. [CrossRef]
35. Tan, X.; Triggs, B. Enhanced local texture feature sets for face recognition under difficult lighting conditions. *IEEE Trans. Image Process.* **2010**, *19*, 1635–1650.
36. Tiong, L.C.O. Multimodal Biometrics Recognition Using Multi-Layer Fusion Convolutional Neural Network with RGB and Texture Descriptor. Ph.D. Thesis, KAIST, Daejeon, Korea, 15 February 2019.
37. Delac, K.; Grgic, M.; Kos, T. Sub-image homomorphic filtering technique for improving facial identification under difficult illumination conditions. In Proceedings of the International Conferences on Systems, Signals and Image Processing, Budapest, Hungary, 21–23 September 2006; pp. 95–98.

38. Martinez W.L.; Martinez, A.R.; Solka, J. Chapter 3 Dimensionality reduction—Nonlinear methods. In *Exploratory Data Analysis with MATLAB*; Martinez W.L., Martinez, A.R., Solka, J., Eds.; CRC Press LLC: Boca Raton, FL, USA, 2005; pp. 61–68.

39. Feichtenhofer, C.; Pinz, A.; Zisserman, A. Convolutional two-stream network fusion for video action recognition. In Proceedings of the IEEE Conference on Computer Vision and Pattern Recognition (CVPR), Las Vegas, NV, USA, 26 June–1 July 2016; pp. 1933–1941.

40. BBC News. Available online: http://www.bbc.com/news (accessed on 10 October 2018).

41. CNN News. Available online: https://edition.cnn.com/ (accessed on 11 October 2018).

42. Naver News. Available online: http://news.naver.com/ (accessed on 11 October 2018).

43. Ng, H.W.; Winkler, S. A data-driven approach to cleaning large face datasets. In Proceedings of the International Conferences on Image Processing (ICIP), CNIT La Défense, Paris, France, 27–30 October 2014; pp. 343–347.

44. Matlab Object Detector. Available online: https://uk.mathworks.com/help/vision/ref/vision.cascadeobjec tdetector-system-object.html (accessed on 10 October 2018).

45. Štruc, V.; Pavešić, N. The complete Gabor-fisher classifier for robust face recognition. *EURASIP J. Adv. Signal Process.* **2010**, 1–26.

46. TensorFlow. Available online: https://tensorflow.org (accessed on 21 November 2018).

47. Schroff, F.; Kalenichenko, D.; Philbin, J. FaceNet: A unified embedding for face recognition and clustering. In Proceedings of the IEEE Conference on Computer Vision and Pattern Recognition (CVPR), Boston, MA, USA, 7–12 June 2015; pp. 815–823.

48. Wu, X.; He, R.; Sun, Z.; Tan, T. A light CNN for deep face representation with noisy labels. *IEEE Trans. Inf. Forensics Secur.* **2018**, *13*, 2884–2896. [CrossRef]

49. Parkhi, O.M.; Vedaldi, A.; Zisserman, A. Deep face recognition. In Proceedings of the British Machine Vision Conference, Swansea, UK, 7–10 September 2015; pp. 1–12.

50. Hernandez-Diaz, K.; Alonso-Fernandez, F.; Bigun, J. Periocular recognition using CNN features off-the-shelf. *arXiv* **2018**, arXiv:1809.06157.

51. Martínez A.; Benavente, R. *The AR Face Database*; CVC Technical Report #24; Robot Vision Lab; Purdue University: Barcelona, Spain, 1998.

 *applied sciences*

Article

Hierarchical Feature Aggregation from Body Parts for Misalignment Robust Person Re-Identification [†]

Yuting Liu [1,2], Hongyu Yang [1,2] and Qijun Zhao [1,2,*]

[1]   College of Computer Science, Sichuan University, Chengdu 610065, China; yuting.liu@stu.scu.edu.cn (Y.L.); yanghongyu@scu.edu.cn (H.Y.)
[2]   National Key Laboratory of Fundamental Science on Synthetic Vision, Sichuan University, Chengdu 610065, China
*   Correspondence: qjzhao@scu.edu.cn; Tel.: +86-028-85417865
†   This paper is an extended version of our paper published in The 2018 IEEE 4th International Conference on Identity, Security, and Behavior Analysis (ISBA).

Received: 8 March 2019; Accepted: 27 May 2019; Published: 31 May 2019

Abstract: In this work, we focus on the misalignment problem in person re-identification. Human body parts commonly contain discriminative local representations relevant with identity recognition. However, the representations are easily affected by misalignment that is due to varying poses or poorly detected bounding boxes. We thus present a two-branch Deep Joint Learning (DJL) network, where the local branch generates misalignment robust representations by pooling the features around the body parts, while the global branch generates representations from a holistic view. A Hierarchical Feature Aggregation mechanism is proposed to aggregate different levels of visual patterns within body part regions. Instead of aggregating each pooled body part features from multi-layers with equal weight, we assign each with the learned optimal weight. This strategy also mitigates the scale differences among multi-layers. By optimizing the global and local features jointly, the DJL network further enhances the discriminative capability of the learned hybrid feature. Experimental results on Market-1501 and CUHK03 datasets show that our method could effectively handle the misalignment induced intra-class variations and yield competitive accuracy particularly on poorly aligned pedestrian images.

Keywords: person re-identification; misalignment; hierarchical feature aggregation

---

1. Introduction

Typical person re-identification (re-ID) systems [1–3] can be broken down into three modules, i.e., person detection, person tracking, and person retrieval. It is generally believed that the first two modules are independent computer vision tasks, thus most re-ID methods focus on the last module, i.e., person retrieval. In this paper, if not specified, person re-ID refers to the person retrieval module. Defined as a classical image retrieval problem, person re-ID is considered as a process of matching identity classes between person-of-interest (query) and detected objects (large galleries) across cameras, which is a fundamental task in several fields such as surveillance, robotics, multimedia and forensics. It has been an area of intense research in the past few years.

Despite years of great efforts, person re-ID remains a challenging task due to the dramatic appearance variations in illumination, human pose, occlusion, and background. The varying poses or poorly detected bounding boxes often lead to misalignment of detected pedestrians (e.g., excessive background and missing or mis-aligned body parts), which is a critical challenge to robust person re-ID systems. The useless background noise and information loss due to misalignment can significantly compromise the feature learning and matching process. Figure 1 shows examples of mis-aligned pedestrian images.

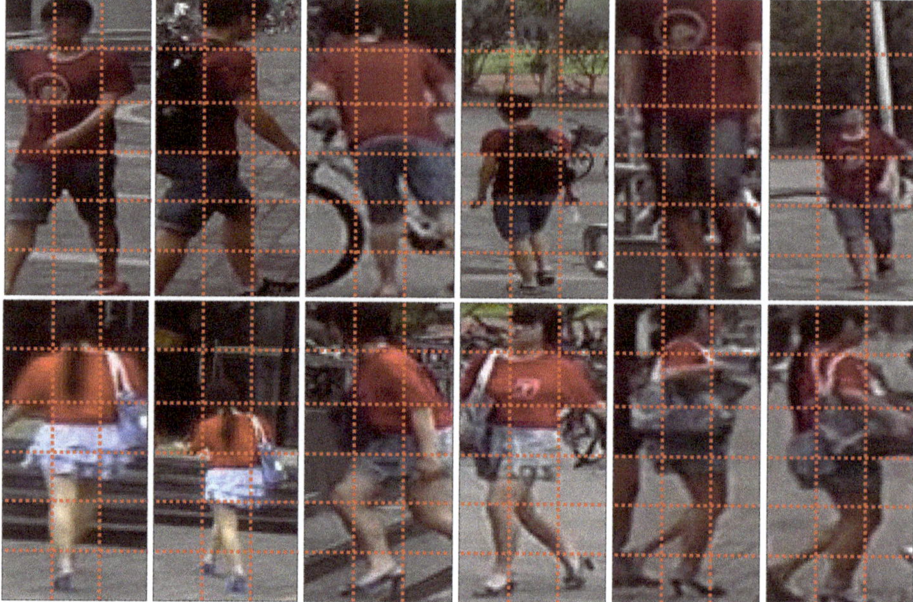

Figure 1. Examples of mis-aligned pedestrian images in Market-1501 dataset caused by pose variations and detection errors. The corresponding image patches of same identity are semantically unmatched (e.g., human head to background).

To handle this problem, early works [4–8] extract features from predefined image patches such as grid cell and horizontal stripes to construct the globally aligned representations for person re-ID. These methods subjectively suppose that every person appears in a similar pose within a tightly surrounded bounding box, ignoring the complex realistic conditions. Thus, they fail to perform well on more difficult databases [5,9]. More reasonable body part partition fashion [10–13] has then been exploited to generate finely aligned representations. With the development of pose estimation techniques [14–18], the above mentioned works have been re-studied. The adapted methods either intuitively perform affine transformation in order to get standard pose-aligned images (PoseBox) [19] or implicitly learn the proper transformation parameters and generate modified pose images with the help of impactful spatial transformer network [20]. However, highly-accurate pose estimation was required to prevent abnormal pose-normalized pedestrian images. To mitigate the problems, we proposed in [21] to apply alignment on feature level by pooling the features around the body parts. Alignment on feature level can not only avoid unnecessary geometric deformation in image but also make full use of the context-aware information encoded in middle convolution layers that can compensate detection errors. Meanwhile, the pooling operation also favors translation and rotation. All these factors make our method more robust to pose estimation errors compared to previous image-level-alignment-based methods. Recent methods [22,23] share similar insights with us in implementing feature level alignment.

Hierarchical-based learning methods are widely used in many tasks. The methods in [24,25] use the hierarchical Hidden Markov Model (HMM) to estimate and synthesize the motion of fingers or full-body while the method in [26] proposes a Bayesian hierarchical model to learn and recognize natural scene categories. These works adopt hierarchies of models to describe the intermediate states or themes of complex motions and scenes. The method in [27] takes advantage of Convolutional Neural Networks to learn hierarchies of features for Scene Labeling. Such hierarchies of features assemble pixel inputs into elements from low-level details to high-level semantic concepts and form good internal representations that are helpful for various visual perception tasks. Similar to these hierarchical-based learning methods, we propose to aggregate features from body parts with different levels of semantics.

Specifically, we construct a deep joint learning (DJL) network to learn misalignment robust feature representations from body parts for person re-ID. We propose to locally align the human bodies based on their landmarks, and pool the features around the body parts on feature maps rather than on original images. This way, our method can effectively handle the misalignment induced intra-class variations even though semantically corresponding body parts are not well aligned on the original images or the detected landmarks deviate from their true positions. As features from multiple layers abstract different level visual patterns of the same pedestrian image, we adopt a Hierarchical Feature Aggregation mechanism to enrich the feature representations for a pedestrian image by aggregating body part features with different levels of semantics. Besides, a Region Re-weighting strategy is applied to learn the importance weight of each body part as well as to mitigate the scale differences [28] among multiple convolution layers. Evaluation experiments on two public benchmark databases prove the effectiveness of our proposed method compared with existing state-of-the-art methods.

This paper is an extended version of our previous conference paper [21] with the following incremental contributions: (i) We further explore the identification performance of multiple layers for re-ID tasks from low-level to semantic-level and propose a Hierarchical Feature Aggregation (HFA) mechanism to take full advantage of different levels of features. (ii) We adopt a Region Re-Weighting (RRW) strategy to learn optimal weight of each body part as well as to mitigate the scale difference of multiple layers. (iii) We get further performance boost, obtaining 88.39% and 85.90% on Market-1501 and CUHK03 datasets. The rest of this paper is organized as follows. Section 2 reviews related work on deep learning based person re-ID methods, global and local features for re-ID and the pedestrian misalignment problem. Section 3 introduces in detail our proposed method, and Section 4 then reports our evaluation experiments. Finally, Section 5 concludes the paper.

2. Related Works

2.1. Deep Learning for Person Re-ID

Early methods solve the person re-ID problem mainly from two aspects, feature extraction and metric learning. Typical features used for person re-ID include color histograms [29–31], color names [9,32], local binary patterns (LBP) [30,33], gabor features [34] and scale invariant local ternary patterns (SILTP) [29,35]. Some researchers apply metric learning methods to seek for effective distance metrics for computing similarity between detected persons [6,29,30,36,37]. The emerging deep learning (DL) technology provides effective approaches for learning both feature representations and distance metrics. These DL-based person re-ID methods are dominating the re-ID community. Recently, attributes [38], transfer learning [39,40], re-ranking [41], mutual learning [42] and different levels of supervision [40,43,44] have also been studied.

2.2. Global and Local Features

Human visual system leverages both global (contextual) and local (saliency) information concurrently [45,46]. This observation supports that global and local features have correlated complementary information in different contexts. Most deep learning methods for person re-ID [47–49] follow the classical image classification mode [50], which favors intrinsically in learning global feature representations. However, these methods ignore the importance of local information. Some methods [5,6,51] utilize local information by decomposing images into horizontal stripes and learning effective local features in each patch. These local stripes in essence globally align the images of detected persons, and are thus still sensitive to misalignment of human bodies in different images.

2.3. Pedestrian Misalignment

Pedestrian misalignment caused by detectors or pose variations is a main challenge for feature matching across images. Most previous works partition pedestrian bounding box into grids or horizontal stripes to handle misaligned pedestrian images [5,9,29,51]. Nevertheless, these methods only work under the assumption of slight vertical misalignment but not for severe misalignment. Some methods [11,12] use the pictorial structure to construct well aligned pedestrian images. However, they only use local body parts while ignoring the global context, which results in suboptimal feature learning.

The recent PIE method [19] proposes a PoseBox fusion (PBF) CNN architecture that takes the original image, the PoseBox, and the pose estimation confidence as input to achieve a globally optimized tradeoff between the global and local feature representations. The PoseBox structure is similar to the pictorial structure [11,12] in enabling well-aligned pedestrian matching. The PDC method [52] first crops part regions and then transforms each part by a Pose Transformation Network (PTN) to automatically learn transformations such as translation, rotation and scale. The PTN outputs the final transformed part images and hence learns partly aligned representations. These methods all attempt to solve the misalignment problem at image level, with few exceptions that directly handle learned features. For example, Zhao et al. [22] followed human body structure to iteratively decompose and fuse features from different semantic region; Li et al. [53] exploited attention models to implicitly learn effective part representations without guidance of body part locations; and Wang et al. [23] encoded human poses in feature maps through bilinear pooling which aggregates appearance and part maps to compute part-aligned representations. Our method differs from them in the following three aspects.

- Our work constructs the "PoseBox" at feature level instead of the image level. We find that the image level PoseBox would lose their discriminative property due to pose estimation errors. In addition, the affine transformation employed by the PIE method may result in unwanted geometric distortion and deteriorating the intrinsic structure of human body. Figure 2 shows some examples of good and bad PoseBox constructed by PIE. Instead of image level affine transformation, we directly pool local body part features on feature maps, and organize them in a fixed order for feature level alignment (concatenate each body part features along channel dimensions). Meanwhile, we propose to model the spatial dependencies between those local body parts through cross-channel convolution computation. Thanks to the capability of CNN feature maps in context-aware semantic information, we suppose that the feature level alignment would be more robust to pose estimation errors.
- We apply max pooling inside local body part regions so as to find the most salient local details. HFA mechanism and RRW strategy are proposed to make the best of multi-level body part features. Our joint optimization of both global and local features further enhances the discriminative capability of learned feature representations for person re-ID.
- By avoiding complicated affine transformation, we can obtain pose aligned features in a simple and efficient way. Moreover, our method can be easily integrated with different person re-ID networks, and effectively enhance their identification accuracy.

Good PoseBox   Bad PoseBox

Figure 2. Examples of good and bad PoseBox constructed by PIE: (Top row) original bounding boxes with detection errors/occlusions; and (Bottom Row) corresponding PoseBoxes.

3. Proposed Method

As shown in Figure 3, our proposed DJL network consists of three main components: the global branch base network, the local branch sub-network, and the multi-loss module. First, the input human body image is segmented into a number of body part regions (Section 3.1). The global branch base network extracts global representations from the original image (Section 3.2). The local branch sub-network then constructs misalignment robust local features according to the segmented body part regions and middle layer feature maps generated by global branch. With three Softmax losses, the multi-loss module optimizes global and local features jointly (Section 3.3). In this section, we introduce first the process of body part segmentation, then the global branch base network, and finally the proposed DJL network.

3.1. Body Part Segmentation

We first segment human body parts through deep pose estimation method CPM [16]. CPM outputs the coordinates of a set of 14 body parts and the corresponding confidence scores, i.e., head, neck, left and right shoulders, left and right elbows, left and right wrists, left and right hips, left and right knees, and left and right ankles. Several previous works [4,6,19] show that the torso and legs make the largest contributions and that integration of the head may introduce noise due to the unstable head detection. In this paper, we thus choose ten of the body parts as region boxes for local feature extraction, including left and right shoulders, left and right elbows, left and right hips, left and right knees, and left and right ankles. Figure 4 shows an illustration of the chosen body parts.

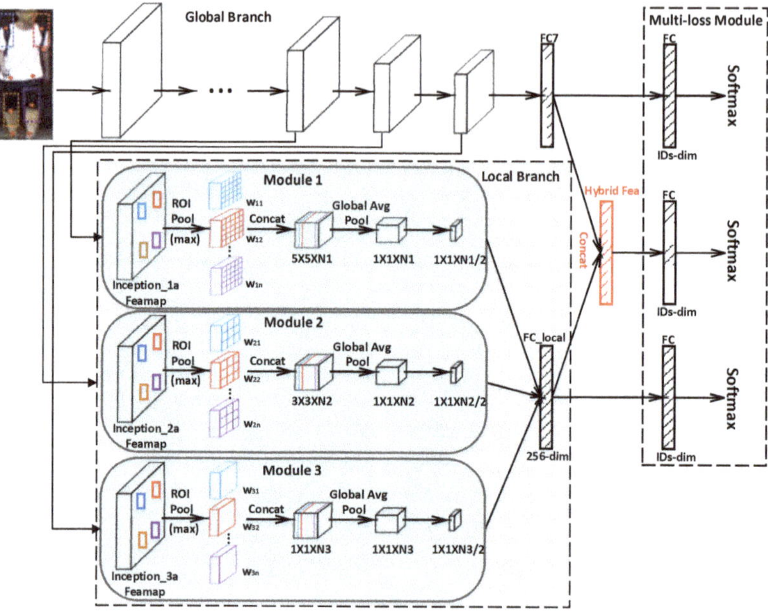

Figure 3. The proposed DJL network with InceptionNet as the base network. The input to DJL includes a pedestrian image and the human body landmarks. We segment ten body part regions according to the landmarks (Section 3.1). A local branch sub-net (Section 3.3) is specially designed in this paper to pool and aggregate multi-level body part representations from the feature maps generated by the global branch base network (Section 3.2). The multi-loss module then optimizes the global and local features jointly.

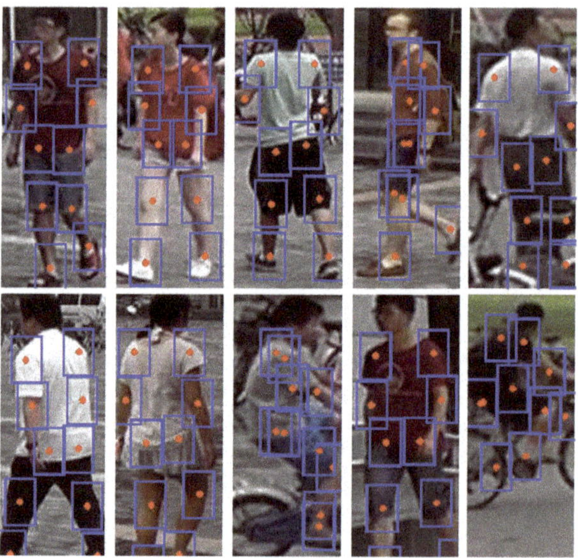

Figure 4. Examples of the segmented ten body parts used in our DJL network.

3.2. Base Networks

We utilize the widely used AlexNet [50], Residual-50 [54] and InceptionNet [48] as the base networks in our proposed method. We refer readers to respective papers for detail network descriptions. We adopt Identification model in this paper and edit the last FC layer to have the same number of neurons as the number of distinct IDs in the training set. As described in [49], the identification model yields superior performance to verification model for the reason that the former makes full use of the re-ID labels while the latter takes limited relationships into consideration, i.e., whether two input images belong to the same person.

3.3. The Deep Joint Learning Network

Two pairs of feature maps extracted by the base network are provided in Figure 5 to give insights into the model design. We observe that high responses are mostly concentrated on the local body parts and they often present attribute-relevant information (e.g., clothing type, color, accessories, etc.), and, when reasonably exploited, those body part features may be helpful to distinguish individuals. Motivated by this, we integrate body part features from low level to semantic level, resulting in misalignment-robust representations for matching.

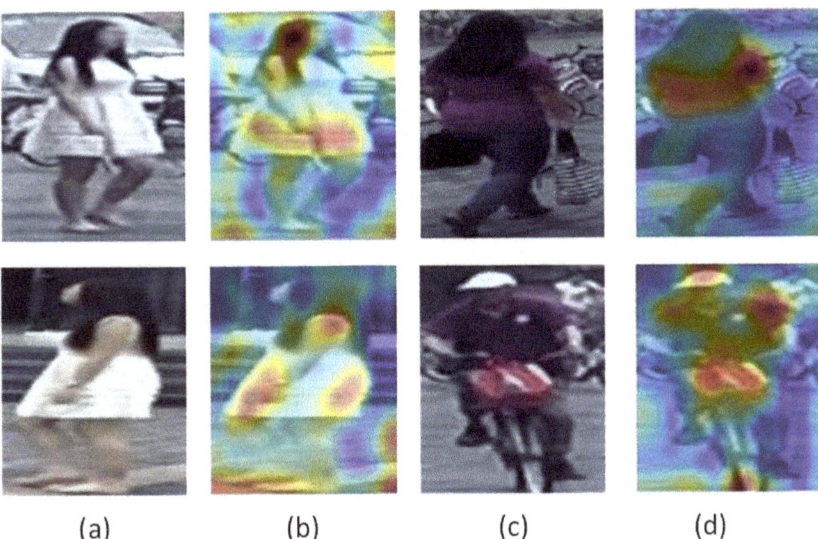

(a)   (b)   (c)   (d)

Figure 5. Two examples to show the effectiveness of the local body part features: (**a**) two images of the same person; (**b**) corresponding feature maps of (**a**); (**c**) two different persons; and (**d**) corresponding feature maps of (**c**).

3.3.1. Network Structure

The input to the DJL network contains a pedestrian image and its ten body parts. Each body part is represented by its position. The global branch of DJL is composed by the base networks, as previously described in Section 3.2. Its objective is to extract global features of pedestrians.

The local branch aims to learn misalignment-robust feature representations from low level to semantic level. It consists of several similar modules, each of which takes as input the output feature maps of a specific middle convolution layer from base network and generates local descriptors of that level. As shown in Figure 3, for a single module, RoI pooling layer [55] is adopted to learn sparse representations of each local body part. The RoI pooling layer uses max pooling to convert the features inside any region of interest window of size $h \times w$ into a small feature map with a fixed spatial extent of $H \times W$, where H and W are layer hyper-parameters. It works by dividing the $h \times w$ RoI window into an $H \times W$ grid of sub-windows of approximate size $h/H \times w/W$ and then max-pooling the values in each sub-window into the corresponding output grid cell. Pooling is applied independently to each sub-window as in standard max pooling. Figure 6 shows an illustration of the RoI pooling operation. Given the middle-layer feature maps and coordinates of body part regions, we perform RoI pooling inside each region to select the most discriminative features. Then, those local body part features are concatenated along channel dimensions in a fixed order, and a global average pooling layer and a convolution layer follow to get the dimension-reduced local descriptors.

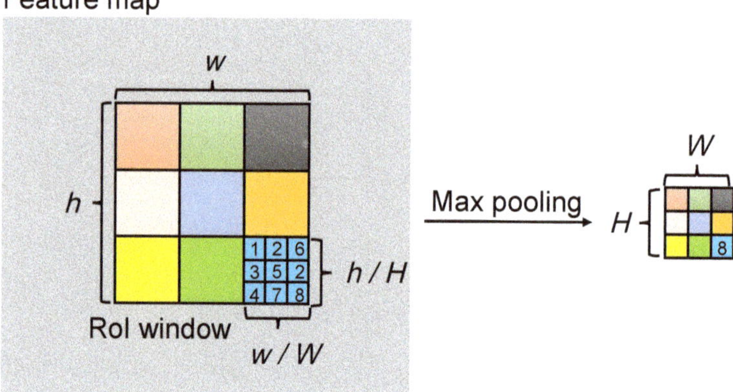

Figure 6. Illustration of the RoI pooling operation.

The multi-loss module consists of three full connection (FC) layers before Softmax loss computation. The sum of the three Softmax losses is used for loss computation. Dimensions of these FC layers are the number of distinct IDs in the training set. In Figure 3, as denoted by the red FC layer, the learned hybrid feature representation for final matching is defined as the concatenated FC7 activations (FC_ local + FC7). The motivation of our multiple loss module is to integrate the discriminative power of global and local features.

3.3.2. Hierarchical Feature Aggregation

Inspired by neuroscience, reasoning across multiple levels of hierarchies has been proven beneficial in some computer vision problems [24,26,27,56,57]. On the one hand, it has been demonstrated that details can be well captured by low-level features from shallow convolution layer rather than by high-level features. On the other hand, high-level features from deeper convolution layer get complementary semantic information as neurons in these layers have lager receptive fields. We thus adopt a Hierarchical Feature Aggregation mechanism to pool features from shallow to deep convolution layers of base network and aggregate the learned local descriptors from detail to semantic. For example, as shown in Figure 3, we perform RoI pooling at Inception_ 3a, Inception_ 2a, Inception_ 1a for InceptionNet with different pooling scales ($H \times W$). The output spatial extents are, respectively, 1×1, 3×3, and 5×5. Here, we adopt coarse spatial division 1×1 in deep layers and fine spatial division 5×5 in shallow layers to capture fine-grained features corresponding to local salient details. Finally, the pose aligned body part features from each module are concatenated to form the final multi-level local descriptors (denoted by FC_ local). We also adopt a Region Re-Weighting strategy (see Section 3.3.3) to make the Hierarchical Feature Aggregation mechanism more effective.

3.3.3. Region Re-Weighting

For the reason that pose estimation method (CPM) may induce ill-positioned body parts and different body part regions may have different importance for person re-identification, we intend to learn the importance weight of each body part region during training procedure. We call this strategy Region Re-Weighting (RRW). RRW performs an element-wise product between body part region features and the corresponding region weights. Formally, for each pooled body part feature of d-dimension $X_i = (x_{i1}, \cdots, x_{id})$, we introduce a weight parameter w_i, which scales per region features as $Y_i = (w_i \cdot x_{i1}, \cdots, w_i \cdot x_{id})$. During training, letting L be the loss we want to minimize, we use back propagation and chain rule to compute derivatives with respect to the weight factor w_i and body part features X_i.

$$\frac{\partial L}{\partial X_i} = \frac{\partial L}{\partial Y_i} \cdot w_i \qquad \frac{\partial L}{\partial w_i} = \sum_{j=1}^{d} \frac{\partial L}{\partial y_{ij}} \cdot x_{ij} \qquad (1 \leq i \leq 10) \tag{1}$$

As mentioned in [28], scales and norms of feature vectors from multiple layers may be quite different, and directly concatenating multi-level features may leads to poor performance as the "larger" features dominate the "smaller" ones. We find that combining RRW with HFA makes the training more stable and enables further performance improvements.

4. Experiments

4.1. Datasets and Protocol

4.1.1. Datasets

This study used CUHK03 [5] and Market-1501 [9] datasets for evaluation. The Market-1501 dataset is featured by 1501 IDs (750 for training and 751 for testing) with 32,668 cropped pedestrian bounding boxes. It contains 3368 query images and 19,732 gallery images (including 2793 distractors). For each query, we aimed to retrieve the ground-truth images from the 19,732 candidate images. This dataset is one of the largest benchmark datasets for person re-identification. Pictures were captured by six cameras: five high-resolution cameras and one low-resolution camera. The CUHK03 dataset contains 13,164 cropped pedestrian bounding boxes of 1360 identities (1160 for training, 100 for validation and 100 for testing) captured by six cameras. Each identity appears in two disjoint camera views (i.e., 4.8 images in each view on average). The bounding boxes of the pedestrians used in this study were generated by the DPM detector [58] instead of human annotated. This was to make the evaluation results more practical as in real-world automatic person re-ID systems.

4.1.2. Protocol

Cumulative Matching Characteristic (CMC) curve and mean average precision (mAP) are commonly used metrics for evaluating person re-ID methods. The CMC curve reflects retrieval precision, while the mAP reflects the recall. On CUHK03, we followed Li et al. [5] to repeat 20 times of random 1160/100 training/test splits and report the results under the single-shot evaluation setting. On Market-1501, the standard training/test split (750/751) was used.

4.2. Implementation Details

This work was implemented using Caffe [59], an open source deep learning framework. Original images were resized to 256×256 (then randomly cropped to 227×227 for AlexNet and 224×224 for Residual-50). As for InceptionNet, original images were resized to 160×64 (then randomly cropped to 144×56). All input images were mirrored randomly for data augmentation. Both AlextNet and Residual-50 were pre-trained on ImageNet dataset [60], while InceptionNet was directly trained from scratch (refer to [48]).

4.2.1. Training Base Networks

We adopted the mini-batch stochastic gradient descent (SGD) algorithm to update the network parameters. The batch size was set to 64 for AlexNet, 16 for Residual-50 and 100 for InceptionNet. The maximum number of training epochs was set to 50, 62, and 232 for AlexNet, Residual-50, and InceptionNet, respectively. AlexNet was trained with an initial learning rate of 0.001 and then reduced by 10 every 20 epochs. Residual-50 was trained with learning rate initialized at 0.001 and reduced by 10 every 25 epochs. For InceptionNet, the initial learning rate was set to 0.1 and was decreased by 4% for every four epochs until it reached 0.0005. The learning rate was then fixed at this value for a few more epochs until convergence.

4.2.2. Training DJL Network

Once the base network was pre-trained, we fine-tuned our Deep Joint Learning network. During training, the coordinates of body parts were transformed along with random image cropping and mirror operation. We set the position of invisible parts as zero. We empirically set the w/h of each body part region as 24/16 for InceptionNet (32/32 for AlexNet and Residual-50). When a body part was invisible, the features corresponding to its region were set to zero. The learning rate policy was changed to decay polynomially from 0.01 with the power parameter set to 0.5 and the whole network was trained for only around 20 epochs.

4.2.3. Testing

Given a pedestrian image of fixed size (227×227 for AlexNet, 224×224 for Residual-50, and 144×56 for InceptionNet), we extracted as features the FC7 activations for AlexNet, Pool5 activations for Residual-50, and FC7 activations for InceptionNet. We measured the similarity between two pedestrian images by the Euclidean distance between the L2-normalized features of them.

4.3. Performance Evaluation

We defined a simple version DJL network (DJL-S) which only contained one module in its local branch and compared it with the complete DJL network (DJL-HFS) with Hierarchical Feature Aggregation mechanism and Region Re-Weighting strategy. We adopted DJL-S structure with different base networks to validate the generalization ability of the proposed method and compared with the PIE method for the sake of fairness. We choose Conv4, Res4a and Inception_3a feature maps to generate the local features for AlexNet, Residual-50 and InceptionNet, respectively. Here, the output spatial extent of the RoI pooling layer was 1×1. To show the effectiveness of the Hierarchical Feature Aggregation as well as Region Re-Weighting strategy, further experiments were designed for InceptionNet based implementation with DJL-HFA structure.

4.3.1. Improvement over Base Networks

We first evaluated the proposed DJL-S network using various base networks on Market-1501 and CUHK03 benchmarks. The overall results are shown in Tables 1 and 2. The improvements over both AlexNet and Residual-50 base networks were significant. When using AlexNet, Rank-1 accuracy on Market-1501 rose from 57.75% to 67.64% and mAP rose from 33.80% to 43.60%. On CUHK03 dataset, Rank-1 accuracy rose by +18.92% for AlexNet. When using Residual-50, Rank-1 accuracy on CUHK03 arrived at 80.83%. On Market1501, consistent improvement could also be observed. Best performance appeared using InceptionNet [48], which obtained Rank-1 accuracy of 85.12% on Market-1501 and 84.25% on CUHK03. These results prove the effectiveness of our DJL-S network.

4.3.2. Comparison with The PIE Method

Our method shares a similar nature with the recent PIE [19] method, which learns pose invariant embedding from both well aligned PoseBox and original image. We compared our method with it under the same experimental settings. Rank-1 accuracy improvement over base networks was used as the measurement criteria here. According to the results in Table 3, our observation was two-fold.

Table 1. Comparison with the three base networks, AlexNet, Residual-50 and InceptionNet on Market-1501 (by adopting the proposed DJL-S structure) in terms of identification accuracy (%) and mAP (%).

Method	Market-1501				
	Rank-1	Rank-5	Rank-10	Rank-20	mAP
AlexNet	57.75	77.52	84.47	89.46	33.80
Residual-50	72.42	86.49	91.03	94.42	48.01
InceptionNet	79.66	91.51	94.54	96.50	56.59
Proposed (AlexNet)	67.64	84.80	89.88	93.53	43.60
Proposed (Residual-50)	78.86	90.38	93.91	96.35	55.49
Proposed (InceptionNet)	**85.12**	**93.91**	**95.69**	**97.51**	**64.82**

Table 2. Comparison with the three base networks, AlexNet, Residual-50 and InceptionNet on CUHK03 (by adopting the proposed DJL-S structure) in terms of identification accuracy (%).

Method	CUHK03			
	Rank-1	Rank-5	Rank-10	Rank-20
AlexNet	53.03	79.53	87.82	94.21
Residual-50	61.79	85.46	92.31	97.86
InceptionNet	80.85	95.90	98.17	99.48
Proposed (AlexNet)	71.95	90.30	94.91	98.16
Proposed (Residual-50)	80.83	95.92	98.66	99.54
Proposed (InceptionNet)	**84.25**	**97.40**	**98.86**	**99.67**

Table 3. Rank-1 accuracy improvement (%) over base networks compared with the PIE method.

Base Network	Market-1501		CUHK03	
	DJL-S	PIE	DJL-S	PIE
AlexNet	**+9.89**	+9.12	**+18.92**	+2.65
Residual-50	**+6.44**	+5.66	**+19.04**	+5.50

First, for both base networks, DJL-S achieved better accuracy than PIE on both databases. This validated the superiority of our proposed local body part features as we did alignment at feature level instead of image level. As for PIE, image level alignment by affine transformation performed worse due to pose estimation errors. The higher accuracy achieved by our proposed method might be owing to two factors. For one thing, we pool body part features on the feature maps that are generated by the middle convolution layers in the base network. These layers have larger receptive fields and thus capture more context-aware information that can compensate misalignment errors of detected persons. For another, discriminative detail information can be learned through max-pooling operation inside local body part regions, which should be helpful to identify individuals with slight difference.

Second, we found that our method obtained significant improvement on CUHK03. We speculate that the higher image resolution in CUHK03 benefited the learned features. We discuss this in detail in Section 4.3.4.

4.3.3. Comparison with More State-of-The-Arts

We compared our DJL with the current state-of-the-art DL-based methods. For ease of comparison, those methods are summarised into two categories: Pose-irrelevant DL-based methods and Pose-relevant DL-based methods. Their results on Market-1501 and CUHK03 are shown in Tables 4 and 5. The proposed DJL-S structure achieved comparable Rank-1 accuracy among the methods, i.e., 85.12% and 84.25% on Market-1501 and CUHK03, respectively. When adopting DJL-HFS structure and combining other re-ranking method (RK) [41], the performance was further boosted, reaching 88.39% on Market-1501. Furthermore, our Deep Joint Learning pipeline can be easily integrated with other state-of-the-art person re-ID networks.

Table 4. Comparison with state-of-the-arts on Market-1501. Rank-1 accuracy (%) and mAP (%) are shown. The best result is marked in bold while the second best in gray.

Methods	Rank-1	mAP
Pose-irrelevant DL-based Methods		
APR [38]	84.29	64.67
DLCE [49]	79.51	59.87
DML [42]	87.73	68.83
Gate-SCNN [7]	65.88	39.55
JLML [51]	85.10	65.50
X-Corr [61]	-	-
Ours		
DJL-S	85.12	64.82
DJL-HFA	85.99	65.65
DJL-HFA(RK)	88.39	**79.97**
Pose-relevant DL-based Methods		
DLPA [53]	81.0	63.4
MSCAN [62]	80.31	57.53
PABP [23]	**88.8**	74.5
PDC [52]	84.14	63.41
PIE [19]	78.65	53.87
PIE + KISSME [19]	79.33	55.95
Spindle [22]	76.90	-

Table 5. Comparison with state-of-the-arts on CUHK03. Rank-1 accuracy (%) is shown. The best result is marked in bold while the second best in gray.

Methods	Rank-1
Pose-irrelevant DL-based Methods	
APR [38]	-
DLCE [49]	83.4
DML [42]	-
Gate-SCNN [7]	68.10
JLML [51]	80.60
X-Corr [61]	72.04
Ours	
DJL-S	84.25
DJL-HFA	85.90
DJL-HFA(RK)	85.12
Pose-relevant DL-based Methods	
DLPA [53]	81.6
MSCAN [62]	67.99
PABP [23]	**88.0**
PDC [52]	78.29
PIE [19]	62.40
PIE + KISSME [19]	67.10
Spindle [22]	-

4.3.4. Further Analysis and Discussion

- **Body part segmentation**

  To evaluate the impact of body part segmentation errors on our method, we randomly disturbed the position of each body part during training. Here, we adopted two settings: small disturbance (Disturb-small) and violent disturbance (Disturb-violent). We translated the coordinates of each body part up to 6% of input image size for small disturbance and 30% for violent disturbance. Tables 6 and 7 show the results of DJL-S on Market-1501 and CUHK03, respectively. Generally, accuracy changed a little under slight disturbances (from 67.64% to 68.82% for AlexNet on Market-1501) while varied dramatically under large disturbances (still better than base networks). This demonstrates that our proposed method can effectively cope with human body misalignment.

- **Low resolution**

  We evaluated the impact of image resolution on our method. Experiments were conducted on CUHK03. We down-sampled all images in CUHK03 to half of their original size and used those low resolution images for training and testing. The results in Table 7 show that low image resolution degrades the performance of DJL-S.

- **RoI pooling effects at different layers**

  An important part of our method is to apply the RoI pooling operation to different middle layers. In Tables 8 and 9, we systematically explore the identification performance of different middle convolution by performing RoI pooling on each of them. We experimented with various network structures (AlexNet, Residual-50 and InceptionNet) and found that pooling at relative deeper layer obtains better performance improvements over the base networks. This observation shows that deeper, semantic CNN features contribute more to person re-ID task.

- **Hierarchical Feature Aggregation and Region Re-Weighting**

  We evaluated the effects of Hierarchical feature aggregation and Region Re-Weighting using the base Inception network with different variants of DJL: DJL-S, DJL-S + RRW, DJL-HFA(w/o RRW), and DJL-HFA. DJL-S denotes pooling body part features from a single convolution layer (Inception_3a). DJL-S + RRW further combines RRW strategy with DJL-S. DJL-HFA (w/o RRW) means DJL-HFA without applying RRW strategy, and DJL-HFA is the full version of our proposed method. As depicted in Tables 10 and 11, the DJL-S + RRW achieves performance gain in Rank-1 accuracy compared with the DJL-S network on both Market-1501 and CUHK03 datasets. When adopting DJL-HFA(w/o RRW), the Rank-1 accuracy improved on CUHK03 dataset while dropped slightly on Market-1501 dataset. We believe the performance drop is due to the inconsistent scale and norm of multiple layers (the "larger" features would dominate the "smaller" ones) [28]. As Region Re-Weighting would automatically learn the scale of features during training procedure, we speculate that integrating RRW with HFA would achieve more performance gain in Rank-1 accuracy. The results in Tables 10 and 11 also demonstrate this: the Rank-1 accuracy arrived at 85.99/85.90 on Market-1501/CUHK03 when using DJL-HFA. Furthermore, we give some illustrations about the learned weight parameters in Table 12, which show the scale and importance differences across multiple layers regions.

- **Complementary effects**

  We evaluated the effects of individual local feature (FC_local), global feature (FC7) as well as their combination on Market-1501 and CUHK03. The results on the two databases are shown in Figure 7. These results demonstrate that, although global and local feature representations alone are competitive for re-ID, further performance gain can be obtained by combining them using our proposed method. This proves that our proposed method can effectively explore the complementary discriminative information in global and local features for more accurate person re-ID. Two example results are shown in Figure 8. As can be seen, even when the probe and gallery pedestrian images have obviously different poses (i.e., they are not well aligned), our proposed method can still correctly retrieve the corresponding gallery images among the first ten ranks.

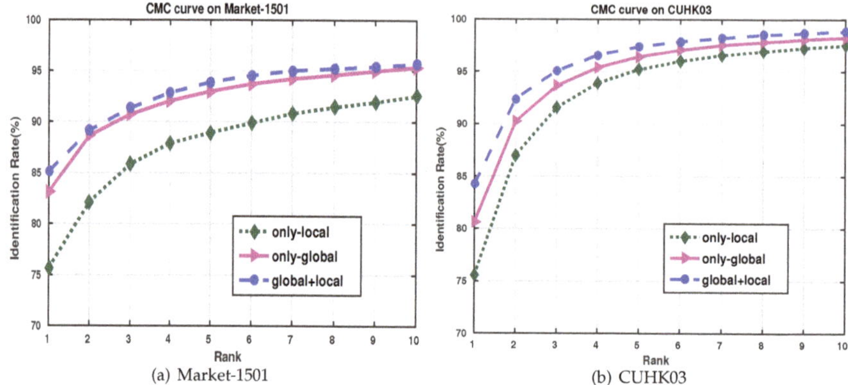

(a) Market-1501   (b) CUHK03

Figure 7. CMC curves on Market-1501 and CUHK03 when using local, global and hybrid features (global+local) extracted by DJL-S (based on InceptionNet).

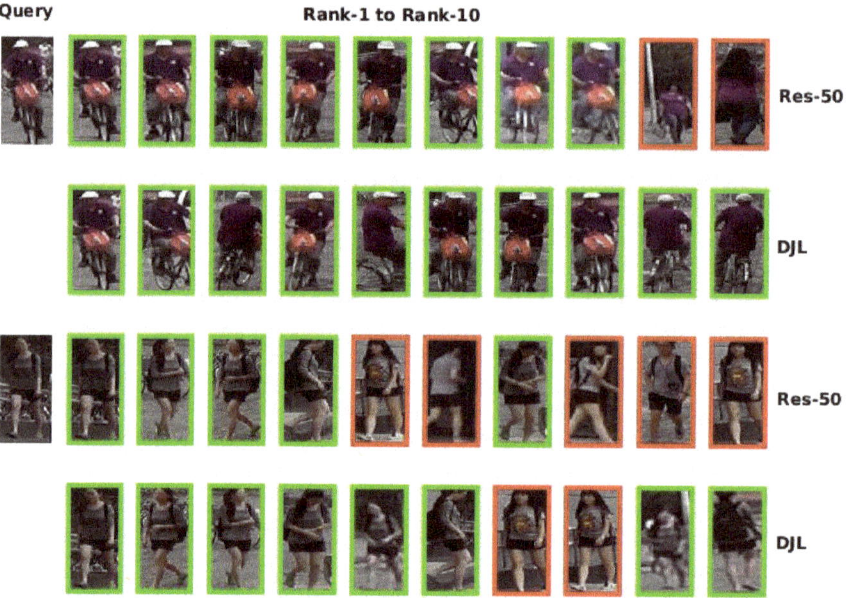

Figure 8. Example person re-ID results by using the base Residual-50 network (Res-50) and the proposed DJL network on Market-1501 database. Correct retrievals are surrounded with green bounding boxes while wrong retrievals are surrounded with red bounding boxes.

Table 6. Identification accuracy (%) and mAP (%) of the proposed method with different base networks on Market-1501 when different disturbances are applied to the segmented body parts. The best results under different settings are marked in bold.

Base Network	Setting	Market-1501				
		Rank-1	Rank-5	Rank-10	Rank-20	mAP
AlexNet	Base	57.75	77.52	84.47	89.46	33.80
	Proposed	67.64	84.80	**89.88**	**93.53**	43.60
	Disturb-small	**68.82**	**84.95**	89.31	93.50	**44.89**
	Disturb-violent	64.79	82.21	88.15	92.22	40.84
Residual-50	Base	72.42	86.49	91.03	94.42	48.01
	Proposed	**78.86**	**90.38**	**93.91**	**96.35**	**55.49**
	Disturb-small	77.76	89.88	92.96	96.02	54.62
	Disturb-violent	75.95	88.60	92.37	95.19	52.71
InceptionNet	Base	79.66	91.51	94.54	96.50	56.59
	Proposed	**85.12**	**93.91**	95.69	97.51	64.82
	Disturb-small	84.53	93.79	95.93	97.54	**64.89**
	Disturb-violent	83.61	93.65	**95.99**	**97.60**	63.44

Table 7. Identification accuracy (%) of the proposed method with different base networks on CUHK03 when different disturbances were applied to the segmented body parts and when low resolution images were used. The best results under different settings are marked in bold.

Base Network	Setting	CUHK03			
		Rank-1	Rank-5	Rank-10	Rank-20
AlexNet	Base	53.03	79.53	87.82	94.21
	proposed	**71.95**	**90.30**	**94.91**	**98.16**
	Disturb-small	68.31	89.19	93.86	97.07
	Disturb-violent	62.07	84.84	91.49	96.03
	Low-resolution	60.35	83.71	90.59	95.49
Residual-50	Base	61.79	85.46	92.31	97.86
	proposed	**80.83**	95.92	98.66	99.54
	Disturb-small	80.53	**96.45**	**99.01**	**99.71**
	Disturb-violent	73.40	93.23	96.75	99.25
	Low-resolution	75.58	93.68	97.28	9 9.15
InceptionNet	Base	80.85	95.90	98.17	99.48
	proposed	**84.25**	97.40	**98.86**	99.67
	Disturb-small	83.38	**97.49**	98.81	99.52
	Disturb-violent	82.41	97.42	98.84	**99.69**
	Low-resolution	82.80	97.12	98.64	99.63

Table 8. Identification accuracy (%) and mAP (%) of the proposed method with performing RoI pooling at different middle layers on Market-1501. The best result over various pooling layers is marked in bold.

Base Network	Pooling Layer	Market-1501				
		Rank-1	Rank-5	Rank-10	Rank-20	mAP
AlexNet	Conv3	66.30	84.29	89.58	93.23	42.75
	Conv4	67.64	84.80	89.88	93.53	43.60
	Conv5	**69.83**	**85.66**	**90.65**	**94.00**	**45.17**
Residual-50	Res3a	77.02	89.88	93.29	95.87	54.17
	Res4a	78.86	90.38	93.91	**96.35**	55.49
	Res5a	**79.48**	**91.30**	**94.51**	96.20	**57.53**
InceptionNet	Inception_1a	82.90	92.99	95.16	97.00	61.24
	Inception_2a	84.59	**94.83**	**96.53**	**98.19**	**65.89**
	Inception_3a	**85.12**	93.91	95.69	97.51	64.82

Table 9. Identification accuracy (%) of the proposed method with performing RoI pooling at different middle layers on CUHK03. The best result over various pooling layers is marked in bold.

Base Network	Pooling Layer	CUHK03			
		Rank-1	Rank-5	Rank-10	Rank-20
AlexNet	Conv3	68.13	89.12	94.84	97.94
	Conv4	71.95	90.30	94.91	98.16
	Conv5	**74.22**	**91.25**	**95.37**	**98.64**
Residual-50	Res3a	77.40	94.63	98.44	99.57
	Res4a	80.83	95.92	98.66	99.54
	Res5a	**83.97**	**96.97**	**98.67**	**99.61**
InceptionNet	Inception_1a	82.66	96.66	98.42	99.33
	Inception_2a	83.01	96.98	98.75	99.53
	nception_3a	**84.25**	**97.40**	**98.86**	**99.67**

Table 10. Effects of Region Re-Weighting and Hierarchical Feature Aggregation using the base Inception network on Market-1501. Identification accuracy (%) and mAP (%) are reported. The best Rank-1 result is marked in bold.

Methods	Market-1501				
	Rank-1	Rank-5	Rank-10	Rank-20	mAP
DJL-S	85.12	93.91	95.69	97.51	64.82
DJL-S + RRW	85.21	93.74	95.78	97.60	65.38
DJL-HFA (**w/o RRW**)	84.95	94.15	96.38	97.74	65.29
DJL-HFA	**85.99**	94.15	96.29	97.77	65.65

Table 11. Effects of Region Re-Weighting and Hierarchical Feature Aggregation using the base Inception network on CUHK03. Identification accuracy (%) is reported. The best Rank-1 result is marked in bold.

Methods	CUHK03			
	Rank-1	Rank-5	Rank-10	Rank-20
DJL-S	84.25	97.40	98.86	99.67
DJL-S + RRW	84.29	97.15	98.80	99.67
DJL-HFA (**w/o RRW**)	84.72	97.17	98.40	99.34
DJL-HFA	**85.90**	97.79	98.90	99.40

Table 12. The learned weights of ten body parts at different pooling layers. The initial weight parameter of each body part region was set to 10.

Body Parts	Pooling Layer		
	Inception_1a	Inception_2a	Inception_3a
Rshoulder(w0)	8.30	7.21	5.56
Lshoulder(w1)	8.35	7.20	5.39
RElbow(w2)	8.96	8.97	6.60
LElbow(w3)	9.97	8.49	6.56
RHip(w4)	8.68	7.24	5.21
LHip(w5)	8.66	7.54	5.76
Rknee(w6)	9.99	7.67	6.07
Lknee(w7)	8.70	7.81	5.46
RAnkle(w8)	10.09	9.24	8.82
LAnkle(w9)	9.92	9.86	8.56

5. Conclusions

This paper proposes a Deep Joint Learning (DJL) network to learn better feature representation from both entire image and local body parts. The local features are pooled from the feature maps generated by the convolution layers, which capture the salient details and are robust to handle pedestrian misalignment. Hierarchical Feature Aggregation mechanism and Region Re-Weighting strategy effectively improve our feature representation by optimally aggregating body parts features from low-level to semantic-level. Multiple Softmax losses are used to integrate the discriminative power of global and local features. Extensive evaluations on Market1501 and CUHK03 benchmarks validated the advantages of the proposed DJL network.

Author Contributions: Conceptualization, Y.L.; Methodology, Y.L.; Supervision, H.Y. and Q.Z.; Validation, Y.L.; Writing—original draft, Y.L.; and Writing—review and editing, H.Y. and Q.Z.

Acknowledgments: This work is supported by the National Key Research and Development Program of China (No. 2017YFB0802300), and the Miaozi Key Project in Science and Technology Innovation Program of Sichuan Province, China (No. 2017RZ0016) and the Shenzhen Fundamental Research fund (JCYJ20180305125822769).

Conflicts of Interest: The authors declare no conflict of interest.

References

1. Zheng, L.; Yang, Y.; Hauptmann, A.G. Person re-identification: Past, present and future. *arXiv* **2016**, arXiv:1610.02984.
2. Gong, S.; Cristani, M.; Yan, S.; Loy, C.C. *Person Re-Identification*; Springer: Berlin, Germany, 2014.
3. Bedagkar-Gala, A.; Shah, S.K. A survey of approaches and trends in person re-identification. *Image Vis. Comput.* **2014**, *32*, 270–286. [CrossRef]
4. Ahmed, E.; Jones, M.; Marks, T.K. An improved deep learning architecture for person re-identification. In Proceedings of the IEEE Conference on Computer Vision and Pattern Recognition, Boston, MA, USA, 7–12 June 2015; pp. 3908–3916.
5. Li, W.; Zhao, R.; Xiao, T.; Wang, X. Deepreid: Deep filter pairing neural network for person re-identification. In Proceedings of the IEEE Conference on Computer Vision and Pattern Recognition, Columbus, OH, USA, 23–28 June 2014; pp. 152–159.
6. Cheng, D.; Gong, Y.; Zhou, S.; Wang, J.; Zheng, N. Person re-identification by multi-channel parts-based cnn with improved triplet loss function. In Proceedings of the IEEE Conference on Computer Vision and Pattern Recognition, Las Vegas, NV, USA, 27–30 June 2016; pp. 1335–1344.

7. Varior, R.R.; Haloi, M.; Wang, G. Gated siamese convolutional neural network architecture for human re-identification. In Proceedings of the European Conference on Computer Vision, Amsterdam, The Netherlands, 11–14 October 2016; pp. 791–808.

8. Yi, D.; Lei, Z.; Liao, S.; Li, S.Z. Deep metric learning for person re-identification. In Proceedings of the International Conference on Pattern Recognition, Stockholm, Swede, 24–28 August 2014; pp. 34–39.

9. Zheng, L.; Shen, L.; Tian, L.; Wang, S.; Wang, J.; Tian, Q. Scalable person re-identification: A benchmark. In Proceedings of the IEEE International Conference on Computer Vision, Las Condes, Chile, 11–18 December 2015; pp. 1116–1124.

10. Corvee, E.; Bremond, F.; Thonnat, M. Person re-identification using spatial covariance regions of human body parts. In Proceedings of the 2010 IEEE 7th International Conference on Advanced Video and Signal Based Surveillance (AVSS), Boston, MA, USA, 29 August–1 September 2010; pp. 435–440.

11. Cheng, D.S.; Cristani, M. Person re-identification by articulated appearance matching. In *Person Re-Identification*; Springer: Berlin, Germany, 2014; pp. 139–160.

12. Cheng, D.S.; Cristani, M.; Stoppa, M.; Bazzani, L.; Murino, V. Custom pictorial structures for re-identification. In Proceedings of the 22nd British Machine Vision Conference, Dundee, UK, 29 August–2 September 2011; Volume 1, p. 6.

13. Farenzena, M.; Bazzani, L.; Perina, A.; Murino, V.; Cristani, M. Person re-identification by symmetry-driven accumulation of local features. In Proceedings of the IEEE Conference on Computer Vision and Pattern Recognition, San Francisco, CA, USA, 13–18 June 2010; pp. 2360–2367.

14. Plagemann, C.; Ganapathi, V.; Koller, D.; Thrun, S. Real-time identification and localization of body parts from depth images. In Proceedings of the 2010 IEEE International Conference on Robotics and Automation, Anchorage, AK, USA, 3–8 May 2010; pp. 3108–3113.

15. Mousas, C.; Anagnostopoulos, C.N. Performance-driven hybrid full-body character control for navigation and interaction in virtual environments. *3D Res.* **2017**, *8*, 18. [CrossRef]

16. Wei, S.E.; Ramakrishna, V.; Kanade, T.; Sheikh, Y. Convolutional pose machines. In Proceedings of the IEEE Conference on Computer Vision and Pattern Recognition, Las Vegas, NV, USA, 27–30 June 2016; pp. 4724–4732.

17. Cao, Z.; Simon, T.; Wei, S.E.; Sheikh, Y. Realtime multi-person 2d pose estimation using part affinity fields. In Proceedings of the IEEE Conference on Computer Vision and Pattern Recognition, Honolulu, HI, USA, 21–26 June 2017; pp. 7291–7299.

18. Fang, H.; Xie, S.; Tai, Y.W.; Lu, C. Rmpe: Regional multi-person pose estimation. In Proceedings of the IEEE International Conference on Computer Vision, Venice, Italy, 22–29 October 2017; pp. 2334–2343.

19. Zheng, L.; Huang, Y.; Lu, H.; Yang, Y. Pose invariant embedding for deep person re-identification. *arXiv* **2017**, arXiv:1701.07732.

20. Jaderberg, M.; Simonyan, K.; Zisserman, A. Spatial transformer networks. In Proceedings of the Advances in Neural Information Processing Systems, Montreal, QC, USA, 7–12 December 2015; pp. 2017–2025.

21. Liu, Y.; Wu, Z.; Zhao, Q. Pooling body parts on feature maps for misalignment robust person re-identification. In Proceedings of the 2018 IEEE 4th International Conference on Identity, Security, and Behavior Analysis (ISBA), Singapore, 11–12 January 2018; pp. 1–8. [CrossRef]

22. Zhao, H.; Tian, M.; Sun, S.; Shao, J.; Yan, J.; Yi, S.; Wang, X.; Tang, X. Spindle net: Person re-identification with human body region guided feature decomposition and fusion. In Proceedings of the IEEE Conference on Computer Vision and Pattern Recognition, Honolulu, HI, USA, 21–26 June 2017; pp. 1077–1085.

23. Suh, Y.; Wang, J.; Tang, S.; Mei, T.; Lee, K.M. Part-Aligned Bilinear Representations for Person Re-identification. In Proceedings of the European Conference on Computer Vision, Munich, Germany, 8–14 September 2018, pp. 402–419.

24. Mousas, C.; Anagnostopoulos, C.N. Real-time performance-driven finger motion synthesis. *Comput. Graph.* **2017**, *65*, 1–11. [CrossRef]

25. Mousas, C. Full-body locomotion reconstruction of virtual characters using a single inertial measurement unit. *Sensors* **2017**, *17*, 2589. [CrossRef] [PubMed]

26. Fei-Fei, L.; Perona, P. A bayesian hierarchical model for learning natural scene categories. In Proceedings of the 2005 IEEE Computer Society Conference on Computer Vision and Pattern Recognition (CVPR'05), San Diego, CA, USA, 20–25 June 2005; Volume 2, pp. 524–531.

27. Farabet, C.; Couprie, C.; Najman, L.; LeCun, Y. Learning hierarchical features for scene labeling. *IEEE Trans. Pattern Anal. Mach. Intell.* **2013**, *35*, 1915–1929. [CrossRef] [PubMed]

28. Liu, W.; Rabinovich, A.; Berg, A.C. Parsenet: Looking wider to see better. *arXiv* **2015**, arXiv:1506.04579.

29. Liao, S.; Hu, Y.; Zhu, X.; Li, S.Z. Person re-identification by Local Maximal Occurrence representation and metric learning. In Proceedings of the IEEE Conference on Computer Vision and Pattern Recognition, Boston, MA, USA, 7–12 June 2015; pp. 2197–2206.

30. Xiong, F.; Gou, M.; Camps, O.; Sznaier, M. Person re-identification using kernel-based metric learning methods. In Proceedings of the European Conference on Computer Vision, Zurich, Switzerland, 6–12 September 2014; pp. 1–16.

31. Zhao, R.; Ouyang, W.; Wang, X. Person re-identification by salience matching. In Proceedings of the IEEE International Conference on Computer Vision, Sydney, Australia, 1–8 December 2013; pp. 2528–2535.

32. Yang, Y.; Yang, J.; Yan, J.; Liao, S.; Yi, D.; Li, S.Z. Salient color names for person re-identification. In Proceedings of the European Conference on Computer Vision, Zurich, Switzerland, 6–12 September 2014; pp. 536–551.

33. Koestinger, M.; Hirzer, M.; Wohlhart, P.; Roth, P.M.; Bischof, H. Large scale metric learning from equivalence constraints. In Proceedings of the IEEE Conference on Computer Vision and Pattern Recognition, Providence, RI, USA, 16–21 June 2012; pp. 2288–2295.

34. Li, W.; Wang, X. Locally aligned feature transforms across views. In Proceedings of the IEEE Conference on Computer Vision and Pattern Recognition, Portland, OR, USA, 23–28 June 2013; pp. 3594–3601.

35. Liao, S.; Zhao, G.; Kellokumpu, V.; Pietikäinen, M.; Li, S.Z. Modeling pixel process with scale invariant local patterns for background subtraction in complex scenes. In Proceedings of the IEEE Conference on Computer Vision and Pattern Recognition, San Francisco, CA, USA, 13–18 June 2010; pp. 1301–1306.

36. Liao, S.; Li, S.Z. Efficient psd constrained asymmetric metric learning for person re-identification. In Proceedings of the IEEE International Conference on Computer Vision, Las Condes, Chile, 11–18 December 2015; pp. 3685–3693.

37. Jose, C.; Fleuret, F. Scalable metric learning via weighted approximate rank component analysis. In Proceedings of the European Conference on Computer Vision, Amsterdam, The Netherlands, 11–14 October 2016; pp. 875–890.

38. Lin, Y.; Zheng, L.; Zheng, Z.; Wu, Y.; Yang, Y. Improving person re-identification by attribute and identity learning. *arXiv* **2017**, arXiv:1703.07220.

39. Shi, Z.; Hospedales, T.M.; Xiang, T. Transferring a semantic representation for person re-identification and search. In Proceedings of the IEEE Conference on Computer Vision and Pattern Recognition, Boston, MA, USA, 7–12 June 2015; pp. 4184–4193.

40. Peng, P.; Xiang, T.; Wang, Y.; Pontil, M.; Gong, S.; Huang, T.; Tian, Y. Unsupervised cross-dataset transfer learning for person re-identification. In Proceedings of the IEEE Conference on Computer Vision and Pattern Recognition, Las Vegas, NV, USA, 27–30 June 2016; pp. 1306–1315.

41. Zhong, Z.; Zheng, L.; Cao, D.; Li, S. Re-ranking person re-identification with k-reciprocal encoding. In Proceedings of the IEEE Conference on Computer Vision and Pattern Recognition, Honolulu, HI, USA, 21–26 June 2017; pp. 3652–3661.

42. Zhang, Y.; Xiang, T.; Hospedales, T.M.; Lu, H. Deep Mutual Learning. In Proceedings of the IEEE Conference on Computer Vision and Pattern Recognition, Salt Lake City, UT, USA, 18–22 June 2018; pp. 4320–4328.

43. Fan, H.; Zheng, L.; Yan, C.; Yang, Y. Unsupervised Person Re-identification: Clustering and Fine-tuning. *ACM Trans. Multimed. Comput. Commun. Appl.* **2018**, *14*, 83:1–83:18. [CrossRef]

44. Wu, Y.; Lin, Y.; Dong, X.; Yan, Y.; Ouyang, W.; Yang, Y. Exploit the unknown gradually: One-shot video-based person re-identification by stepwise learning. In Proceedings of the IEEE Conference on Computer Vision and Pattern Recognition, Salt Lake City, UT, USA, 18–22 June 2018; pp. 5177–5186.

45. Navon, D. Forest before trees: The precedence of global features in visual perception. *Cogn. Psychol.* **1977**, *9*, 353–383. [CrossRef]

46. Torralba, A.; Oliva, A.; Castelhano, M.S.; Henderson, J.M. Contextual guidance of eye movements and attention in real-world scenes: The role of global features in object search. *Psychol. Rev.* **2006**, *113*, 766–786. [CrossRef] [PubMed]

47. Chen, W.; Chen, X.; Zhang, J.; Huang, K. Beyond triplet loss: A deep quadruplet network for person re-identification. In Proceedings of the IEEE Conference on Computer Vision and Pattern Recognition, Honolulu, HI, USA, 21–26 June 2017; pp. 403–412.

48. Xiao, T.; Li, H.; Ouyang, W.; Wang, X. Learning deep feature representations with domain guided dropout for person re-identification. In Proceedings of the IEEE Conference on Computer Vision and Pattern Recognition, Las Vegas, NV, USA, 27–30 June 2016; pp. 1249–1258.

49. Zheng, Z.; Zheng, L.; Yang, Y. A Discriminatively Learned CNN Embedding for Person Reidentification. *ACM Trans. Multimed. Comput. Commun. Appl.* **2017**, *14*, 13:1–13:20. [CrossRef]

50. Krizhevsky, A.; Sutskever, I.; Hinton, G.E. Imagenet classification with deep convolutional neural networks. In Proceedings of the Advances in Neural Information Processing Systems, Lake Tahoe, NV, USA, 3–8 December 2012; pp. 1097–1105.

51. Li, W.; Zhu, X.; Gong, S. Person re-identification by deep joint learning of multi-loss classification. *arXiv* **2017**, arXiv:1705.04724.

52. Su, C.; Li, J.; Zhang, S.; Xing, J.; Gao, W.; Tian, Q. Pose-driven deep convolutional model for person re-identification. In Proceedings of the IEEE International Conference on Computer Vision, Venice, Italy, 22–29 October 2017; pp. 3960–3969.

53. Zhao, L.; Li, X.; Zhuang, Y.; Wang, J. Deeply-Learned Part-Aligned Representations for Person Re-identification. In Proceedings of the IEEE Conference on Computer Vision and Pattern Recognition, Honolulu, HI, USA, 21–26 June 2017; pp. 3239–3248.

54. He, K.; Zhang, X.; Ren, S.; Sun, J. Deep residual learning for image recognition. In Proceedings of the IEEE Conference on Computer Vision and Pattern Recognition, Las Vegas, NV, USA, 27–30 June 2016; pp. 770–778.

55. Girshick, R. Fast r-cnn. In Proceedings of the IEEE International Conference on Computer Vision, Las Condes, Chile, 11–18 December 2015; pp. 1440–1448.

56. He, K.; Zhang, X.; Ren, S.; Sun, J. Spatial pyramid pooling in deep convolutional networks for visual recognition. In Proceedings of the European Conference on Computer Vision, Zurich, Switzerland, 6–12 September 2014; pp. 346–361.

57. Hariharan, B.; Arbeláez, P.; Girshick, R.; Malik, J. Hypercolumns for object segmentation and fine-grained localization. In Proceedings of the IEEE Conference on Computer Vision and Pattern Recognition, Boston, MA, USA, 7–12 June 2015; pp. 447–456.

58. Felzenszwalb, P.; McAllester, D.; Ramanan, D. A discriminatively trained, multiscale, deformable part model. In Proceedings of the IEEE Conference on Computer Vision and Pattern Recognition, Anchorage, AK, USA, 23–28 June 2008; pp. 1–8.

59. Jia, Y.; Shelhamer, E.; Donahue, J.; Karayev, S.; Long, J.; Girshick, R.; Guadarrama, S.; Darrell, T. Caffe: Convolutional architecture for fast feature embedding. In Proceedings of the 22nd ACM international conference on Multimedia, Orlando, FL, USA, 3–7 November 2014; pp. 675–678.

60. Deng, J.; Dong, W.; Socher, R.; Li, L.J.; Li, K.; Fei-Fei, L. Imagenet: A large-scale hierarchical image database. In Proceedings of the IEEE Conference on Computer Vision and Pattern Recognition, Miami, FL, USA, 20–25 June 2009; pp. 248–255.

61. Subramaniam, A.; Chatterjee, M.; Mittal, A. Deep neural networks with inexact matching for person re-identification. In Proceedings of the Advances in Neural Information Processing Systems, Barcelona, Spain, 5–10 December 2016; pp. 2667–2675.

62. Li, D.; Chen, X.; Zhang, Z.; Huang, K. Learning deep context-aware features over body and latent parts for person re-identification. In Proceedings of the IEEE Conference on Computer Vision and Pattern Recognition, Honolulu, HI, USA, 21–26 June 2017; pp. 384–393.

Article

Finger-Vein Verification Based on LSTM Recurrent Neural Networks

Huafeng Qin [1,*] and Peng Wang [2]

[1] Chongqing Engineering Laboratory of Detection Control and Integrated System, School of Computer Science and Information Engineering, Chongqing Technology and Business University, Chongqing 400067, China

[2] National Research Base of Intelligent Manufacturing Service, Chongqing Technology and Business University, Chongqing 400067, China; 2017658007@email.ctbu.edu.cn

* Correspondence: qinhuafengfeng@163.com; Tel.: +86-138-8375-8680

Received: 27 February 2019; Accepted: 8 April 2019; Published: 24 April 2019

Abstract: Finger-vein biometrics has been extensively investigated for personal verification. A challenge is that the finger-vein acquisition is affected by many factors, which results in many ambiguous regions in the finger-vein image. Generally, the separability between vein and background is poor in such regions. Despite recent advances in finger-vein pattern segmentation, current solutions still lack the robustness to extract finger-vein features from raw images because they do not take into account the complex spatial dependencies of vein pattern. This paper proposes a deep learning model to extract vein features by combining the Convolutional Neural Networks (CNN) model and Long Short-Term Memory (LSTM) model. Firstly, we automatically assign the label based on a combination of known state of the art handcrafted finger-vein image segmentation techniques, and generate various sequences for each labeled pixel along different directions. Secondly, several Stacked Convolutional Neural Networks and Long Short-Term Memory (SCNN-LSTM) models are independently trained on the resulting sequences. The outputs of various SCNN-LSTMs form a complementary and over-complete representation and are conjointly put into Probabilistic Support Vector Machine (P-SVM) to predict the probability of each pixel of being foreground (i.e., vein pixel) given several sequences centered on it. Thirdly, we propose a supervised encoding scheme to extract the binary vein texture. A threshold is automatically computed by taking into account the maximal separation between the inter-class distance and the intra-class distance. In our approach, the CNN learns robust features for vein texture pattern representation and LSTM stores the complex spatial dependencies of vein patterns. So, the pixels in any region of a test image can then be classified effectively. In addition, the supervised information is employed to encode the vein patterns, so the resulting encoding images contain more discriminating features. The experimental results on one public finger-vein database show that the proposed approach significantly improves the finger-vein verification accuracy.

Keywords: biometrics; finger-vein verification; deep learning; convolutional neural network; representation learning

---

1. Introduction

With the wide application of internal and increasing risk of terrorist attacks, information security became a hot topic and received more and more attention. A key point is how to recognize a person to protect personal poverty and privacy. Biometrics as an authentication method of recognizing a person has been widely investigated in past years. Currently, various biometric characteristics such as fingerprints [1], palm-print [2], finger-vein [3,4], hand-vein [5], palm-vein [6], face [7], iris [8], voice [9], signature [10] have been employed for verification and can be broadly classified into two categories. (1) Extrinsic characteristics (e.g., fingerprints, palm-print, face, voice, signature); (2) Intrinsic characteristics (e.g., finger-vein, palm-vein, hand-vein). The extrinsic characteristics are prone to be attacked because faked face and fingerprint can successfully cheat the verification system [11]. As the intrinsic characteristics such as finger-vein conceal the skin and not easily copied and forged, they show high security and privacy in practical application.

However, vein verification faces serious challenges. In practical applications, various factors such as environmental illumination [12–14], ambient temperature [3,14,15], light scattering [16,17], and user behavior [12,13] affect the finger-vein image quality. Generally, these factors are not controlled, so many capturing images not only contain vein patterns but also noise and irregular shadowing. Generally, the separability between the vein and non-vein patterns is poor in the regions associated with noise and irregular shadowing. Performing matching from such regions degrades the verification accuracy. To solve this problem, many segmentation-based methods are proposed to segment robust vein network for finger-vein recognition. Broadly, they can be categorized into two groups.

(1) Handcraft-based segmentation approaches. In this category, researchers employed the existing mathematical models to detect vein features based on attribute assumptions such as valleys and straight-lines. For example, they assume that the vein patterns can be approximated to line-like texture in a predefined neighborhood region and the descriptors such as Gabor filters are proposed to extract the vein pattern. The representative works include wide line detector (WLD) [13], Gabor filters [3,18–21], and matched filters [22]. Some researchers observe that the cross-sectional profile of a vein pattern shows the attribute of valley shape. Therefore, many models are built to detect the valley for vein pattern extraction. For instance, the curvature is sensitive to valley, so various approaches are proposed to enhance the vein patterns by computing mean curvature [14], difference curvature [23], and maxim curvature [15] of pixels in an image. In [24–27], the vein patterns are detected by computing the depth of the valley. In the region growth approach [27], both depth and symmetry of valley are combined to extract vein pattern. Recently, according to the anatomical knowledge, some characteristics of finger-vein structure, e.g., directionality, continuity, width variability, smoothness, and solidness are taken into account for finger-vein texture extraction in [28].

(2) Deep learning-based segmentation approaches. Unlike handcrafted approaches, the deep learning-based approaches are capable of extracting the vein patters from a raw image without the manual attribute distribution assumption and have shown promising performance in medical image segmentation such as neuronal membrane segmentation [29], prostate segmentation [30], retinal blood vessels [31], and brain image segmentation [32]. In work [33], the Convolutional Neural Network (CNN) model is firstly employed for finger-vein segmentation, and outperforms handcrafted feature-based approaches in terms of verification errors improvement. In their work, the pixels are automatically labeled and a patch-based dataset is built for CNN training. For testing, an image is split into various patches and each patch is put into the CNN to predict the probability of its center point being a vein pattern.

The approaches described above achieve good performance on some finger-vein recognition tasks, but they suffer from the following problems. For example, existing handcrafted approaches segment vein pattern based on assumptions. However, these assumptions are not always effective to detect the finger-vein patterns because some vein pixels may be created by more complicated distributions than valleys or straight lines. Also, they explicitly extract some vein features by an image processing method, which might discard relevant information about the finger-vein pattern. In addition, they do not get any prior knowledge from the different images as they segment each image independently from the others. For the deep learning-based approach [33], these problems are alleviated to an extent because it directly uncovers hierarchical features from raw images to minimize its decision errors on vein patterns without the attribute distribution assumptions. Meanwhile, rich prior knowledge is harnessed by training it on a huge patch-based training data from different images. However, these approaches, including CNN in [33], segment each pixel independently based on a predefined neighborhood region (e.g., patch) instead of considering spatial dependencies among the closed pixels. Factually, finger-vein vein patterns extend from finger root to fingertip, and show clear direction and good connectivity [34]. Therefore, there exists spatial dependencies among the closed vein pixels. So, the performances of these existing approaches are still limited for finger-vein texture pattern extraction.

Recurrent neural networks have shown powerful capacity for the representation of long-term dependency information and have been successfully applied to human activity [35,36], speech recognition [37], and handwriting recognition [38]. In recent years, LSTM networks [39] as the most successful extensions of recurrent neural networks have received more and more attention. The Long Short-Term Memory (LSTM) model adopts a gating mechanism controlling the contents of an internal memory cell so that it is capable of learning a better and more complex representation of long-term dependencies in the input sequential data. Consequently, LSTM networks work well for feature learning over time series data. Some researches employ it to learn the complex spatial dependencies for scene labeling and action recognition [40–42].

Inspired by this idea, in this paper we proposed a stacked Convolutional Neural Networks and Long Short-Term Memory (SCNN-LSTM) for finger-vein texture segmentation by combining the CNN model and LSTM model. Compared to existing segmentation-based methods, our approach not only predicts the probability of a pixel based only on its pixels and their correlations in a local region, but it does so by relying also on the spatial dependencies in its neighboring contexts, through a feature representation learned by LSTM from a large sequence training set. The main paper contributions are summarized as follows:

(1) We proposed a stacked Convolutional Neural Network and Long Short-Term Memory model to automatically learn features from raw pixels for finger-vein verification. First, the vein and background pixels are automatically labeled based on several baselines. For each labeled pixel, we generated four sequences along different directions. As a result, there are various sequence-based training sets, on which several SCNN-LSTMs are independently trained to form a complementary and an over-complete representation. Secondly, for a testing image, the probability of each sequence being to vein pattern is predicted and the scores from patch-based sequences are conjointly input to P-SVM to segment the vein patterns. As the CNN model has the capacity for representation of vein texture features in a local region (i.e., patch) and the LSTM model captures the spatial dependencies among the closed regions, the proposed SCNN-LSTM model is capable of predicting the probability of belonging to a vein pattern. The rigorous experimental results on a public finger-vein database imply that the proposed approach is able to extract vein pattern, which results in a significant improvement for finger-vein verification accuracy.

(2) This paper investigates a new approach to encode the finger-vein for verification. Generally, the existing finger-vein segmentation approaches encode an image to extract binary vein patterns based on one or more thresholds, which are not related to verification error reduction. Different from them, an effective supervised scheme is employed to automatically select the threshold for vein pattern encoding. We search for a robust threshold to encode image by maximizing the inter-class distance and minimizing intra-class distance, which is not based on human domain knowledge. So the proposed scheme directly targets biometrics verification performance instead of human perception. We analyze the experimental results and estimate the verification performance.

2. The Proposed Approach

To learn compositional representations of the texture feature and spatial dependencies information, a SCNN-LSTM model is proposed for finger-vein feature extraction. First, we employed seven baselines to label the pixel from a training set and validation set. Secondly, for each labeled pixel, different sequences are created along different orientations. Thirdly, each sequence is forwarded to SCNN-LSTM to predict its probability of belonging to a vein patten. As a result, there are several labeled scores for different orientations, which are taken out of the input of SVM to extract a vein feature. Applying the proposed SCNN-LSTM model to the whole image in this way, the vein images are enhanced. To achieve verification, the resulting enhancement image is encoded by a supervised encoding scheme. The framework of the proposed approach is illustrated in Figure 1.

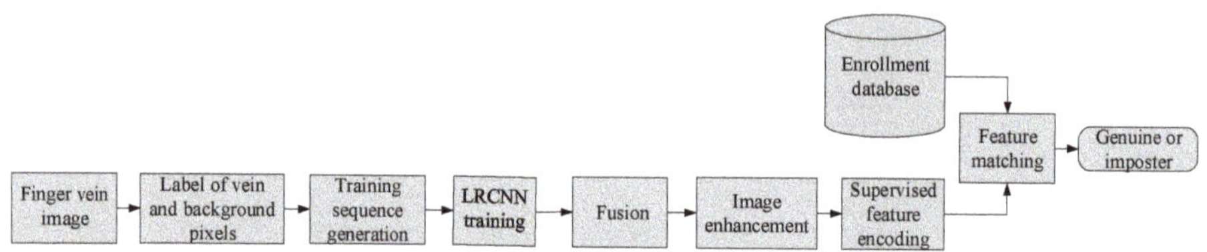

Figure 1. The framework of the proposed approach.

2.1. Label Vein Sequences

2.1.1. Label Vein Patterns

Similar to work [33], for each input finger-vein image, seven baselines, i.e., Repeated line tracking [24], Maximum Curvature points [15], Mean curvature [14], Different Curvature [43], Region growth [27], Wide line detector [13], and Gabor filters [3] are employed to segment vein pattern, resulting in seven binary images (as shown from Figure 2a–i). The values in each binary image (0 and 1 denote background and vein pixels, respectively) are treated as labels of corresponding pixels in the input image. We compute the average of seven binary images and obtain an average image F (Fin.3(i)). For a pixel (x, y), it is labeled as vein pattern if $F(x, y) = 1$ white region in Figure 2f), and it is labeled as vein for $F(x, y) = 0$ (black region in Figure 2j). We do not label the pixels in the remaining region (the color region in Figure 2j).

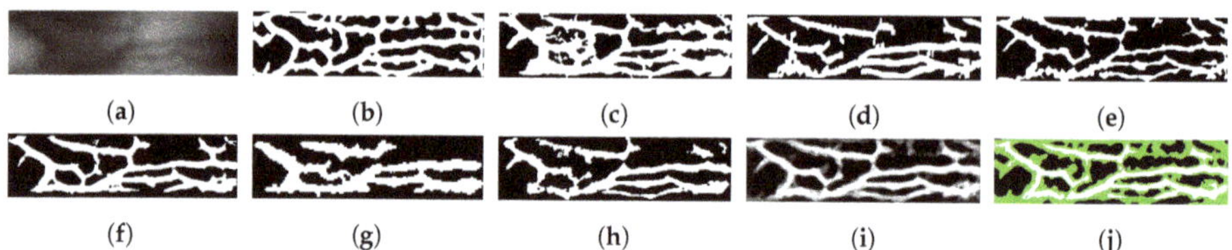

Figure 2. Segmented results. (**a**) Original finger-vein image; (**b**) Gabor filter; (**c**) Difference curvature; (**d**) Maximun curvature point; (**e**) Mean curvature; (**f**) Region growth; (**g**) Repeated line tracking; (**h**) Wide line detector; (**i**) Probability map; (**j**) Labeled pixels (wight region and black region denote vein and background, respectively).

2.1.2. Labeling Vein Sequences

In this section, the training sequences are produced based on labeled pixel (as shown in Figure 2j) for SCNN-LSTM training. Firstly, we select a labeled pixel as a current point c_0 and determine its $K - 1$ adjacent pixels along a given orientation θ. This results in a set of K pixels $\{c_{-(K-1)/2,\theta}, ..., c_0, ...c_{(K-1)/2,\theta}\}$ for orientation θ, where $0 \leq \theta < \pi$ and K is the odd number to enforce symmetry. Then, we create K patches of $s \times s$ centered on c_0 and its $K - 1$ adjacent pixels from image in training, and the resulting patches construct a sequence $S_\theta = \{P_{-(K-1)/2,\theta}, ..., P_0, ...P_{(K-1)/2,\theta}\}$. Similarly, the labels of K pixels create a labeled sequence $L_\theta = \{l_{-(K-1)/2,\theta}, ..., l_0, ...l_{(K-1)/2,\theta}\}$ for S_θ. Here, we quantize all the possible vein orientations θ into a set of C values by

$$\theta_i = \frac{i\pi}{C} \tag{1}$$

where $i = 1, 2, ..., C$ and C is heuristically set as 4. Namely, $\theta \in \{0°, 45°, 90°, 135°\}$, as shown in Figure 3a. Therefore, there are four sequences for each labeled pixel c_0. Figure 3 shows a sequence $S_{0°} = \{P_{-(K-1)/2,0°}, ..., P_0, ...P_{(K-1)/2,0°}\}$ of current pixel c_0 along the $0°$ orientation.

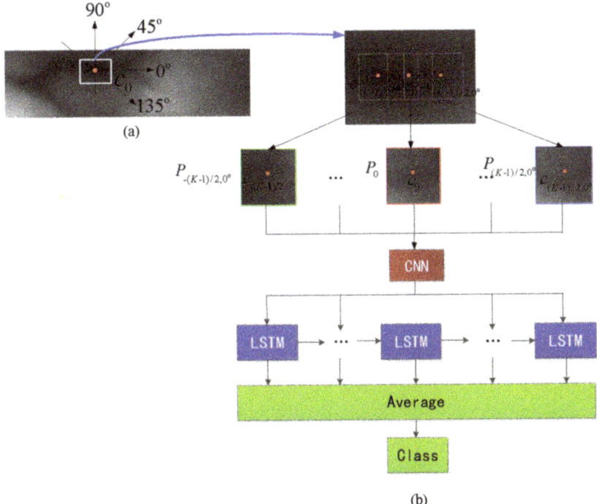

Figure 3. Illustration of the SCNN-LSTM model. (**a**) The four orientations for a pixel c_0; (**b**) SCNN-LSTM for prediction. A sequence sampled along $0°$ orientation is taken as an input of SCNN-LSTM to predict the probability of the centered point c_0 being to vein pattern. The LSTMs share same weights.

2.2. Stacked Convolutional Neural Networks and Long Short-Term Memory

The proposed stacked Convolutional Neural Networks and Long Short-Term Memory (SCNN-LSTM) consist of a CNN model and LSTM model (as shown in Figure 3) and are trained to learn the joint texture and spatial dependency representations for finger-vein texture segmentation. Our SCNN-LSTM takes a sequence associated with K patches as its input. In SCNN-LSTM, a deep CNN model is built by removing the output layer of an existing CNN model [33] for the vein texture representation. Then we take any patch as an input of the CNN model and it outputs a fixed-length vector representation which is further forwarded to a recurrent sequence learning module (LSTM) to learn the compositional representations in space, as shown Figure 3b.

Figure 4 shows the architecture of the proposed SCNN-LSTM. As shown in Figure 4, our approach consists of a CNN model and LSTM model. This CNN model (as shown in the red box in Figure 4) consists of three convolutional layers and one fully connected layer. There are 24 kernels of 5×5 in the first convolutional layer, 48 kernels of 5×5 in the second convolutional layer, and 100 kernels in the fully connected layer. The LSTM model (the blue region in Figure 4) includes 128 kernels. For SCNN-LSTM training, its input is a sequence of 7 patches with size of 11×11. Each patch in the sequence is forwarded to CNN model to obtain a 100 dimensional vector. As a result, there are 7 vectors for an input sequence with length of 7. The resulting vectors are taken as an input of LSTM model to obtain a 100 dimensional representative vector. Finally, the output of LSTM model is put into the last layer for classification. The output of last layer is a 2 dimensional vector because there are two classes (vein and background) for vein segmentation. When the input size changes, the width and height of the map in each convolutional layer changes accordingly. Along the forward direction, a patch-based sequence is represented effectively.

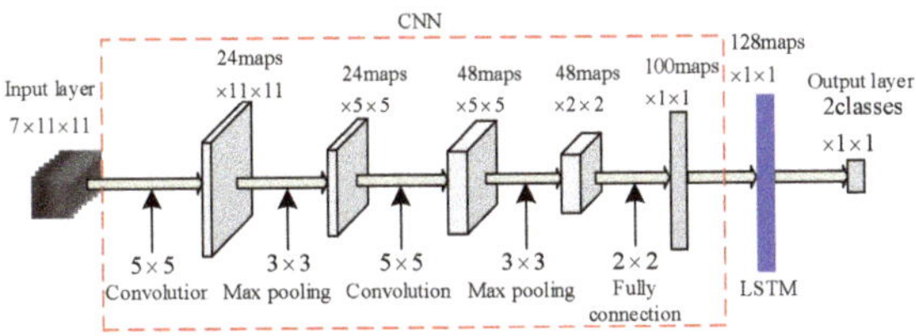

Figure 4. Architecture of SCNN-LSTM.

2.2.1. CNN Module

As the existing CNN model with three layers described in [33] has achieved promising performance for vein feature segmentation, we create a CNN module for feature representation of vein or background patch by removing the output layer of CNN in work [33]. During the training stage, our CNN is initialized using weights of an existing CNN [33]. Our CNN model consists of one input layer, three convolutional layers, two max-pooling layers, and one full-connection layer, respectively. The number of kernels in the three layers are 24, 48, and 100 respectively, and the sizes of kernels in both convolutional layers are 5. Each layer is detailed as follows.

Convolutional layer: The concept of Rectified Linear Units (y = max(0, x)) is used to active the hidden neurons.

Pooling alyer: The max-pooling is employed to extract location information by ensuring robustness to translation.

$$R_{i,j}^k = \max_{0 \leq m,n < s} (r_{i.s+m,j.s+n}^k) \tag{2}$$

where r_k denotes as k-th output map obtained by the k-th filter; The value $R_{i,j}^k$ pools over non-overlapping $r \times r$ local regions in I_k to extract the compact feature.

Dropout: The drop-out technique [44] is applied in three fully connected layers. The overfitting can be greatly prevented by randomly omitting half of the hidden units.

2.2.2. LSTM Module

The LSTM module is a subnet of our SCNN-LSTM which allows to easily memorize the context information for long periods of time in sequence data. In general, LSTM is proposed to model the temporal dependencies. In images, this temporal dependency learning is converted to the spatial domain [41]. Therefore, we employ a LSTM unit as described in [39] to model spatial dependencies by mapping the deep feature sequences produced from CNN to hidden states. To predict a distribution over spatial step, the softmax is employed in output layer. Finally, we average the outputs of the LSTM network's softmax layer to compute the predicted distribution, as shown in Figure 3b. Given inputs x_t, h_{t-1}, and c_{t-1}, the LSTM updates at the position t are

$$i_t = \sigma(W_{xi}x_t + W_{hi}x_{t-1} + b_i) \tag{3}$$
$$f_t = \sigma(W_{xf}x_t + W_{hf}x_{t-1} + b_f) \tag{4}$$
$$o_t = \sigma(W_{xo}x_t + W_{ho}x_{t-1} + b_o) \tag{5}$$
$$g_t = tanh(W_{xc}x_t + W_{hc}h_{t-1} + b_c) \tag{6}$$
$$c_t = f_t * c_{t-1it} * g_t \tag{7}$$
$$h_t = o_t * tanh(c_t) \tag{8}$$

where σ and *tanh* are logistic sigmoid (sigm) and hyperbolic tangent (*tanh*), which are defined as

$$\sigma(x) = (1 + e^{-x})^{-1}, \tag{9}$$

$$tanh(x) = \frac{e^x - e^{-x}}{e^x + e^{-x}} \tag{10}$$

and $*$ is the element-wise product. In addition, h_t, i_t, f_t, o_t, g_t, and c_t denote hidden unit, input gate, forget gate, output gate, input modulation gate, and memory cell, respectively, at the position t.

Output layer: The outputs from the last hidden layer are normalized with the softmax function:

$$y_m = \frac{\exp(z_m)}{\sum\limits_{n=1}^{N} \exp(z_n)} \tag{11}$$

where z_n is a linear combination of outputs in LSTM hidden states.

2.3. Multi-SCNN-LSTM Feature Representation

For a pixel with a label $l \in \{0,1\}$ from a given finger-vein image F, we produce a sequence $S_{\theta^*} \in S^{s \times s \times K}$ along an orientation θ^* and label it as $L_{\theta^*} \in \Re^{K \times 1}$ using the scheme described in Section 2.1, where 0 and 1 denote respectively background and vein. The training set used for vein segmentation is represented as $\{(S_{\theta^*,1}, L_{\theta^*,1})\}, \{(S_{\theta^*,2}, L_{\theta^*,2})\}, \ldots, \{(S_{\theta^*,N}, L_{\theta^*,N})\}$, where N is the number of sequences from finger-vein images in the training database. As we quantize all the possible vein orientations into four orientations, we in this way obtain 4 training datasets. Let $\{(S_{\theta_i,1}, L_{\theta_i,1})\}, \{(S_{\theta_i,2}, L_{\theta_i,2})\}, \ldots, \{(S_{\theta_i,N}, L_{\theta_i,N})\}$ be the i-th training dataset ($i = 1,2,\ldots,4$). A different SCNN-LSTM for each dataset is then trained independently, and each SCNN-LSTM produces a score from a particular sequence. We combine the outputs of the 4 SCNN-LSTMs to generate a 4-dimensional vector $v = [v_1, v_2, v_3, v_4]$, which is taken as an input of P-SVM to predict the probability of the pixel (Figure 5).

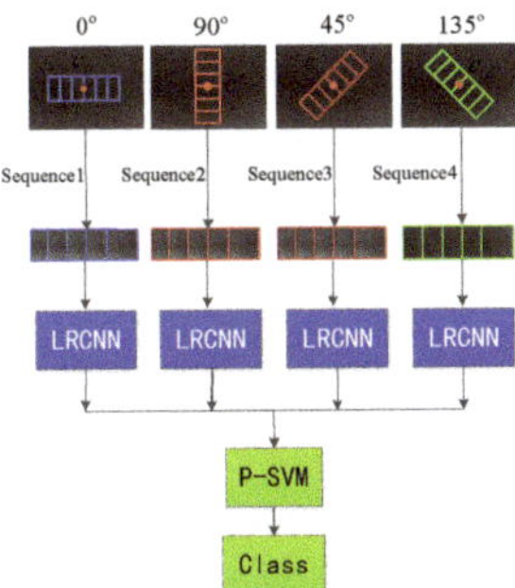

Figure 5. The framework of the Muilti-SCNN-LSTM. The prediction scores of a pixel are computed from four sequences along four orientations $(0°, 45°, 90°, 135°)$ and combined to generate a complementary score vector, taken as input of P-SVM to jointly predict the probability of centered pixel c_0 being to vein pattern.

2.4. Generating Score

A SVM model is employed to compute the probability of a pixel belonging to vein pattern based on its predicted distribution along four orientations. In this work, we employ the P-SVM model [45], which requires a set of vectors for training, to combine all features from all orientations (shows in Figure 5). Let v be a vector extracted from four sequences of a pixel with a label $l \in \{0,1\}$. The P-SVM is trained to provide a probabilistic value p (0 to 1: from background to vein)

$$p(q = 1|\varepsilon(v)) = \frac{1}{1 + \exp(w \cdot \varepsilon(v) + \gamma)} \tag{12}$$

where $\varepsilon(v)$ is the output of a general two-class SVM [46] with v as the input feature vector, and w and γ as fitting parameters trained by P-SVM. After training, we are able to compute the probability of any pixel based on its feature vector v and Equation (12).

2.5. Supervised Feature Encoding

In this section, we propose a scheme to obtain the threshold for vein feature encoding. After applying SCNN-LSTM for all pixels, an enhancing vein image is obtained and then we encode it for matching. In existing works [3,13–15,27,33,43], the vein patterns are encoded by one or more thresholds. For example, the probability of 0.5 is employed to obtain vein patterns in [33]. In [3], the vein image is enhanced by Gabor and then subject to binarzition using threshold of 0. In the classic repeated line tracking approach [24], two global thresholds (i.e., 85 and 175) are used to divide a image into three regions for matching. Some curvature-based approaches [14,15,23] enhanced vein patterns by computing the curvature of all pixels and an empirical threshold is employed to encode resulting enhancement image. For the finger-vein verification, the primary target of feature encoding is to improve performance, mainly verification error rates. However, the approaches determine the threshold based on human perception instead of minimizing the verification error, so the resulting binary code (vein texture features) may not be robust for finger-vein verification. To overcome this problem, in this section, a supervised scheme is proposed to encode vein pattern. Our approach decides the threshold by maximizing the distance between intra-class score set and inter-class score set computed from a training set, such that the resulting threshold is directly related to verification performance. The robust thresholds T are computed as follows.

Assume that there are N classes in the training set and each class provides M samples. Using the proposed SCNN-LSTM model (Figure 5), all finger-vein images are enhanced and we denote the mth enhancement image in the nth class as $x_{m,n}$, where $m = 1, 2, ..., M$ and $n = 1, 2, ...N$. We aim to find a function to map and quantize each enhancement image into a binary image $b_{m,n} \in \{0,1\}^{I \times J}$ which encodes a more discriminative information for verification error minimization. In our work, the binary code (vein texture pattern) $b_{m,n}$ of $x_{m,n}$ is computed by

$$b_{m,n} = 0.5 \times (sgn(x_{m,n} - T) + 1) \tag{13}$$

where $sgn(z)$ is equal to -1 if $z \leq 0$ and 1 otherwise and $T \in [0 \quad 1]$ is a parameter which is determined as follows.

Based on the Equation (13), all training samples are mapped into Hamming space, so a Hamming distance in [47] is employed to match two images for verification. We match the binary codes from same class to generate intra-class scores while the inter-class scores are produced by matching the binary codes from different class. So there are $a_1 = N \times C_2^M$ genuine matching scores $\Omega_1 = \{d_1(T), d_2(T), ..., d_{a_1}(T)\}$ and $a_2 = N \times (N-1) \times M \times M/2$ impostor scores $\Omega_2 = \{d'_1(T), d'_2(T), ..., d'_{a_2}(T)\}$. To make $b_{m,n}$ discriminative, we enforce an important criterion to encode the enhancement images that the resulting binary codes should maximize the distance between two sets Ω_1 and Ω_2. Therefore, we formulate the following optimization objective function:

$$\max_T J(T) = \frac{|u_1(T) - u_2(T)|}{D_1(T) + D_2(T)} \tag{14}$$

where $|\cdot|$ represents the absolute value. $u_1(T)$ and $u_2(T)$ are the means of the scores in Ω_1 and Ω_2, and $D_1(T)$ and $D_2(T)$ are the variances of the scores in the sets Ω_1 and Ω_2

To facilitate to search the threshold T, all enhancement images are converted to gray-scale images with integer values between 0 and 255. The parameter T is assigned from 0 to 255 to transform the enhancement image into a binary code map according to Equation (13). So, 256 different values $J(T)$ ($T = 1, 2, ..., 256$) are computed using Equation (14). The parameter T_*, which can maximize Equation (14), are selected to encode the vein pattern. The binary code of $x_{m,n}$ is computed by

$$b_{m,n} = 0.5 \times (sgn(x_{m,n} - T_*/255) + 1) \tag{15}$$

2.6. Feature Matching

After all training images are mapped into Hamming space, the Hamming distance is employed to match two images. In general, the capturing images are subject to translation and rotation normalization, but there are still some variations due to inaccurate localization and normalization. However, Hamming distance is not robust enough to reduce these variations. So, an enhanced Hamming distance is employed to compute the non-overlapping region between two images with possible spatial shifts for finger-vein matching. Assuming Q and B are enrolment and test binarized feature codes with size of $I \times J$, respectively (as shown in Figure 6), the height and width of Q are extended to $2E + I$ and $2H + J$, and then its expanded image $\bar{Q}$ is obtained and expressed as:

$$\bar{Q}(i,j) = \begin{cases} Q(i - E, j - H) & \text{if} \quad 1 + E \leq i \leq I + E, \\ & \qquad\qquad 1 + H \leq j \leq J + H \\ -1 & \text{otherwise} \end{cases} \tag{16}$$

Figure 6b illustrates the extended image $\bar{Q}$ of a template Q and the extend region with values of -1 is marked in color. The matching distance between Q and B is obtained by

$$d(Q, B) = \min_{0 \leq e \leq 2E, 0 \leq h \leq 2H} \frac{Hamdistance(\bar{Q}^{e,h}, B) - \Phi(\bar{Q}^{e,h})}{Hamdistance(\bar{Q}^{e,h}, U)} \tag{17}$$

In Equation (17), $\Phi(V)$ is the amount of -1 values in matrix V. U is a matrix with size of $I \times J$ and the values of its elements are -1. *Hamdistance* represents the hamming distance between two encoding images, i.e., summation of the number of positions that are different. $\bar{Q}^{(e,h)}$ is a matrix (the red rectangle box in Figure 6b) when the translation distances are e and h over horizontal and vertical directions. $d(Q, B)$ basically computes the minimal amount of non-overlap between Q and B at different spatial shifts, excluding the pixels located in the expanded region (e.g., the green region in Figure 6b). The parameters E and H control the translation distance in horizontal and vertical directions and are heuristically set to 20 and 60.

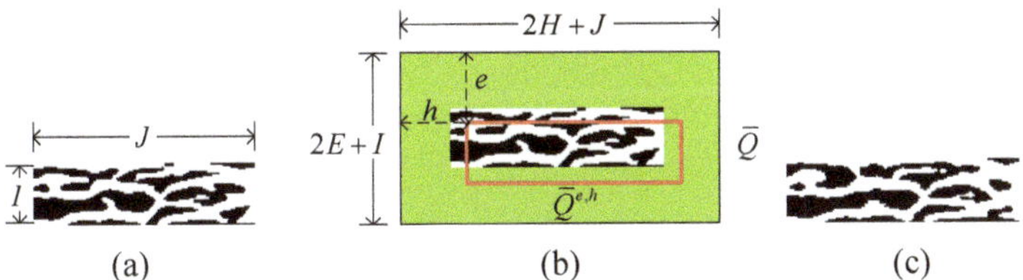

Figure 6. Matching sample. (**a**) A finger-vein template; (**b**) The extended image from (**a**); (**c**) A testing image. The values in green region are -1 in (**b**). The red rectangle box translates in the extended images from top left corner to lower right corner, and $\bar{Q}^{e,h}$ is a map in the red rectangle box when the translation distances are e and h over horizontal and vertical directions.

3. Experiments and Results

To estimate the performance of our approach, we compare various approaches with respect to verification performance improvement. In our experiments, we repeat the experimental results of classic approaches, i.e., Repeated line tracking [24], Maximum Curvature points [15], and recent approaches, i.e., Mean curvature [14], Different Curvature [43], Region growth [27], Wide line detector [13], and Gabor filter [3] for comparison. Also, we show the performance of the deep-based segmentation approach [33] to estimate the verification performance of our approach. In addition, based on the supervised encoding scheme in Equation (15), we can extract the finger-vein patterns from the probability map which is computed by the proposed SCNN-LSTM approach. To simplify the description, we denote them as the SCNN-LSTM + Supervised encoding. To test our encoding approach, we also encode the resulting probability map using a probability threshold of 0.5. This scheme is presented as SCNN-LSTM + Unsupervised encoding. The corresponding performance is shown in the following experiments. We compare all finger-vein extraction approaches mentioned above with the proposed one to get more insights into the problem of finger-vein verification. All experiments are carried out on one public database, namely the PolyU [3] finger-vein database, which is described below.

3.1. HKPU Database

The Hong Kong Polytechnic University (HKPU) finger-vein image database [3] includes 3132 images with a resolution of 513×256 pixels. All images are collected from 156 subjects using an open and contactless imaging device. The first 105 subjects provided 2520 finger images (105 subjects $\times$ 2 fingers $\times$ 6 images $\times$ 2 sessions) in two separate sessions with a minimum interval of one month and a maximum of over six months, with an average of 66.8 days. In each session, each subject provided 2 fingers (index finger and middle finger) and each finger provided 6 image samples. The remaining 51 subjects only provided image data in one session. To verify our approach, the 2520 finger images captured in two sessions are employed in our experiment because it is closer to a practical captured environment. A pre-processing method [3] is employed to extract the region of interest (ROI) image and carry out translation and orientation alignment. In addition, the image background is cropped because it contributes matching errors and computation cost. As a result, all images are normalized to 39×146.

3.2. Experimental Setting

To test our approach, we split the database into three data sets: training set associated with 660 (55 fingers $\times$ 12) images, validation associated with 600 (50 fingers $\times$ 12) images, and testing set associated with 1260 (105 fingers $\times$ 12) images. Based on the label scheme described in Section 2.1, we label vein and background pixels from the training set and validation set. To train our model, we select the sequences centered on vein pixel as positive samples and sequences centered on background pixels as negative ones. For each image in training set, we only employ about 80 positive sequences and negative sequences, respectively. As the length of sequences is fixed to 11 using next experiments in Section 3.3, there are about 1760 (80 sequences $\times$ 11 (length of sequences) $\times$ 2 (positive sequences and negative sequences)) patches for an image. This results in a total of 100,000 training sequences (50,000 positive sequences and 50,000 negative sequences) from 660 images. In the testing phase, we generate a patch for each pixel in a test image. So, for an image with size of 39×146, there are 5694 (39×146) patches, based on which a sequence is created for each pixel along a given orientation. In our work, the length of the sequence is 11. Therefore, for a pixel, the patches centered on its 11 adjacent pixels form a sequence along a given orientation (shown in Figure 3), which results 5694 sequences for a test image with size of 39×146. Then, the sequence of each pixel is put into our model, the output of which is taken as the probability of this pixel to belong to vein pattern.

3.3. Parameter Estimation

As described in Section 2.1, each sequence from images in training set consists of K patches with size of $s \times s$. The CNN module in our SCNN-LSTM is trained by fine-tuning the CNN with an input of 11×11 patch in [33]. Such a size has also shown good performance in work [33], so the patch size s for SCNN-LSTM is fixed to 11. The length of the sequence is important to achieve high verification accuracy. If K is too small, more detailed vein patterns are extracted but including more noise. Matching pixels in noisy regions can create errors which result in verification accuracy reduction. On the contrary, sequences with large K will suppress vein feature details, leading to smooth vein features, which also degrades the verification accuracy. Therefore, we determine the appropriate size of sequence for SCNN-LSTM experimentally. Firstly, we train the proposed SCNN-LSTM model to extract the vein feature of the finger-vein images in the training and validation at different lengths of sequence. To reduce the redundant information, we obtain patches with sampling intervals of one pixel to create training sequences. Secondly, the first 6 images acquired at the first session are employed as registration templates and the remaining as testing images. Therefore, there are 300 (50 × 6) genuine scores and 14,700 (50 × 49 × 6) impostor scores. The False Rejection Rate (FRR) is computed by the genuine scores and the False Acceptance Rate is computed by impostor scores. The Equal Error Rate (EER) is the error rate when FAR is equal to FRR. Figure 7 illustrates the relationships between length of sequence and EER, and the results are obtained by using only the validation data. From Figure 7, we can see that a smaller equal error rate is achieved at a sequence of length 11 and 13. With increasing the length K, the computation time will be increased. Therefore, we fix the length of a sequence to 11 in our experiments.

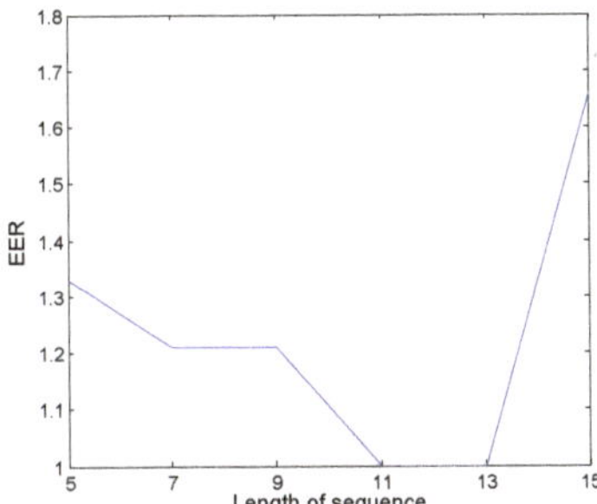

Figure 7. Relationship between the length of sequence and EER.

To verify over-fitting of our model, we shows learning curves in Figure 8. Figure 8a,b show the accuracy on the validation dataset and loss on the training dataset. From Figure 8, we can observe that the accuracy of validation dataset increases to about 65% and the loss decreases slowly after 2000 backpropagations. When the number of iteration steps is between 5000 and 10,000, the accuracy increases to more than 90% and the loss dramatically reduces. After 10,000 iterative steps, the loss fluctuates but it still decreases slowly. Therefore, our SCNN-LSTM model has good convergence for finger-vein segmentation.

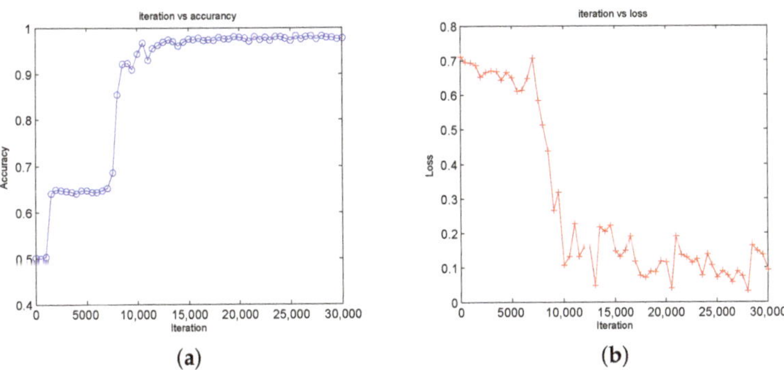

Figure 8. Training curves (**a**) Accuracy vs. iteration and (**b**) Loss vs. iteration.

3.4. Visual Assessment

In this experiment, we visually analyze the extracted finger-vein patterns from various approaches to get more insights into the proposed approach. The seven baselines and a state of the art [33] are employed to segment the vein texture, respectively. Also, the vein patterns encoded by a threshold of 0.5 and supervised threshold are reported in our experiment. Figure 9 shows the extracted results of various approaches. We can see from Figure 9 that the deep learning-based approaches suppresses the noise, and extract more connective and smoothness vein texture compared to the seven baselines. Observed the experiments in Figure 9i,j,f, it sees that the SCNN-LSTM-based approaches outperform the CNN in terms of extracting the connective vein patterns.

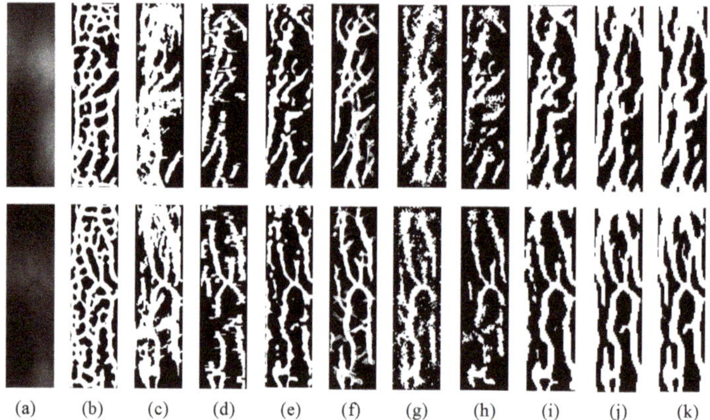

(a)   (b)   (c)   (d)   (e)   (f)   (g)   (h)   (i)   (j)   (k)

Figure 9. Experimental results from various approaches. (**a**) Original image; (**b**) Gabor filters; (**c**) Difference curvature; (**d**) Maximun curvature point; (**e**) Mean curvature; (**f**) Region growth; (**g**) Repeated line tracking; (**h**) Wide line detector; (**i**) CNN; (**j**) SCNN-LSTM + Unsupervised encoding; (**k**) SCNN-LSTM + Supervised encoding.

3.5. Verification Results Based on Image Dataset from One Session

In this section, we evaluate the performance of various approaches on the HKPU finger-vein dataset by considering vein images collected in each of the two sessions. First, the performance is evaluated in each session, individually. In one session, there are 630 images from 105 fingers. Therefore, the total number of genuine scores and impostor scores is 1575 ($105 \times C_2^6$) and 196,560 ($105 \times 104 \times 36/2$). To compute the impostor score, the symmetric matches are not executed. Second, the performance of combining scores from two sessions is reported. So, there are 3150 (1575×2 sessions) genuine scores and 393,120 ($196,560 \times 2$ sessions) impostor scores. Table 1 lists the verification error of various approaches for each session taken separately, and then for the two sessions, mixed. The receiver operating characteristics (ROC) curve for the corresponding performances is illustrated in Figure 10. The experimental results from Table 1 imply that the proposed SCNN-LSTM approach outperforms existing approaches including CNN [33] and achieves low errors, e.g., 1.12%, 0.62%, and 1.01% for data in the first session, second session, and two mixed sessions, respectively. The ERRs are further reduced to 1.08%, 0.58%, and 0.95% using the proposed encoding approach. We also observe from Figure 10 that the SCNN-LSTM-based approaches significantly improve FRR when the FAR is lower than 0.01%, which implies that our system achieve lower verification error than the methods considered in our work at high security level system.

Table 1. EER of various approaches on image dataset from one session.

Methods	First Session	Second Session	Two Sessions
Repeated line tracking [24]	4.76	5.67	5.21
Maximum curvature point [15]	3.91	3.27	3.59
Region growth [27]	2.32	1.24	1.75
Wide line detector [13]	3.68	3.11	3.39
Gabor filters [3]	2.10	1.84	1.95
Mean curvature [14]	2.06	1.50	1.73
Difference curvature [43]	3.61	3.64	3.64
CNN [33]	1.21	0.86	1.12
SCNN-LSTM + Unsupervised encoding	1.12	0.62	1.01
SCNN-LSTM + Supervised encoding	1.01	0.58	0.95

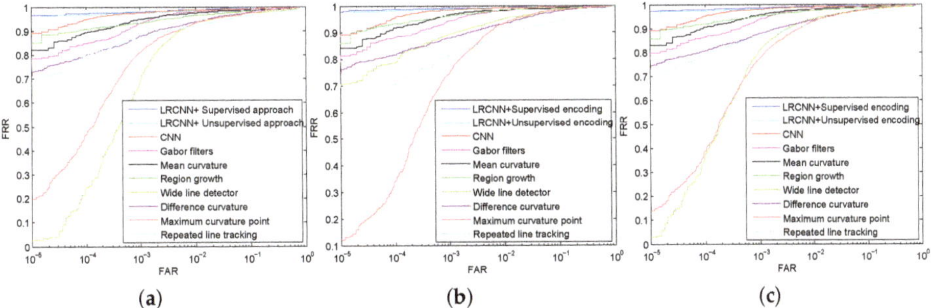

(a)   (b)   (c)

Figure 10. Receiver operating characteristics from image data collected at (**a**) first session, (**b**) second session, and (**c**) two mixed sessions.

3.6. Verification Results Based on Image Dataset from Two Sessions

This experiment aims at estimating the effectiveness and robustness of various algorithms on the finger-vein image data from both sessions. In the testing dataset, there are 1260 (105 fingers $\times$ 6 images $\times$ 2 sessions) images, acquired at two sessions. For each finger, we select the 6 images captured at the first session as enrollment samples and the remaining 6 images captured at the second session as testing samples. The genuine matching scores are produced by matching samples from same finger, while the impostor scores are produced by matching samples from different fingers. This results in a total of 630 (105 $\times$ 6) genuine scores and (105 $\times$ 104 $\times$ 6/2) impostor scores, based on which we the compute FRR and FAR. In addition, we computed the sensitive index(d') [48] by $d' = Z(hit\ rate) - Z(false\ alarm\ rate)$ to estimate the performance of various approaches.

The experimental results from various approaches are summarized in Table 2. The ROC curves for the corresponding performances are illustrated in Figure 11. The experimental results summarized in Table 2 show consistent trends with the those from experiments in each session. The proposed SCNN-LSTM-based approaches (e.g., SCNN-LSTM + Unsupervised encoding and SCNN-LSTM + Supervised encoding) get the best results, especially at the lower FAR. The lowest EER of 2.38% is achieved using the supervised encoding approach. Similarly, the proposed method achieves higher d' (e.g., 3.89 and 3.95) compared to existing approaches, which implies that the lowest verification error is achieved using our SCNN-LSTM model.

Table 2. EER of various approaches on image dataset from two different sessions.

Methods	EER (%)	Sensitive Index (d′)
Repeated line tracking [24]	12.85	2.26
Maximum curvature point [15]	8.30	2.76
Region growth [27]	5.71	3.15
Wide line detector [13]	7.62	2.86
Gabor filters [3]	5.08	2.97
Mean curvature [14]	4.20	3.45
Difference curvature [43]	7.90	2.92
CNN [33]	3.02	3.72
SCNN-LSTM+Unsupervised encoding	2.59	3.89
SCNN-LSTM+Supervised encoding	2.38	3.95

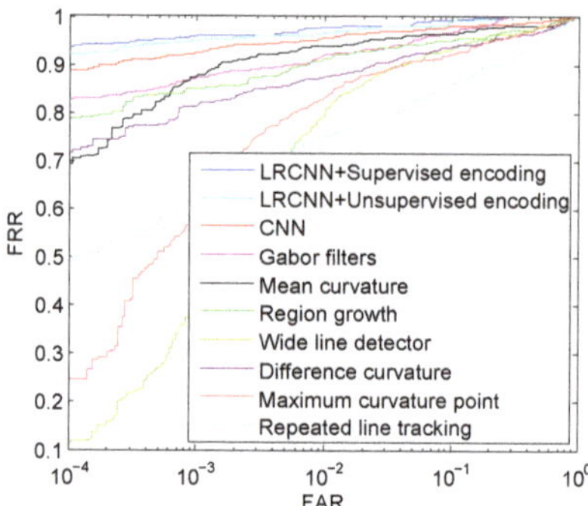

Figure 11. Receiver operating characteristics on image data from two separate sessions.

4. Discussion

The experiments depicted in Tables 1 and 2, Figures 10 and 11 show that the proposed SCNN-LSTM-based models achieve best performance among the all approaches considered in our work, including seven baselines and the CNN-based model. For example, the EER achieved by the best one (CNN) among existing approaches is reduced to 2.53% using the proposed SCNN-LSTM model with unsupervised encoding scheme on the data set acquired from two sessions. The verification accuracy may be further improved by combing the features of sequences along more directions or enlarging the training set. The good performance can be explained by the following fact. The existing handcrafted approaches (seven baselines) explicitly extract some features by image processing method, which might discard relevant information about finger-vein pattern. Also, they do not get any prior knowledge from the different images as they segment each image independently from the others. In addition, all approaches, including CNN, independently process each pixel based on a predefined neighborhood region or cross-sectional profile during the segmentation procedure, and ignore the spatial dependencies among different vein pixels. By contrast, the proposed approach uncovers hierarchical features for vein texture representation by training its CNN module and harnesses rich dependency information by training its LSTM module on a huge sequence set from different images. Therefore, it is capable of predicting the probability of a pixel belonging to a vein pattern.

We can also observe from the experimental results (Tables 1 and 2, Figures 10 and 11), the performance is improved after adopting a supervised encoding scheme. For instance, the EER is reduced to 0.95% (about 6% relative error reduction) on the data from two mixed sessions. When we employ the images in the first session as templates and the remaining images captured at the second session as testing samples, a EER, namely 2.38% (about 8.1% relative error reduction) is achieved by the SCNN-LSTM + Supervised encoding. The experimental results are explained by this fact. The existing finger-vein encoding approaches do not infer any prior knowledge from the different images because they compute the threshold from each image independently from others or employ some empirical threshold values such as 0.5 and 0. By contrast, the proposed encoding approach harnesses a rich prior knowledge acquired by maximizing the distance between the genuine score set and impostor score set (as shown in Equation (14)) and the resulting threshold is directly related to verification error reduction. Therefore, our approach can extract the discriminative vein texture for verification. Also, the experimental results show that the supervised encoding shows more significant improvement on the data acquired in two sessions. The reason is that there is not large room for improvement because it is easier to distinguish the images from one session compared to those from two sessions. Actually, the 2-sessions scenario is more realistic so the supervised encoding scheme is effective to reduce the verification error.

Compared to the experimental results in Section 3.5 (Table 1 and Figure 10) and in Section 3.6 (Table 2 and Figure 11), we see that all approaches achieve significant improvement in terms of verification accuracy on image datasets acquired in one session. Such a good performance can be attributed to the fact that there exist smaller within-class variations in the images captured at the same session because the imaging environment is similar and the subjects increase familiarity in the finger presentations during finger-vein image acquisition within a short duration. On the contrary, there are the larger within-class variations for the data acquired in two different sessions, which causes more mismatching errors.

In addition, we also compare our approach with existing approaches with respect to the computational cost. All experiments are carried out in Matlab 2014a and conducted on a high performance computer with 8 Core E3-1270v3 3.5 GHz processor, 16 GB of RAM, and a NVIDIA Quadro GTX1070 graphics cards. For our approach and CNN [33], they are trained with Caffe package [49] on the graphics cards, and tested with Matlab on the central processing unit (CPU). To improve the time cost, we optimize SCNN-LSTM to extract the vein feature of a test image. First, as described in Section 3.2, a test image with size of 39×146 is divided into 39×146 overlapping patches, based on which 39×146 sequences are generated for all pixels along a given orientation using the scheme in Section 2.1.2. Therefore, there are same patches in the sequences of adjacent pixels. If we input the sequence for each pixel into SCNN-LSTM for feature extraction, it results in a lot of repeated feature extraction operations in the CNN model. To further reduce the computation time, 39×146 patches from an image are separately input into CNN model of SCNN-LSTM and we take its output (a 100 dimensional vector) as the feature vector of the input. Then, for each pixel, we arrange resulting vectors along a given orientation to form a sequence, which is forwarded to the LSTM model to extract its spatial dependence feature. Therefore, as each patch is only subject to one feature extraction operation using CNN model, the computational time for our model is significantly reduced in this way. Second, the four SCNN-LSTMs (shown in Figure 5) for four orientations are implemented in parallel to further reduce time cost. For the remaining approaches mentioned in our work, all experiments are implemented in Matlab on CPU. The average verification time of an image using various methods is listed in Table 3. We can see from Table 1 that the proposed method, CNN, and Repeated line tracking approach require more than two seconds to verify a finger-vein image, e.g., 3.25 s, 2.13 s, and 2.53 s, respectively, which are more than those achieved by the remaining approaches. This can be explained by the following fact. The proposed approach and CNN process the patch centered on each pixel and predict its probability of belonging to a vein pattern. When the size of test image is large, it is computationally expensive. The Repeated line tracking approach starts at a seed point and then tracks all vein patterns pixel by pixel by detecting the local dark line. When a dark line is not detectable, a new tracking operation starts at another position. The local line tracking operation is repeatedly performed and the tracking number for each pixel is recorded in a tracking matrix for segmentation. The larger tracking number will enhance the vein pattern and result in high verification accuracy, but the computational cost increases. Overall, our approach shows high time cost, but it can achieve best performance for finger-vein verification (as shown in experimental results in Tables 1 and 2 and Figures 10 and 11). Moreover, these time costs are expected to be significantly reduced after code optimization. For example, implementing these algorithms in C++ can also improve the computation speed. With development of parallel computing technologies such as CUDA, the computing performance can be dramatically improved by harnessing the power of the graphics processing unit (GPU). Therefore, our approach can achieve computational requirement for practical application after accelerating using GPU.

Table 3. Average computational time of various approaches.

Methods	Time (s)
Repeated line tracking [24]	2.53
Maximum curvature point [15]	1.01
Region growth [27]	0.54
Wide line detector [13]	0.04
Gabor filters [3]	1.96
Mean curvature [14]	0.14
Difference curvature [43]	1.16
CNN [33]	2.13
The proposed approach	3.25

5. Conclusions

In this paper, we proposed an approach to extract the finger-vein pattern for verification. First, a SCNN-LSTM is proposed to predict the probability of a vein pixel belonging to a vein patten. As SCNN-LSTM combines recurrent models such as LSTMs with deep convolutional networks, it can be jointly trained to learn the complex spatial dependencies and convolutional perceptual representations. Second, to improve the performance, we proposed a supervised scheme to encode the vein patterns. As the threshold for encoding is related to verification performance, it can extract robust vein texture features for verification. Experimental results show that the proposed approach extracts robust vein features and significantly improves the verification error rate with respect to state of the art.

As our model can learn the complex spatial dependencies, it extract continuous vein network for verification. Also, our approach is employed to extract the hand-vein and palm-vein for recognition. In medical image analysis, some images such as retinal image, brain segmentation, and neuronal membranes contain continuous texture patterns, so the proposed approach can be applied to segment such texture patterns for disease diagnosis. In addition, if the patterns in vision image show the similar connectivity to vein pattern (as shown in Figure 1), our approach can be used to process vision image. In future work, we will extend the application of our approach to further verify its generalization.

Author Contributions: Conceptualization, H.Q.; methodology, H.Q.; software, P.W.; validation, P.W. and P.W.; formal analysis, H.Q.; investigation, H.Q.; resources, P.W.; data curation, P.W.; writing—original draft preparation, P.W.; writing—review and editing, H.Q.; visualization, P.W.; supervision, H.Q.; project administration, H.Q.; funding acquisition, H.Q.

Funding: This work is supported by the National Natural Science Foundation of China (Grant No. 61402063), the Natural Science Foundation Project of Chongqing (Grant No.cstc2017jcyjAX0002, Grant No.cstc2018jcyjAX0095, Grant No.cstc2013kjrc-qnrc40013,Grant No.cstc2017zdcy-zdyfX0067).

Conflicts of Interest: The authors declare no conflict of interest.

References

1. Jain, A.; Hong, L.; Bolle, R. On-line fingerprint verification. *IEEE Trans. Pattern Anal. Mach. Intell.* **1997**, *19*, 302–314. [CrossRef]

2. Zhang, D.D.; Kong, W.; You, J.; Wong, M. Online palmprint identification. *IEEE Trans. Pattern Anal. Mach. Intell.* **2003**, *25*, 1041–1050. [CrossRef]

3. Kumar, A.; Zhou, Y. Human identification using finger images. *IEEE Trans. Image Process.* **2012**, *21*, 2228–2244. [CrossRef] [PubMed]

4. Yang, L.; Yang, G.; Xi, X.; Su, K.; Chen, Q.; Yin, Y. Finger Vein Code: From Indexing to Matching. *IEEE Trans. Inf. Forensics Secur.* **2019**, *14*, 1210–1223. [CrossRef]

5. Kumar, A.; Prathyusha, K.V. Personal authentication using hand vein triangulation and knuckle shape. *IEEE Trans. Image Process.* **2009**, *18*, 2127–2136. [CrossRef]

6. Zhou, Y.; Kumar, A. Human identification using palm-vein images. *IEEE Trans. Inf. Forensics Secur.* **2011**, *6*, 1259–1274. [CrossRef]

7. Turk, M.A.; Pentland, A.P. Face recognition using eigenfaces. In Proceedings of the 1991 IEEE Computer Society Conference on Computer Vision and Pattern Recognitio, Maui, HI, USA, 3–6 June 1991; pp. 586–591.

8. Daugman, J. How iris recognition works. *IEEE Trans. Circuits Syst. Video Technol.* **2004**, *14*, 21–30. [CrossRef]

9. Ramírez, J.; Segura, J.C.; Górriz, J.M.; García, L. Improved voice activity detection using contextual multiple hypothesis testing for robust speech recognition. *IEEE Trans. Audio Speech Lang. Process.* **2007**, *15*, 2177–2189. [CrossRef]

10. El-Yacoubi, M.A.; Gilloux, M.; Bertille, J.M. A statistical approach for phrase location and recognition within a text line: An application to street name recognition. *IEEE Trans. Pattern Anal. Mach. Intell.* **2002**, *24*, 172–188. [CrossRef]

11. Lu, Y.; Xie, S.J.; Yoon, S.; Yang, J.; Park, D.S. Robust finger vein ROI localization based on flexible segmentation. *Sensors* **2013**, *13*, 14339–14366. [CrossRef] [PubMed]

12. Hashimoto, J. Finger vein authentication technology and its future. In Proceedings of the 2006 Symposium on VLSI Circuits, Honolulu, HI, USA, 15–17 June 2006; pp. 5–8.

13. Huang, B.; Dai, Y.; Li, R.; Tang, D.; Li, W. Finger-vein authentication based on wide line detector and pattern normalization. In Proceedings of the 2010 20th International Conference on Pattern Recognition, Istanbul, Turkey, 23–26 August 2010; pp. 1269–1272.

14. Song, W.; Kim, T.; Kim, H.C.; Choi, J.H.; Kong, H.J.; Lee, S.R. A finger-vein verification system using mean curvature. *Pattern Recognit. Lett.* **2011**, *32*, 1541–1547. [CrossRef]

15. Miura, N.; Nagasaka, A.; Miyatake, T. Extraction of finger-vein patterns using maximum curvature points in image profiles. *IEICE Trans. Inf. Syst.* **2007**, *90*, 1185–1194. [CrossRef]

16. Yang, J.; Shi, Y. Towards finger-vein image restoration and enhancement for finger-vein recognition. *Inf. Sci.* **2014**, *268*, 33–52. [CrossRef]

17. Lee, E.C.; Park, K.R. Image restoration of skin scattering and optical blurring for finger vein recognition. *Opt. Lasers Eng.* **2011**, *49*, 816–828. [CrossRef]

18. Yang, W.; Huang, X.; Zhou, F.; Liao, Q. Comparative competitive coding for personal identification by using finger vein and finger dorsal texture fusion. *Inf. Sci.* **2014**, *268*, 20–32. [CrossRef]

19. Yang, J.; Shi, Y. Finger–vein ROI localization and vein ridge enhancement. *Pattern Recognit. Lett.* **2012**, *33*, 1569–1579. [CrossRef]

20. Yang, J.; Shi, Y.; Yang, J. Finger-vein recognition based on a bank of Gabor filters. In Proceedings of the Asian Conference on Computer Vision, Xi'an, China, 23–27 September 2009; pp. 374–383.

21. Yu, C.B.; Qin, H.F.; Cui, Y.Z.; Hu, X.Q. Finger-vein image recognition combining modified hausdorff distance with minutiae feature matching. *Interdiscip. Sci. Comput. Life Sci.* **2009**, *1*, 280–289. [CrossRef]

22. Chaudhuri, S.; Chatterjee, S.; Katz, N.; Nelson, M.; Goldbaum, M. Detection of blood vessels in retinal images using two-dimensional matched filters. *IEEE Trans. Med. Imaging* **1989**, *8*, 263–269. [CrossRef]

23. Qin, H.; He, X.; Yao, X.; Li, H. Finger-vein verification based on the curvature in Radon space. *Expert Syst. Appl.* **2017**, *82*, 151–161. [CrossRef]

24. Miura, N.; Nagasaka, A.; Miyatake, T. Feature extraction of finger-vein patterns based on repeated line tracking and its application to personal identification. *Mach. Vis. Appl.* **2004**, *15*, 194–203. [CrossRef]

25. Liu, T.; Xie, J.; Yan, W.; Li, P.; Lu, H. An algorithm for finger-vein segmentation based on modified repeated line tracking. *Imaging Sci. J.* **2013**, *61*, 491–502. [CrossRef]

26. Gupta, P.; Gupta, P. An accurate finger vein based verification system. *Digit. Signal Process.* **2015**, *38*, 43–52. [CrossRef]

27. Qin, H.; Qin, L.; Yu, C. Region growth–based feature extraction method for finger-vein recognition. *Opt. Eng.* **2011**, *50*, 057208. [CrossRef]

28. Yang, L.; Yang, G.; Yin, Y.; Xi, X. Finger vein recognition with anatomy structure analysis. *IEEE Trans. Circuits Syst. Video Technol.* **2018**, *28*, 1892–1905. [CrossRef]

29. Ciresan, D.; Giusti, A.; Gambardella, L.M.; Juergen, S. Deep neural networks segment neuronal membranes in electron microscopy images. In Proceedings of the 25th International Conference on Advances in Neural Information Processing Systems, Lake Tahoe, NV, USA, 3–6 December 2012; pp. 2843–2851.

30. Guo, Y.; Gao, Y.; Shen, D. Deformable MR Prostate Segmentation via Deep Feature Learning and Sparse Patch Matching. *IEEE Trans. Med. Imaging* **2016**, *35*, 1077–1089. [CrossRef] [PubMed]
31. Liskowski, P.; Krawiec, K. Segmenting Retinal Blood Vessels with Deep Neural Networks. *IEEE Trans. Med. Imaging* **2016**, *35*, 2369–2380. [CrossRef]
32. Zhang, W.; Li, R.; Deng, H.; Wang, L.; Lin, W.; Ji, S.; Shen, D. Deep convolutional neural networks for multi-modality isointense infant brain image segmentation. *Neuro Image* **2015**, *108*, 214–224. [CrossRef]
33. Qin, H.; El-Yacoubi, M.A. Deep representation-based feature extraction and recovering for finger-vein verification. *IEEE Trans. Inf. Forensics Secur.* **2017**, *12*, 1816–1829. [CrossRef]
34. Han, T.Z.; Yu, X. Anatomical study and clinical application of superficial palmar digital veins in finger replantation. *Chin. J. Clin. Anat.* **1997**, *15*, 39–41.
35. Donahue, J.; Anne Hendricks, L.; Guadarrama, S.; Rohrbach, M.; Venugopalan, S.; Saenko, K.; Darrell, T. Long-term recurrent convolutional networks for visual recognition and description. In Proceedings of the IEEE Conference on Computer Vision and Pattern Recognition, Boston, MA, USA, 7–12 June 2015. pp. 2625–2634.
36. Du, Y.; Wang, W.; Wang, L. Hierarchical recurrent neural network for skeleton based action recognition. In Proceedings of the IEEE Conference on Computer Vision and Pattern Recognition, Boston, MA, USA, 7–12 June 2015; pp. 1110–1118.
37. Graves, A.; Mohamed, A.R.; Hinton, G. Speech recognition with deep recurrent neural networks. In Proceedings of the 2013 IEEE International Conference on Acoustics, Speech and Signal Processing (ICASSP), Vancouver, BC, Canada, 26–31 May 2013; pp. 6645–6649.
38. Graves, A.; Schmidhuber, J. Offline handwriting recognition with multidimensional recurrent neural networks. In Proceedings of the 21st International Conference on Neural Information Processing Systems, Vancouver, BC, Canada, 7–10 December 2009; pp. 545–552.
39. Hochreiter, S.; Schmidhuber, J. Long short-term memory. *Neural Comput.* **1997**, *9*, 1735–1780. [CrossRef] [PubMed]
40. Zhu, W.; Lan, C.; Xing, J.; Zeng, W.; Li, Y.; Shen, L.; Xie, X. Co-Occurrence Feature Learning for Skeleton Based Action Recognition Using Regularized Deep LSTM Networks. In Proceedings of the Thirtieth AAAI Conference on Artificial Intelligence, Phoenix, AZ, USA, 12–17 February 2016; Volume 2, p. 6.
41. Byeon, W.; Breuel, T.M.; Raue, F.; Liwicki, M. Scene labeling with lstm recurrent neural networks. In Proceedings of the 2015 IEEE Conference on Computer Vision and Pattern Recognition, Boston, MA, USA, 7–12 June 2015; pp. 3547–3555.
42. Jain, A.; Zamir, A.R.; Savarese, S.; Saxena, A. Structural-RNN: Deep learning on spatio-temporal graphs. In Proceedings of the IEEE Conference on Computer Vision and Pattern Recognition, Las Vegas, NV, USA, 27–30 June 2016; pp. 5308–5317.
43. Qin, H.; Qin, L.; Xue, L.; He, X.; Yu, C.; Liang, X. Finger-vein verification based on multi-features fusion. *Sensors* **2013**, *13*, 15048–15067. [CrossRef]
44. Krizhevsky, A.; Sutskever, I.; Hinton, G.E. Imagenet classification with deep convolutional neural networks. *NIPS* **2012**, *25*, 1097–1105. [CrossRef]
45. Platt, J. Probabilistic outputs for support vector machines and comparisons to regularized likelihood methods. *Adv. Large Margin Classif.* **1999**, *10*, 61–74.
46. Taigman, Y.; Yang, M.; Ranzato, M.; Wolf, L. Deepface: Closing the gap to human-level performance in face verification. In Proceedings of the IEEE Conference on Computer Vision and Pattern Recognition, Columbus, OH, USA, 23–28 June 2014; pp. 1701–1708.
47. Hamming, R.W. Error detecting and error correcting codes. *Bell Syst. Tech. J.* **1950**, *29*, 147–160. [CrossRef]
48. Macmillan, N.; Creelman, C. *Detection Theory: A User's Guide*; Lawrence Erlbaum: Mahwah, NJ, USA, 2005.
49. Jia, Y.; Shelhamer, E.; Donahue, J.; Karayev, S.; Long, J.; Girshick, R.; Guadarrama, S.; Darrell, T. Caffe: Convolutional architecture for fast feature embedding. In Proceedings of the 22nd ACM International Conference on Multimedia, Orlando, FL, USA, 3–7 November 2014; pp. 675–678.

MDPI
St. Alban-Anlage 66
4052 Basel
Switzerland
Tel. +41 61 683 77 34
Fax +41 61 302 89 18
www.mdpi.com

Applied Sciences Editorial Office
E-mail: applsci@mdpi.com
www.mdpi.com/journal/applsci